Experimental Methods in the Physical Sciences

VOLUME 32

VACUUM ULTRAVIOLET SPECTROSCOPY II

EXPERIMENTAL METHODS IN THE PHYSICAL SCIENCES

Robert Celotta and Thomas Lucatorto, *Editors in Chief*

Founding Editors

L. MARTON
C. MARTON

Volume 32

Vacuum Ultraviolet Spectroscopy II

Edited by

J. A. R. Samson
Department of Physics and Astronomy
University of Nebraska
Lincoln, Nebraska

and

D. L. Ederer
Physics Department
Tulane University
New Orleans, Louisiana

ACADEMIC PRESS

San Diego   London   Boston   New York   Sydney   Tokyo   Toronto

627422

This book is printed on acid-free paper. ∞

Copyright © 1998 by Academic Press

All Rights Reserved.
No part of this publication may be reproduced or transmitted in any form or by any means, electronic or mechanical, including photocopy, recording, or any information storage and retrieval system, without permission in writing from the Publisher.
The appearance of the code at the bottom of the first page of a chapter in this book indicates the Publisher's consent that copies of the chapter may be made for personal or internal use of specific clients. This consent is given on the condition, however, that the copier pay the stated per-copy fee through the Copyright Clearance Center, Inc. (222 Rosewood Drive, Danvers, Massachusetts 01923) for copying beyond that permitted by Sections 107 or 108 of the U.S. Copyright Law. This consent does not extend to other kinds of copying, such as copying for general distribution, for advertising or promotional purposes, for creating new collective works, or for resale. Copy fees for pre-1998 chapters are as shown on the title pages. If no fee code appears on the title page, the copy fee is the same as for current chapters.
1079-4042/98 $25.00

Academic Press
a division of Harcourt Brace & Company
525 B Street, Suite 1900, San Diego, CA 92101-4495, USA
http://www.apnet.com

Academic Press
24-28 Oval Road, London NW1 7DX, UK
http://www.hbuk.co.uk/ap/

International Standard Serial Number: 1079-4042/98

International Standard Book Number: 0-12-475979-3

PRINTED IN THE UNITED STATES OF AMERICA
98 99 00 01 02 EB 9 8 7 6 5 4 3 2 1

CONTENTS

Contributors — xi
Volumes in Series — xiii

1. **Normal-Incidence Monochromators and Spectrometers**
   MASATO KOIKE

   1.1. Concave Grating Monochromators and Spectrometers — 1
   1.2. Plane Grating Monochromators and Spectrometers — 17
   References — 19

2. **Grazing-Incidence Monochromators for Third-Generation Synchrotron Radiation Sources**
   H. A. PADMORE, M. R. HOWELLS, and W. R. McKINNEY

   2.1. Introduction — 21
   2.2. Grating Theory — 23
   2.3. Application of Aberration Theory to the Design of an Undulator-Based SGM — 33
   2.4. Focusing in Variable-Included-Angle Monochromators — 43
   2.5. Diffraction Efficiency — 46
   2.6. An Optimized Beam Line for Microscopy by Photoelectron Emission Microscopy and Micro-X-Ray Photoelectron Spectroscopy — 47
   References — 52

3. Spectrographs and Monochromators Using Varied Line Spacing Gratings
JAMES H. UNDERWOOD

3.1. Limitations of Uniformly Spaced Grating Instruments	55
3.2. Paraxial Focusing Equations for VLS Gratings	56
3.3. Harada-Style Focusing	58
3.4. Hettrick-Style Focusing	60
3.5. Flat Field Spectrographs and Spectrometers	62
3.6. Light Path Function for a System of a Mirror and a VLS Grating	65
3.7. Monochromators for Synchrotron Radiation	66
3.8. Holographically Recorded Gratings	70
3.9. Conclusions	70
References	70

4. Interferometric Spectrometers
ANNE P. THORNE and MALCOLM R. HOWELLS

4.1. Introduction	73
4.2. Fabry–Perot Interferometry	77
4.3. Fourier Transform Spectrometry: Principal Features	80
4.4. Fourier Transform Spectrometry in Practice in the VUV	87
4.5. Spatially Heterodyned, Nonscanning Interferometers for FTS	93
4.6. All-Reflection FT Spectrometers	97
4.7. Soft X-Ray FTS by Grazing Reflection	99
References	105

5. Gas Detectors
J. B. WEST

5.1. Ionization Chambers	107
5.2. Proportional Counters	112
References	115

6. Photodiode Detectors
L. R. CANFIELD

6.1. Introduction	117
6.2. Radiometric Standards	118
6.3. Photodiode Types	119
6.4. Proper Use of Photodiodes in the VUV	134
References	137

7. Amplifying and Position Sensitive Detectors
OSWALD H. W. SIEGMUND

7.1. Photon Detection	139
7.2. Amplifying Detectors	148
7.3. Position Sensing Techniques	162
References	172

8. Absolute Flux Measurements
S. V. BOBASHEV

8.1. Introduction	177
8.2. Primary Radiator Radiometry	178
8.3. Primary Detector Radiometry	179
8.4. Transfer Detector Standards for Absolute Flux Measurement	186
8.5. Conclusion	189
References	190

9. Vacuum Techniques
ROGER L. STOCKBAUER

9.1. Introduction	193
9.2. Design of the Vacuum Environment	193

10. Lithography
YULI VLADIMIRSKY

9.3. Leak Detection	200
9.4. Optics Cleaning	202
References	203
10.1. Integrated Circuit Fabrication and Lithographic Process	205
10.2. Photoresist in Lithography	206
10.3. Optical Lithography	208
10.4. X-Ray Lithography	214
References	221

11. X-Ray Spectromicroscopy
HARALD ADE

11.1. Introduction	225
11.2. X-Ray Spectromicroscopy Approaches	230
11.3. Applications	239
11.4. Discussion	256
11.5. Conclusions	257
References	258

12. Optical Spectroscopy in the VUV Region
MARSHALL L. GINTER and KOUICHI YOSHINO

12.1. Introduction	263
12.2. Wavelength Measurements and Energy Levels	265
12.3. Intensity Measurements and Cross Sections	270
References	275

13. Soft X-Ray Fluorescence Spectroscopy
THOMAS A. CALLCOTT

13.1. Introduction	279
13.2. Characteristics of the Soft X-Ray Fluorescence Spectra of Solids	282
13.3. Instrumentation for SXF Spectroscopy	293
13.4. Survey of Recent SXF Spectroscopy Research	297
References	298

INDEX 301

CONTRIBUTORS

Numbers in parentheses indicate the pages on which the authors' contributions begin.

HARALD ADE (225), *Department of Physics, North Carolina State University, Raleigh, North Carolina 27695-8202*

S. V. BOBASHEV (177), *A. F. Ioffe Physico-Technical Institute RAS, St. Petersburg 194021, Russia*

THOMAS A. CALLCOTT (279), *Department of Physics and Astronomy, University of Tennessee, Knoxville, Tennessee 37996*

L. R. CANFIELD (117), *Physics Laboratory, National Institute of Standards and Technology, Gaithersburg, Maryland 20899*

MARSHALL L. GINTER (263), *Institute for Physical Science and Technology, University of Maryland, College Park, Maryland 20742*

MALCOLM R. HOWELLS (21, 73), *Advanced Light Source, Lawrence Berkeley National Laboratory, Berkeley, California 94720*

MASATO KOIKE (1), *Advanced Photon Research Center, Japan Atomic Energy Research Institute, Osaka 572-0019, Japan*

WAYNE R. MCKINNEY (21), *Advanced Light Source, Lawrence Berkeley National Laboratory, Berkeley, California 94720*

HOWARD A. PADMORE (21), *Advanced Light Source, Lawrence Berkeley National Laboratory, Berkeley, California 94720*

OSWALD H. W. SIEGMUND (139), *Space Sciences Laboratory, University of California, Berkeley, California 94720*

ROGER L. STOCKBAUER (193), *Department of Physics and Astronomy, Louisiana State University, Baton Rouge, Louisiana 70803*

ANNE P. THORNE (73), *Blackett Laboratory, Imperial College, London SW7 2BZ, United Kingdom*

JAMES H. UNDERWOOD (55), *Center for X-ray Optics, Lawrence Berkeley National Laboratory, Berkeley, California 94720*

YULI VLADIMIRSKY (205), *Center for X-ray Lithography, University of Wisconsin–Madison, Stoughton, Wisconsin 53589-3097*

J. B. WEST (107), *Daresbury Laboratory, Warrington WA4 4AD, United Kingdom*

KOUICHI YOSHINO (263), *Harvard Smithsonian Center for Astrophysics, Cambridge, Massachusetts 02138*

VOLUMES IN SERIES
EXPERIMENTAL METHODS IN THE PHYSICAL SCIENCES

(formerly Methods of Experimental Physics)

Editors-in-Chief
Robert Celotta and Thomas Lucatorto

Volume 1. Classical Methods
Edited by Immanuel Estermann

Volume 2. Electronic Methods, Second Edition (in two parts)
Edited by E. Bleuler and R. O. Haxby

Volume 3. Molecular Physics, Second Edition (in two parts)
Edited by Dudley Williams

Volume 4. Atomic and Electron Physics—Part A: Atomic Sources and Detectors; Part B: Free Atoms
Edited by Vernon W. Hughes and Howard L. Schultz

Volume 5. Nuclear Physics (in two parts)
Edited by Luke C. L. Yuan and Chien-Shiung Wu

Volume 6. Solid State Physics—Part A: Preparation, Structure, Mechanical and Thermal Properties; Part B: Electrical, Magnetic and Optical Properties
Edited by K. Lark-Horovitz and Vivian A. Johnson

Volume 7. Atomic and Electron Physics—Atomic Interactions (in two parts)
Edited by Benjamin Bederson and Wade L. Fite

Volume 8. Problems and Solutions for Students
Edited by L. Marton and W. F. Hornyak

Volume 9. Plasma Physics (in two parts)
Edited by Hans R. Griem and Ralph H. Lovberg

Volume 10. Physical Principles of Far-Infrared Radiation
By L. C. Robinson

Volume 11. Solid State Physics
Edited by R. V. Coleman

Volume 12. Astrophysics—Part A: Optical and Infrared Astronomy
Edited by N. Carleton

Part B: Radio Telescopes; Part C: Radio Observations
Edited by M. L. Meeks

Volume 13. Spectroscopy (in two parts)
Edited by Dudley Williams

Volume 14. Vacuum Physics and Technology
Edited by G. L. Weissler and R. W. Carlson

Volume 15. Quantum Electronics (in two parts)
Edited by C. L. Tang

Volume 16. Polymers—Part A: Molecular Structure and Dynamics; Part B: Crystal Structure and Morphology; Part C: Physical Properties
Edited by R. A. Fava

Volume 17. Accelerators in Atomic Physics
Edited by P. Richard

Volume 18. Fluid Dynamics (in two parts)
Edited by R. J. Emrich

Volume 19. Ultrasonics
Edited by Peter D. Edmonds

Volume 20. Biophysics
Edited by Gerald Ehrenstein and Harold Lecar

Volume 21. Solid State: Nuclear Methods
Edited by J. N. Mundy, S. J. Rothman, M. J. Fluss, and L. C. Smedskjaer

Volume 22. Solid State Physics: Surfaces
Edited by Robert L. Park and Max G. Lagally

Volume 23. Neutron Scattering (in three parts)
Edited by K. Sköld and D. L. Price

Volume 24. Geophysics—Part A: Laboratory Measurements; Part B: Field Measurements
Edited by C. G. Sammis and T. L. Henyey

Volume 25. Geometrical and Instrumental Optics
Edited by Daniel Malacara

Volume 26. Physical Optics and Light Measurements
Edited by Daniel Malacara

Volume 27. Scanning Tunneling Microscopy
Edited by Joseph Stroscio and William Kaiser

Volume 28. Statistical Methods for Physical Science
Edited by John L. Stanford and Stephen B. Vardaman

Volume 29. Atomic, Molecular, and Optical Physics—Part A: Charged Particles; Part B: Atoms and Molecules; Part C: Electromagnetic Radiation
Edited by F. B. Dunning and Randall G. Hulet

Volume 30. Laser Ablation and Desorption
Edited by John C. Miller and Richard F. Haglund, Jr.

Volume 31. Vacuum Ultraviolet Spectroscopy I
Edited by J. A. R. Samson and D. L. Ederer

Volume 32. Vacuum Ultraviolet Spectroscopy II
Edited by J. A. R. Samson and D. L. Ederer

1. NORMAL-INCIDENCE MONOCHROMATORS AND SPECTROMETERS

Masato Koike
Advanced Photon Research Center
Kansai Research Establishment
Japan Atomic Energy Research Institute
Osaka, Japan

1.1 Concave Grating Monochromators and Spectrometers

1.1.1 Seya–Namioka Monochromator

The concept of a Seya–Namioka monochromator was conceived by Seya [1] in the course of his exploration to design a vacuum ultraviolet (VUV) spherical grating monochromator without using the Rowland circle, and the monochromator was analyzed and constructed by Namioka [2, 3]. In this system, the entrance and exit slits are fixed, and the wavelength scanning is performed by rotation of the grating around its axis tangent to the zeroth groove at the grating center. Therefore, the directions of incidence and exiting beams remain unchanged while wavelengths are scanned. The unique included-angle ~70° makes it possible to satisfy the horizontal focal condition very closely over a wide wavelength range (e.g., 0–700 nm with a 600 grooves/mm grating).

For a Seya–Namioka monochromator (Fig. 1),

$$r, r' = \text{const.}, \quad 2K = \alpha - \beta_0 = \text{const.}, \quad \alpha = K + \theta, \quad \beta_0 = \theta - K \quad (1)$$

where r and r' are the distances from the centers of the entrance and exit slits to the grating centers, respectively, $2K$ is the included angle between the principal incident and exiting rays, θ is the angle of grating rotation measured from the bisector of the angle $2K$ and has the same sign as the spectral order m, and α and β_0 are the angles of incidence and diffraction, respectively. The relation between wavelength λ and θ is given by

$$\lambda = (2\sigma/m) \cos K \sin \theta \quad (2)$$

where σ is the distance between the grating grooves and is called the grating constant.

Values of r, r', and $2K$ in Eq. (1) that fulfill very closely the horizontal focal condition $F_{200} = 0$, over the given scanning range from θ_1 to θ_2 can be obtained by solving the following equations [2, 4]:

$$\partial I_{200}/\partial r = 0, \quad \partial I_{200}/\partial r' = 0, \quad \partial I_{200}/\partial K = 0. \quad (3)$$

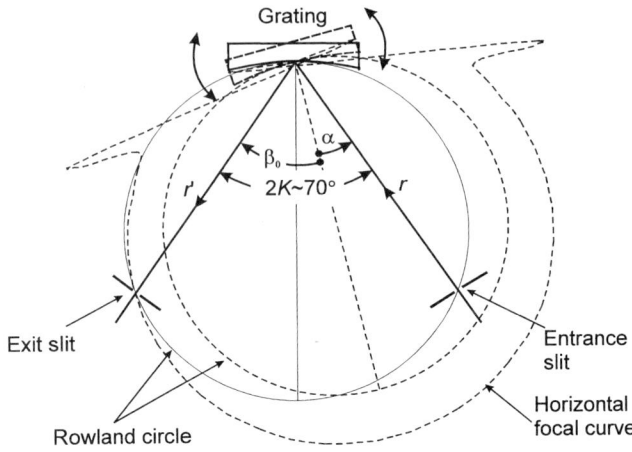

FIG. 1. A Seya–Namioka monochromator.

where

$$I_{200} \equiv \int_{\theta_1}^{\theta_2} F_{200}^2 \, d\theta \qquad (4)$$

and θ_1 and θ_2 are the angles of the grating rotation for the minimum and maximum wavelengths through λ_{min} and λ_{max}, respectively. For λ_{min} = zeroth order and $\lambda_{max} = 0.42\sigma$ nm in the negative first order ($m = -1$), for example, the instrumental constants r, r', and $2K$ are from Eq. (3) as

$$R/r = 1.220527, \qquad R/r' = 1.216931, \qquad 2K = 69°44' \qquad (5)$$

where R is the radius of curvature of the spherical grating. Figure 1 shows the Rowland circles of the grating at the scanning wavelength $\lambda = 0$ and 0.36σ nm, and the horizontal focal curve at $\lambda = 0.36\sigma$ nm (the Rowland circle at the zeroth order is indicated by the solid line and the Rowland circle and the horizontal focal curve at $\lambda = 0.36\sigma$ nm are shown by dotted lines). It can be shown also for other wavelengths that the horizontal focal curves always go through the exit slit almost exactly. Note that the Rowland circle is no longer the horizontal focal curve in this case, except for the zeroth order.

As seen in Eqs. (3)–(5) the optimum values of R/r, R/r', and $2K$ depend on the choice of a scanning range and spectral order. However, small variations in the instrumental constants have no practical importance because residual higher-order aberrations limit the resolution of the monochromator.

The disadvantage of the Seya–Namioka monochromator is its rather large astigmatism. This disadvantage can be eliminated either by placing an additional mirror in between the light source and the entrance slit [5, 6], or by using an aspheric grating [7, 8]. However, the former method reduces the luminosity of the monochromator, especially in the VUV region, and the latter method is still not easily accessible.

Recently, the mechanically ruled varied-line-spacing (VLS) gratings [9] and various holographic gratings [4, 10] have remarkably improved the resolution and reduction of astigmatism in the Seya–Namioka monochromator. Some design examples are given in the following paragraphs to show the improvement in the resolution of a monochromator equipped with a holographic grating recorded with spherical wave fronts, a holographic grating recorded with a spherical wave front and an aspheric wave front, or a mechanically ruled VLS grating over that with a conventional grating.

The instrumental constants assumed are wavelength range 0–233 nm, grating constant $\sigma = 1/1800$ mm, radius of curvature $R = 1000$ mm, and grating size 100 (W) × 60 (H) mm^2. The optimization was made by the analytical design method [11, 12] with the weighting factor $\mu = 0$ and the design wavelengths $\lambda_i = 50, 100, 150,$ and 200 nm. The parameters r and $2K$ were optimized for the conventional grating, and their optimum values were used also in the case of the holographic gratings and the mechanically ruled VLS grating. On the other hand, r' was optimized in individual cases.

The parameters thus obtained are:

a. conventional grating: $r = 819.32$ mm, $r' = 821.74$ mm, and $2K = 69°44'$,
b. holographic grating recorded with spherical wave fronts: $r' = 821.40$ mm, $\lambda_0 = 457.93$ nm, $r_C = 846.96$ mm, $r_D = 1271.27$ mm, $\gamma = -65.8703°$, and $\delta = -5.0717°$,
c. holographic grating recorded with a spherical wave front and an aspheric wave front: $r' = 821.70$ mm, $\lambda_0 = 457.93$ nm, $R_D = 1000$ mm, $r_C = 877.64$ mm, $p_D = 309.16$ mm, $q_D = 453.60$ mm, $\gamma = -64.9445°$, $\delta = -4.6819°$, and $\eta_D = 8.0499°$,
d. mechanically ruled VLS grating with straight grooves: $r' = 822.62$ mm, $2K = 69°44'$, $2a = 1.735396 \times 10^{-14}$ mm, $6b = -1.189319 \times 10^{-16}$ mm, and $4c = -1.771678 \times 10^{-24}$ mm, where λ_0 is the wavelength of the recording laser. For the definition of other parameters refer to Volume 31, Sections 17.1.2 through 17.1.4.

Figure 2 shows the spot diagrams constructed for the monochromator with the instrumental parameters and the gratings given, (a)–(d). In Fig. 2, 1000 rays of wavelength λ were generated with the source being 10 mm long on the entrance slit and traced through the monochromator assuming a ruled area of 100 (W) × 60 (H) mm^2 and $\lambda = 50, 100, 150,$ and 200 nm. Each diagram shows the value of σ_λ, which is defined by the product of the standard deviation σ_Y of the ray-traced spots in the direction of dispersion and the reciprocal linear dispersion at λ. It is seen in Figs. 2c and 2d that coma-type aberrations are well

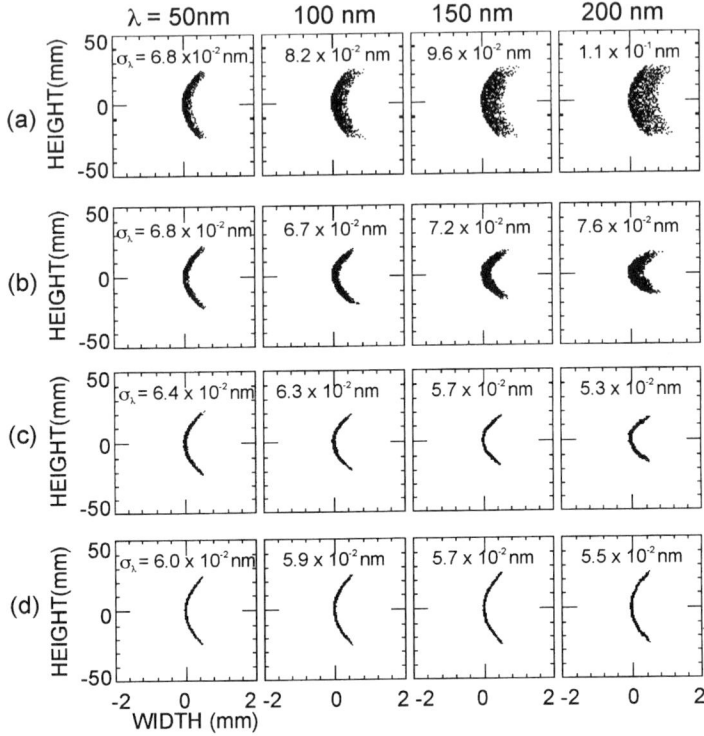

FIG. 2. Spot diagrams constructed for the Seya–Namioka monochromator with (a) conventional grating, (b) holographic grating with spherical wave fronts, (c) holographic grating with a spherical wave front and an aspheric wave front, and (d) mechanically ruled VLS grating with straight grooves. The standard deviation σ_λ of the spectral spread is also indicated in the respective diagrams.

corrected for the holographic grating recorded with a spherical wave front and an aspheric wave front and the mechanically ruled VLS grating with straight grooves. For the holographic gratings, astigmatism is also reduced at longer wavelengths, as is seen in Figs. 2b and 2c, although no attempt was made to reduce astigmatism at the design stages (i.e., weighting factor μ was set to zero). This feature is caused by the curvature of the grooves of the gratings. However, this is by no means unique to the holographic grating. Ruled grating with curved grooves can also reduce astigmatism [13].

Next, a monochromatized beam emerging from the exit slit of a limited length is considered. Figure 3 shows the line profiles constructed from the spot diagrams of Fig. 2 by taking only the spots that fell within the exit slit 10 mm in length. In each diagram are shown the standard deviation σ_λ of the spectral spread for the rays of λ and the throughput T. These diagrams clearly show the

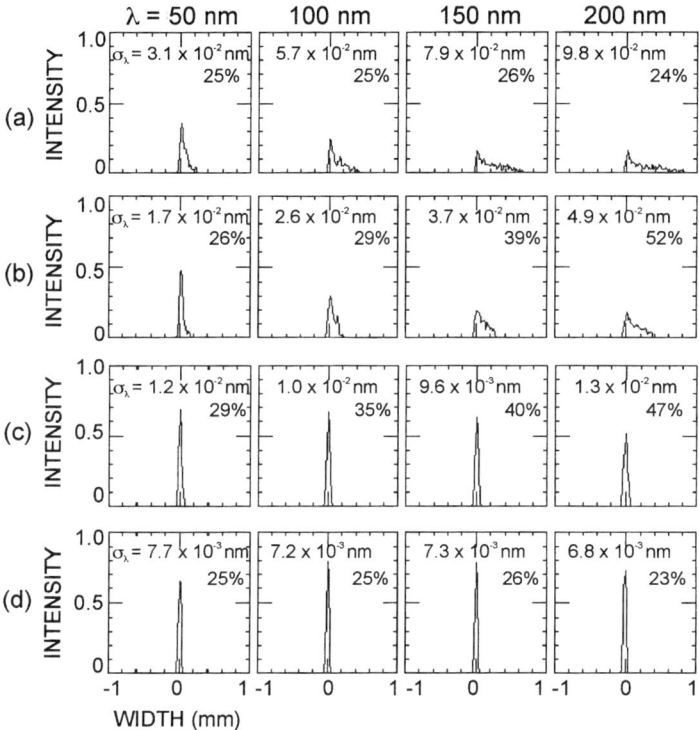

FIG. 3. Line profiles for the valid rays through the exit slit having a 10-mm length for the Seya–Namioka monochromator with (a) conventional grating, (b) holographic grating with spherical wave fronts, (c) holographic grating with a spherical wave front and an aspheric wave front, and (d) mechanically ruled VLS grating with straight grooves. The standard deviation σ_λ of the spectral spread and throughput T are also indicated in the respective diagrams.

advantages of the holographic gratings and the mechanically ruled VLS grating over the conventional grating in obtaining high resolution and throughput.

In the Seya–Namioka monochromator equipped with a conventional grating, the astigmatism is calculated from Eq. (56) in Volume 31, Section 17.6.7:

$$L_{ast} = H\frac{r'}{r} + Lr'\left(\frac{1}{r} + \frac{1}{r'} - \frac{\cos \alpha + \cos \beta_0}{R}\right) \qquad (6)$$

where H and L are the total illuminated length of the entrance slit and that of the grooves, respectively. For the conventional grating, $r'/r \cong 1.003$, and the second term in Eq. (6) varies from $0.655L$ to $0.687L$ as wavelengths are scanned from the zeroth order to 200 nm. The curvature in the spectral line of the point source, called the astigmatic curvature, is expressed as

$$\rho_1 = \frac{r' g_{010}^2 \cos \beta_0}{2 f_{020}} \quad (7)$$

where f_{020} and g_{010} are given in Volume 31, Section 17.6.4. For the conventional grating, ρ_1 varies from $0.571R$ to $0.611R$ as wavelengths are scanned from the zeroth order to 200 nm. Another type of curvature appears in the spectral lines. This curvature, known as the enveloping curvature, is observed in the spectral line formed by the central rays from every point on the infinite narrow slit and expressed as

$$\rho_2 = \frac{r' \cos \beta_0}{\sin \alpha + \sin \beta_0}. \quad (8)$$

1.1.2 Wadsworth Monochromator and Spectrometer

One common drawback of the Seya–Namioka and Rowland circle mounts is the presence of astigmatism. The result is loss of intensity and spatial resolution in the direction parallel to the entrance slit

Wadsworth [14] created his new spherical grating mount, in 1896, and showed the stigmatic nature of its spectral images. This mount, known as the Wadsworth mount, uses parallel light to illuminate the grating and uses the normal spectrum (see Volume 31, Section 17.6.6). This is illustrated in Fig. 4. Light from the entrance slit is rendered parallel by a concave mirror and is reflected onto the spherical grating. The diffracted rays are then focused on the exit slit. Assuming that the grating is illuminated by a parallel beam both horizontally and vertically, the horizontal and the vertical focal distances, r'_h and

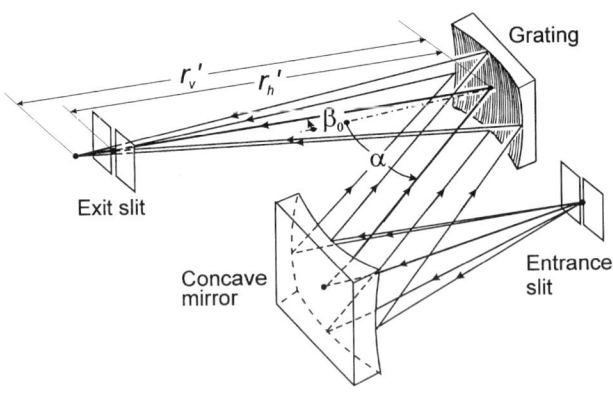

FIG. 4. A Wadsworth mount.

r'_v, are obtained from Eqs. (51) and (52) in Volume 31, Section 17.6.6:

$$r'_h = \frac{R \cos^2 \beta_0}{\cos \alpha + \cos \beta_0}, \quad r'_v = \frac{R}{\cos \alpha + \cos \beta_0} \quad (9)$$

At $\beta_0 \sim 0$, both the horizontal and vertical focal curves are reduced to

$$r'_h \cong r'_v \cong \frac{R}{1 + \cos \alpha} \quad (10)$$

producing nearly anastigmatic spectral images. For a small angle of incidence, the focal distance becomes almost $R/2$, providing the possibility of realizing a fast spectrograph. In spite of these merits, this mount has not achieved the popularity of the Rowland circle mount because its focal curve is parabola-like.

A modified Wadsworth monochromator [15] is used on synchrotron radiation (SR) beamlines for providing high luminosity and moderate resolution by taking advantage of the simple optical configuration without a collimating mirror and an entrance slit.

1.1.3 Eagle and Eagle-Type Monochromators and Spectrometers

The Eagle mount [16, 17] is the spherical grating version of the Littrow mount of a prism or a plane grating in the sense that the angles of incidence and diffraction are approximately equal, that is, $\alpha \cong \beta_0$. The advantages of this mount are compactness and the lowest level of aberrations among Rowland circle mounts. Figures 5 and 6 are schematic diagrams of the in-plane and off-plane spectrographs, respectively.

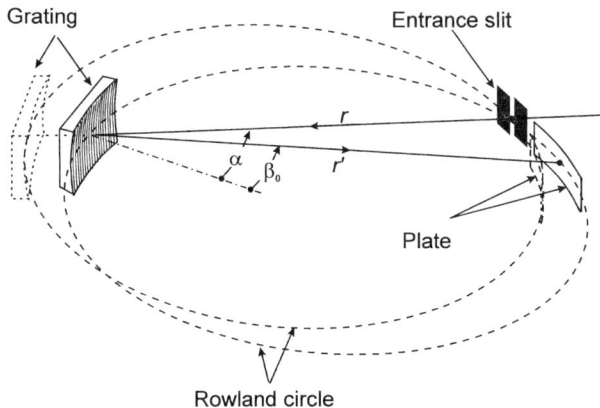

FIG. 5. A spectrograph using an in-plane Eagle mount.

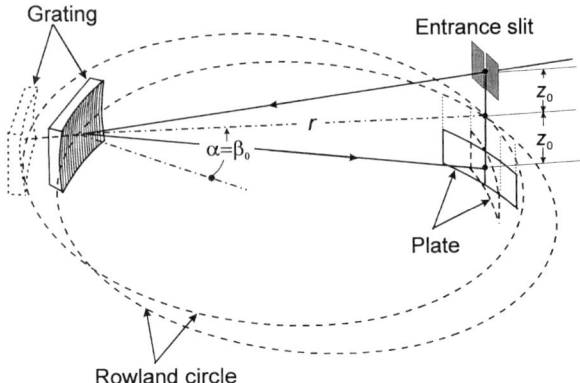

FIG. 6. A spectrograph using an off-plane Eagle mount.

In an in-plane Eagle spectrograph the grating and the plate holder should be translated and rotated in a way so as to remain on the Rowland circle while changing the wavelength. On the other hand, for the off-plane Eagle spectrograph, the positions of the entrance slit and plate holder can be fixed independent of wavelength. However, the grating must be translated and rotated and the plate holder must be rotated as the wavelength is changed. These features are attractive and should be kept in mind when designing a monochromator.

For the off-plane mount, $\alpha = \beta_0$. Using the notation defined in Volume 31, Section 17.6.2, the wavelength is expressed as [18]

$$\lambda = \frac{2\sigma}{m}\left(1 + \frac{z_0^2}{R^2 \cos^2 \alpha}\right)^{-1/2} \sin \alpha \qquad (11)$$

where z_0 is the height of the center of the entrance slit from the Rowland circle plane.

The aberrations present in an off-plane Eagle spectrograph are corrected by a slight rotation of the entrance slit with respect to the direction of the ruling of the grating [18, 19].

$$\varphi = \frac{z_0}{R} \tan \alpha \sec \alpha. \qquad (12)$$

In an off-plane mount the angle subtending the incidence and diffraction beams for the grating is smaller than 2° to prevent a degradation in the resolution [17]. This requirement implies the radius of curvature of the grating of at least ~3 m. The application of off-plane Eagle spectrometers on SR beamlines is discussed in Sections 1.1.4 and 1.1.5.

Various in-plane normal incidence monochromators such as the in-plane Eagle mount but without using the Rowland circle were developed [20–22]. In most mounts the positions of the entrance and exit slits are fixed, and the spherical grating is simultaneously rotated and translated along the line of the grating normal at the position corresponding to zeroth order. The deviation from the horizontal focus condition is kept very small with a simple mechanism for wavelength scanning. Furthermore, the adoption of mechanically ruled VLS gratings and holographic gratings will make it possible to reduce astigmatism and the amount of the grating translation of the normal incidence high-resolution spherical grating monochromators [11, 13, 23].

1.1.4 High-Resolution Monochromator on an Undulator Beamline

At the Advanced Light Source, an undulator beamline was constructed for the study of chemical dynamics [24, 25]. This beamline consists of two branch lines; one is the high-resolution branch line and the other the high-flux branch line. A vertical dispersion 6.65-m off-plane Eagle monochromator is used on the high-resolution branch line. The radiation source is a 4.5-m long undulator having a 10-cm period, and the fundamental in the energy range of 6 to 30 eV is used in the experiments. Figure 7 is a schematic of the beamline, and the parameters for the optical elements are listed in Table I.

The spherical mirror M1 and the following toroidal mirror M2 accepts the beam from the undulator (not shown in Fig. 7) and produces a convergent beam toward the center of the gas filter GF. The silicon-coated M2 mirror cuts off high-energy photons greater than 70 eV. A 406 W power load is reduced to 6 W after the reflections at the M1 and M2 mirrors. The substrates of M1 and M2 are

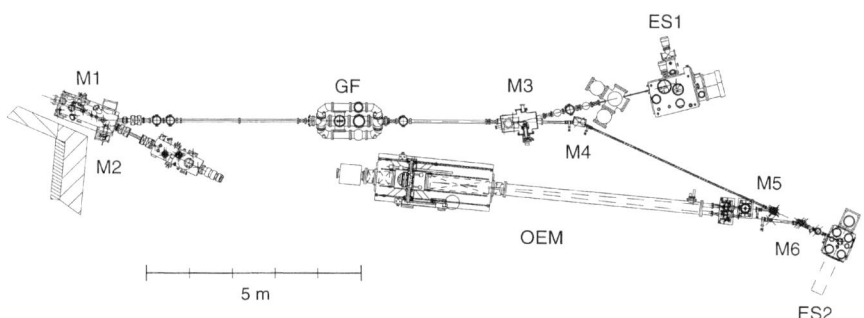

FIG. 7. A VUV high-resolution and high-flux undulator beamline at the Advanced Light Source. M1, spherical mirror; M2, retractable toroidal mirror; GF, gas filter; M3, retractable toroidal mirror; M4 and M5, cylindrical mirrors; OEM, 6.65-m off-plane Eagle monochromator; M6, toroidal mirror; ES1, end station 1; ES2, end station 2. The horizontal plane view is shown.

TABLE I. Optical Elements of the Off-Plane Eagle Monochromator Beamline at the Advanced Light Source

Type	Coating and blank material	Dimensions (mm) foot print (mm²)	Radius of curvature (m)	Incidence angle (°)	Groove density (grooves/mm)	Blaze wavelength nm/ blaze angle (°)
M1 Spherical mirror	Ni- and C-coated GlidCop™	381 (W) × 76 (H) 252 (W) × 13 (H)	302.300	87.0	—	—
M2 Toroidal mirror	Si-coated GlidCop™	210 (W) × 76 (H) 60 (W) × 13 (H)	66.288 (h) 1.728 (v)	77.0	—	—
M3 Toroidal mirror	Si-coated GlidCop™	257 (W) × 76 (H) 119 (W) × 11 (H)	29.175 (h) 0.433 (v)	83.0	—	—
M4 Bendable cylindrical mirror	Si-coated GlidCop™	150 (W) × 40 (H) 66 (W) × 13 (H)	24.900–28.514 (h) ∞ (v)	77.9	—	—
M5 Cylindrical mirror	SiC-coated quartz	50 (W) × 50 (H) 3 (W) × 25 (H)	∞ (h) 1782.904 (v)	9.8	—	—
G1 Spherical grating (ruled, replica)	Al/MgF$_2$-coated quartz	210ϕ 186 (W) × 14 (H) (at 150 nm)	6.650	5.2 (at 150 nm)	1200	150/5.17
G2 Spherical grating (ruled, replica)	Os-coated quartz	180ϕ 174 (W) × 13 (H) (at 150 nm)	6.650	21.1 (at 150 nm)	4800	100/13.88
G3 Spherical grating (holographic, master)	SiC-coated quartz	180ϕ 180 (W) × 13 (H) (at 150 nm)	6.100	10.4 (at 150 nm)	2400[a]	100/7.00
M6 Toroidal mirror	Si-coated quartz	80 (W) × 50 (H) 37 (W) × 37 (H)	15.781 (h) 0.182 (v)	81.0	—	—

Source: Ref. 24.
[a] Effective groove density. Recording parameters for the holographic grating are given in Ref. 23.

made of GlidCop AL-15,* which is an alloy of OFHC copper that contains 0.15% by weight of aluminum oxide. The maximum absorbed power densities for M1 and M2 are found to be 10.4 and 7.6 W/cm^2, respectively. The toroidal mirror M3 focuses the beam in the end station ES1, which is devoted to the study of photochemistry and chemical reactivity. When M3 is retracted, the beam is focused, both horizontally and vertically, onto the entrance slit of the 6.65-m off-plane Eagle monochromator (OEM) by the cylindrical mirrors M4 and M5. The toroidal mirror M6 focuses the emerging beam from the exit slit at the end station ES2.

To compensate for the astigmatism of OEM and obtain a small spot size at ES2, a bendable cylindrical mirror is adopted for M4. The OEM is equipped with two concave gratings having 1200 and 4800 grooves/mm. To scan over a range of 50 to 200 nm, the radius of M4 is required to vary from 28 to 25 m. The anticipated spot size at ES2 is 360 μm (h) × 240 μm (v) at 200 nm.

The monochromator is normally operated in the first order and is tuned to the fundamental wavelength of the undulator radiation. The undulator radiation operating at a high K-value delivers a substantial radiant power into the higher harmonics. This would necessitate the use of cooled optics in the monochromator. Furthermore, higher harmonics of the undulator going to the higher orders of the grating overlap with the first-order spectrum of the fundamental radiation, causing deterioration of the spectral purity. To address these problems, a windowless rare-gas harmonic filter has successfully been used [26]. Figure 8 shows the harmonics filter with a three-stage differential pumping system. This system has proved to be capable of maintaining the beamline vacuum better than 5×10^{-9} torr.

Figure 9 shows the performance of the harmonics filter measured by a transmission grating spectrometer (grating constant = 1/5000 mm) [27]. The 8-cm period undulator was operated at $K = 5.24$, and the spectra were measured without a filter gas (solid line) and with 30.6 torr neon gas in the gas cell (dotted line). In the absence of gas, a number of peaks originate from the undulator and higher orders of the transmission grating. Comparison of the spectra measured with and without neon gas shows the suppression of higher harmonics greater than four orders of magnitude. It also showed insignificant (<5%) attenuation of the fundamental radiation.

Figure 10 shows an elevation view of the OEM. The entrance slit is behind the exit slit, and both slits are oriented horizontally. The grating, which can be rotated through a 30° angle, disperses the radiation vertically. The grating chamber is mounted on a granite table. The grating carriage and vacuum chamber independently ride on pairs of cross-roller bearing slides allowing a

---

* GlidCop™ is a proprietary product manufactured by SCM Metal Products, Research Triangle Park, NC.

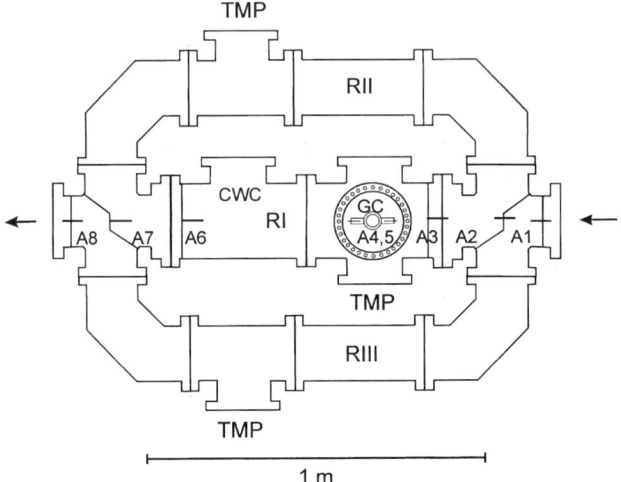

Fig. 8. A harmonics (gas) filter. RI, RII, RIII, Regions I, II, III; apertures: A1, A8, 3-mm diameter; A2, A6, A7, 2-mm diameter; A3, A4, A5, 1-mm diameter; GC, gas cell; CWC, chopper wheel chamber; TMP, turbo molecular pump.

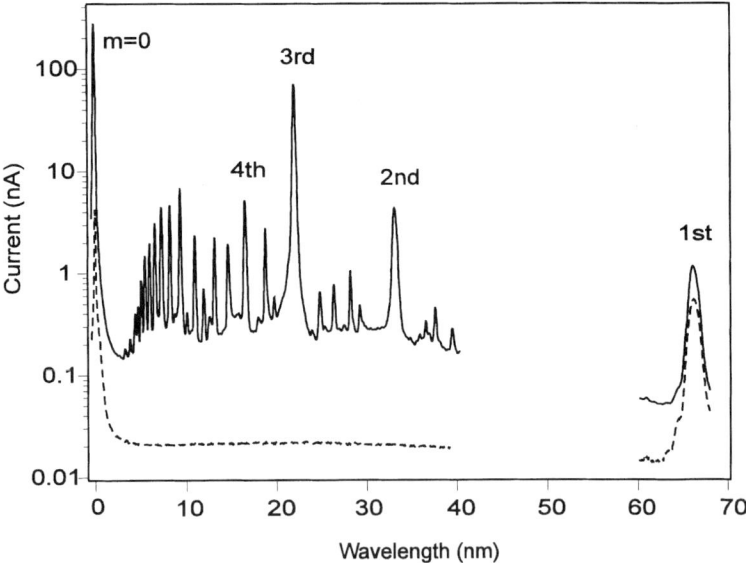

Fig. 9. Advanced Light Source U8 undulator spectrum at $K = 5.24$ measured by a transmission grating ($\sigma = 1/5000$ mm) spectrometer without Ne (solid line) and with 30.6 torr neon gas in the gas cell (dotted line).

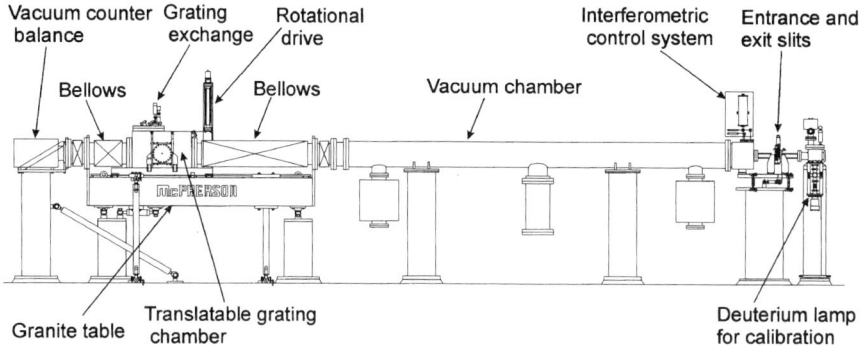

FIG. 10. An elevation view of the vertical-dispersion 6.65-m off-plane Eagle monochromator (courtesy of McPherson, Inc.).

translation of 950 mm at a resolution of 1 μm and an accuracy of 5 μm using a linear encoder. The ranges of grating rotation and translation permit a scan from zeroth order to 200 nm in the first order with the 4800 lines/mm grating. Two separate stepping motors drive the grating rotation and translation. This provides the adjustability required for gratings of somewhat different radius of curvature and/or focal properties.

An interferometric control system is used to control the grating's angular position and its translational linearity (Fig. 11). In-vacuum operation of the interferometer removes any interference produced by operation at atmospheric pressure. A laser beam from a two-frequency He–Ne laser, which is mounted on the same base plate as the slits, is split into two beams by a polarized beam splitter to provide the interferometric control. Rotating corner cubes are attached to the rotation table of the grating carriage for linearity measurements. Using these techniques, the grating rotation is encoded to a resolution of 0.02 arcsec.

The OEM has achieved a resolving power of $\sim 7 \times 10^4$ at ~ 57 nm in the first order using a 4800 lines/mm grating and a maximum flux of 3×10^{11} photons and a resolving power of 2.5×10^4 at ~ 160 nm using a 1200 lines/mm grating [25].

1.1.5 High-Resolution Spectrometer/Monochromator Beamline on a Bending Magnet Beamline

A high-resolution VUV spectroscopic facility was installed at the Photon Factory [28, 29]. This facility uses the high-order spectrum of a 6.65-m vertical dispersion off-plane Eagle spectrometer/monochromator called 6VOPE and a simple and efficient predisperser system. The facility can be operated in three

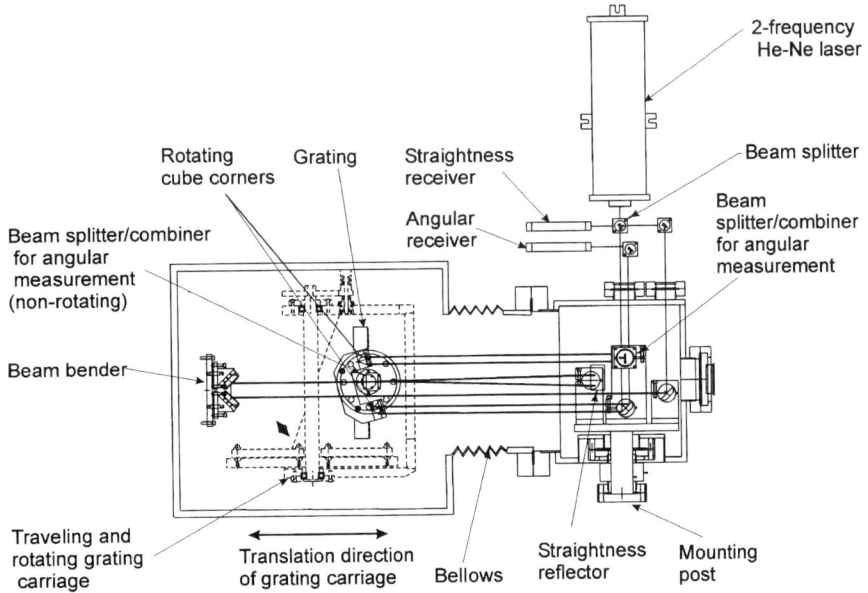

FIG. 11. An interferometric straightness and angular control system (courtesy of McPherson, Inc.).

different modes: spectrograph mode, focal plane scanning mode, and monochromator mode. The first and second modes are for high-resolution photographic and photoelectric measurements, respectively, whereas the third mode is used to provide monochromatic radiation.

Figure 12 shows a schematic diagram of the optical system. The bending magnet source point, the plane mirror M, the first VLS spherical grating G1, and the intermediate slit S1 constitute a band-pass filter type monochromator with a constant-deviation angle of 70°. The slit S1, the second VLS spherical grating G2, and the main entrance slit S2 form another constant-deviation monochromator with an included angle of 90°. The second monochromator cancels the dispersion of the filtered beam produced by the first monochromator. Thus, simple rotation of G1 and G2, together with adjustment of the width of S1, provides a properly filtered SR beam of various bandwidths to the 6VOPE [30]. Table II summarizes the specifications of the optical elements of the facility.

Figure 13 shows a schematic diagram of the unique mechanism capable of switching between the spectrograph mode and the focal plane scanning mode without breaking the vacuum [31]. The cassette holder and exit slit/photodetector unit are rigidly fixed to the subframe. Mode switching is accomplished by moving the subframe along the guide rods attached to the main frame so that the cassette holder or the exit slit/photodetector unit is brought to the focal

TABLE II. Optical Elements of the 6VOPE Facility at the Photo Factory

	Type	Coating and blank material	Dimensions (mm) effective area (mm)^2	Radius of curvature (m)	Groove density (grooves/mm)	Blaze wavelength (nm)/ blaze angle (°)
M	Plane mirror	CVD-SiC on sintered SiC	100 (W) × 280 (H) 80 (W) × 280 (H)	—	—	—
G1[a]	Spherical grating (ruled, replica)	Pt-coated Pyrex	110ϕ 70 (W) × 75 (H)	4.321	200[b]	74/0.52
G2[a]	Spherical grating (ruled, replica)	Pt-coated Pyrex	110ϕ 70 (W) × 60 (H)	2.189	204[b]	70/0.58
G3	Spherical grating (ruled, replica)	Os-coated BK-7	210ϕ 175 (W) × 100 (H)	6.650	1200	550/19.27
G3'	Spherical grating (ruled, replica)	Os-coated BK-7	210ϕ 175 (W) × 100 (H)	6.650	1200	150/5.17
G3"	Spherical grating (ruled, replica)	Os-coated quartz	180ϕ 125 (W) × 110 (H)	6.650	4800	90

Source: Ref. 28.
[a] In the monochromator mode, G1 and G2 are replaced by spherical mirrors M1 and M2 having the same radii of curvature and dimension as those of G1 and G2.
[b] Effective groove density. Ruling parameters for varied line spacing gratings are given in Ref. 30.

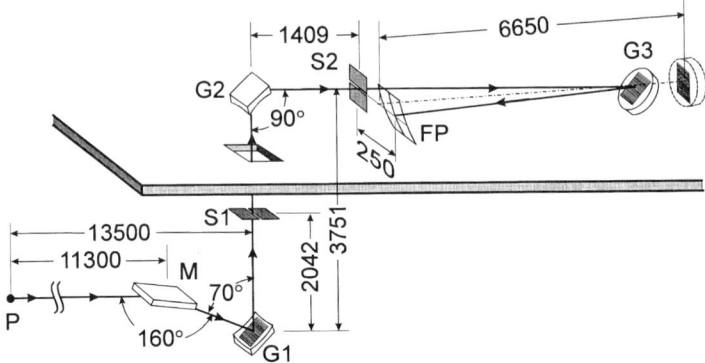

FIG. 12. A 6VOPE optical system at the Photon Factory. P, bending magnet source point; M, SiC plane mirror; G1, first VLS spherical grating; S1, intermediate slit; G2, second VLS spherical grating; S2, entrance slit; G3, main grating; FP, focal plane.

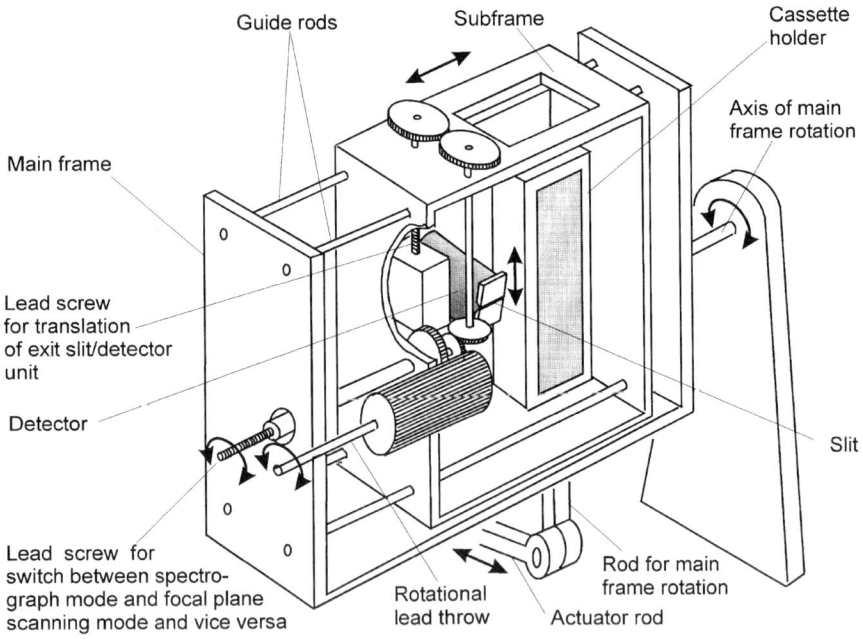

FIG. 13. Schematic diagram of the mechanism that facilitates spectrograph mode and focal plane scanning mode of 6VOPE (courtesy of Shimadzu Corp.).

position. In the spectrograph mode, the main frame is rotated by means of the actuator rod, under computer control, until the plane of the photographic plate coincides with the Rowland circle. The main frame can be rotated to cover the wavelength range of 0–640 nm in the first order of 1200 grooves/mm main grating. In the focal plane scanning mode, the exit slit/photodetector unit is moved, under computer control, along the Rowland circle over a distance of 250 mm at a rate of 2.5×10^{-4} nm/step in terms of wavelength. Scanning motion is accomplished by synchronous translation of the exit slit/photodetector unit with rotation of the main frame.

The 6VOPE facility has been operated successfully for over ten years. It has achieved a resolving power of better than 2.5×10^5 at ~79 nm in the seventh order in the photographic mode and better than 1.5×10^5 at ~88 nm in the eighth order in the focal plane scanning mode using a 1200 lines/mm grating [29, 31].

1.2 Plane Grating Monochromators and Spectrometers

Plane grating normal incidence monochromators and spectrometers have been used extensively in the visible and ultraviolet region. By contrast, they are not popular, at present, in the VUV region. However, introduction of high-quality aspheric mirrors and modern gratings such as mechanically ruled VLS gratings and holographic gratings with aspheric wave fronts may give a new life in the VUV region to some of the existing plane grating normal incidence instruments. Note that plane grating grazing incidence monochromators are becoming popular in the soft x-ray region. Therefore, some of the plane grating mounts will be described briefly as follows.

The Czerny–Turner mount [32] is a mount in which a plane grating is illuminated by light collimated with a concave mirror, and the diffracted light is focused into the exit slit by another concave mirror (Fig. 14a). The wavelength scanning is performed by rotation of the grating on its axis parallel to the ruling. When spherical mirrors are used, the coma-type aberration is canceled at one wavelength by an asymmetric layout of the mirrors [33]. Furthermore, the usage of off-axis parabolic mirrors in the monochromator essentially reduces the aberrations over a wide wavelength range. This mount can also be used as a spectrograph. The condition necessary to obtain a nearly flat field is discussed in Ref. 24.

The Ebert–Fastie mount [35, 36] uses one large mirror for collimating and focusing. The off-plane monochromator is used for a high-resolution spectrograph (Fig. 14b). Improved resolution over a wide wavelength can be attained for the in-plane monochromator by the use of the curved entrance and exit slits

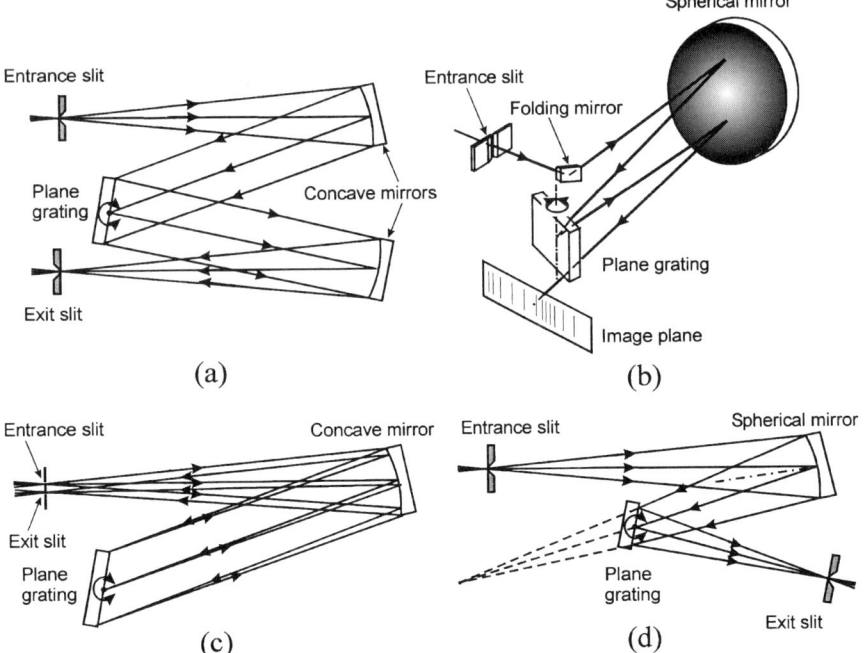

FIG. 14. Plane grating mounts. (a) The Czerny–Turner monochromator, (b) the off-plane Ebert–Fastie spectrograph, (c) the Littrow monochromator, and (d) the Monk–Gillieson monochromator.

lying on the same circle [37, 38]. This mount has small astigmatism and is sometimes used as an imaging spectrograph.

The Littrow mount also uses one concave mirror for both collimating and focusing (Fig. 14c). The entrance and exit slits are placed very closely side by side in the in-plane or on top of each other in the off-plane mount. Therefore, it has both the advantage of compactness and the disadvantage of a less workable space around both slits. The use of an off-axis parabolic mirror is effective to reduce astigmatism and coma-type aberrations [39].

The Monk–Gillieson mount [40, 41] is a mount in which the grating is illuminated with a convergent beam produced by a spherical mirror (Fig. 14d). The diffracted beam is focused into the exit slit directly. The wavelength scanning is performed by rotation of the grating on its axis parallel to the ruling. The drawback of this mount is the relatively large defocus in conjunction with wavelength scanning [42]. However, this drawback is overcome by introducing the scanning system, which is a simple rotation of the grating on an axis parallel to the ruling that is displaced from the grating surface by the proper amount

[43]. Another approach to aberration correction using modern gratings is to modify the grating focal curve so as to minimize the variation of the focal distance. The first experiment had been performed in the visible region using a holographic grating recorded with spherical wave fronts [44], and high resolution grazing incidence spectrometers equipped with mechanically ruled VLS gratings have worked successfully in the soft x-ray region [45]. This subject is described in more detail in Chapter 3.

References

1. M. Seya, *Sci. Light* **2**, 8–17 (1952).
2. T. Namioka, *Sci. Light* **3**, 15–24 (1954).
3. T. Namioka, *J. Opt. Soc. Am.* **49**, 951–961 (1959).
4. H. Noda, T. Namioka, and M. Seya, *J. Opt. Soc. Am.* **64**, 1043–1048 (1974).
5. W. A. Rense and T. Violett, *J. Opt. Soc. Am.* **49**, 139–141 (1959).
6. T. Namioka, *J. Quant. Spectrosc. Radiat. Transfer* **2**, 697–704 (1962).
7. E. Schönheit, *Optik* **23**, 304–312 (1966).
8. S. A. Strezhnev and A. I. Andreeva, *Opt. Spectr.* **28**, 426–428 (1970).
9. T. Harada and T. Kita, *Appl. Opt.* **19**, 3987–3993 (1980).
10. H. Noda, Y. Harada, and M. Koike, *Appl. Opt.* **28**, 4375–4380 (1989).
11. T. Namioka and M. Koike, *Nucl. Instrum. Methods* **A139**, 219–227 (1992).
12. T. Namioka, M. Koike, and D. Content, *Appl. Opt.* **33**, 7261–7274 (1994).
13. M. Itou, T. Harada, T. Kita, K. Hasumi, I. Koyano, and K. Tanaka, *Appl. Opt.* **25**, 2240–2242 (1980).
14. F. L. O. Wadsworth, *Astrophys. J.* **3**, 47–62 (1896).
15. M. Skibowski and W. Steinmann, *J. Opt. Soc. Am.* **57**, 112–113 (1967).
16. A. Eagle, *Proc. Phys. Soc.* **23**, 233 (1911).
17. P. J. Wilkinson, *J. Mol. Spectroscopy* **1**, 288–305 (1957).
18. T. Namioka, *J. Opt. Soc. Am.* **49**, 460–465 (1959).
19. T. Namioka, *J. Opt. Soc. Am.* **49**, 961–965 (1959).
20. B. Vodar, *Rev. Optique* **21**, 97–113 (1942).
21. S. Robin, *J. Phys. Radium.* **14**, 551–552 (1953).
22. J. A. R. Samson, "Techniques of Vacuum Ultraviolet Spectroscopy," p. 64. Pied Publication, Lincoln, 1980.
23. M. Koike, P. A. Heimann, and T. Namioka, *Proc. SPIE, Denver 1996*, **2856**, 300–306 (1996).
24. M. Koike, P. Heimann, A. Kung, T. Namioka, R. DiGennaro, B. Gee, and N. Yu, *Nucl. Instr. Methods* **A347**, 282–286 (1994).
25. P. A. Heimann, M. Koike, C. W. Hsu, M. Evans, C. Y. Ng, D. Blank, X. M. Yang, C. Flaim, A. G. Suits, and Y. T. Lee, *Proc. SPIE, Denver 1996*, **2856**, 90–99 (1996).
26. A. G. Suits, P. Heimann, X. Yang, M. Evans, C. W. Hsu, K. Lu, Y. T. Lee, and A. H. Kung, *Rev. Sci. Instrum.* **66**, 4841–4844 (1995).
27. D. A. Mossessian, P. A. Heimann, E. Gullikson, R. K. Kaza, J. Chin, and J. Akre, *Nucl. Instrum. Methods* **A347**, 244–248 (1994).
28. K. Ito, T. Namioka, Y. Morioka, T. Sasaki, H. Noda, K. Goto, T. Katayama, and M. Koike, *Appl. Opt.* **25**, 837–847 (1986).
29. K. Ito and T. Namioka, *Rev. Sci. Instrum.* **60**, 1573–1578 (1989).

30. T. Namioka, H. Noda, K. Goto, and T. Katayama, *Nucl. Instr. Methods* **208**, 215–222 (1983).
31. K. Ito, K. Maeda, Y. Morioka, and T. Namioka, *Appl. Opt.* **28**, 1813–1817 (1989).
32. M. Czerny and A. F. Turner, *Z. Physik* **61**, 792 (1930).
33. A. B. Shafer, L. R. Megill, and L. Droppleman, *J. Opt. Soc. Am.* **54**, 879–887 (1964).
34. S. A. Khrshanovskii, *Opt. Spectr.* **9**, 207–210 (1960).
35. H. Ebert, *Wiedemann's Annalen* **38**, 489 (1889).
36. W. G. Fastie, *J. Opt. Soc. Am.* **42**, 641–647 (1952).
37. W. G. Fastie, *J. Opt. Soc. Am.* **42**, 647–651 (1952).
38. W. T. Welford, "Aberration theory of gratings and grating mountings," in *Progress in Optics* (E. Wolf, ed.), Vol. IV, pp. 243–279. North-Holland, Amsterdam (1965).
39. H. Yoshinaga, B. Okazaki, and S. Tatsuoka, *J. Opt. Soc. Am.* **50**, 437–445 (1960).
40. G. S. Monk, *J. Opt. Soc. Am.* **17**, 358–364 (1928).
41. A. H. C. P. Gillieson, *J. Sci. Instr.* **26**, 335–339 (1949).
42. T. Namioka and M. Seya, *Appl. Opt.* **9**, 459–464 (1970).
43. T. Kaneko, T. Namioka, and M. Seya, *Appl. Opt.* **10**, 367–381 (1971).
44. H. Nagata, *Ohyobuturi* **47**, 992–996 (1978).
45. M. C. Hettrick, J. H. Underwood, P.J. Batson, and M. J. Eckart, *Appl. Opt.* **27**, 200–202 (1988).

2. GRAZING-INCIDENCE MONOCHROMATORS FOR THIRD-GENERATION SYNCHROTRON RADIATION SOURCES

H. A. Padmore, M. R. Howells, and W. R. McKinney

Advanced Light Source
Lawrence Berkeley National Laboratory
Berkeley, California

2.1 Introduction

In this chapter, we will discuss the foundations of modern vacuum ultraviolet (VUV) and soft x-ray (SXR) monochromator design based on the well-established theories of geometrical aberrations in grazing-incidence optical systems and grating diffraction efficiency. We will restrict our treatment to topics of relevance to the construction of optical systems for third-generation synchrotron radiation sources. These design considerations will be illustrated by examples from the Advanced Light Source (ALS) and will show clearly how monochromator design is evolving, both for undulator and bending magnet sources and for several different applications in spectroscopy and microscopy.

It is clear from the many types of monochromators used today that a variety of solutions are possible within the prevailing optical design rules, and it is important to understand the historical development of these designs to predict where the future will take us and to see what the driving forces are. The evolution of the field is best seen through reviews and the many instrumentation conferences held over the years. In particular, the period of instrumentation for first-generation machines can be seen in the reviews of Haensel and Kunz in 1967 [10] and by Madden in 1974 [24] as well as in the conference proceedings of the International Symposium for Synchrotron Radiation Users, edited by Marr and Munro in 1973 [25], and in the proceedings of the 4th International Conference on Vacuum Ultraviolet Radiation Physics in 1974 [20]. The first international conference on synchrotron radiation instrumentation (SRI) in 1977, chaired by Wuilleumier and Farge [59], and the first U.S. national conference on synchrotron radiation, chaired by Ederer and West in 1979 [7], heralded the era of the second-generation machines. The international conferences on SRI in Hamburg (1982) [51], Stanford (1985) [52], Tsukuba (1988) [53], and Chester (1991) [54] cover the period of operation of the second-generation machines, whereas the operation of the third-generation machines began roughly with the Stony Brook SRI meeting in 1994. During this period, all the monochromator designs used today emerged: the Petersen plane-grating monochromator (PGM)

[38–40] and its derivatives, the spherical-grating monochromator (SGM) [5, 6, 14, 31] and the variable-angle SGM [32, 36, 37]. In Sections 2.3, 2.4, and 2.6 these designs are discussed, particularly in relation to how considerations of focusing drive the efficiency of the optical system. The reviews of optical systems by Gudat and Kunz [8], Johnson [19], Howells [15], West and Padmore [57], Peatman and Senf [37], and Padmore and Warwick [34] cover the operational periods of the first- to the third-generation machines.

The constraints on optical system designs go well beyond those traditionally imposed by strict considerations of resolving power. Although still important, the fact that modern spherical-grating monochromators can routinely achieve a resolving power of 10^4 (one at the ALS has achieved $>6.5 \times 10^4$), means that this aspect of the design is well understood and works as expected within limits set by optical manufacturing. However, other aspects are much less well documented. We would highlight three main areas:

Optical matching. The high collimation of undulator radiation and the small source size of bending magnets on third-generation machines offer important opportunities in optical design for increasing light collection and efficiency. This is especially significant for beam lines where some form of microscopy is to be practiced at the endstation. Incorrect matching of the microscope optics to the source can lead to an enormous loss of flux.

Fexibility. The high cost of undulators and their associated high-power beamlines initially led to beamline designs in which the experimental needs of rather different communities—for example, high-resolution spectroscopists and high-spatial-resolution microscopists—were accommodated in a single instrument. This was by necessity a compromise, and it is now clear that beamlines with a single function can be designed with better performance and at less cost.

Applications. New applications of synchrotron radiation are changing the way that we must approach beamline design. In some experiments, access to an increased energy range is important. For example, in solid-state photoemission it is sometimes desirable to go to a low photon energy (<30 eV) to perform band mapping studies while being able to go to high photon energies (>1500 eV) to measure core-level shifts. In addition, it is often important to tune rapidly between the low- and high-energy regions, while still preserving precise energy calibration. Although significant advances have been made in this area, several challenges remain. A new area of application is that of x-ray microscopy to problems in materials science. A particular thrust of work at the ALS is in microscope systems directed toward the needs of the local microelectronics industry. These include microfocused x-ray photoelectron spectroscopy (micro-XPS) and x-ray absorption-based photoelectron emission microscopy (X-PEEM). These have very different optical design parameters from spectroscopic beamlines. Moreover, the competitive environment of analytical instrumentation drives many design decisions based on "return on investment." The need is for systems that are dedicated full time to analytical measurements, functioning rather like a standard instrument operating with a laboratory radiation source. This means that the performance, cost, and sample throughput capability must be assessed by comparison to commercial instruments.

In the examples discussed later, we describe an SGM undulator beamline and two "application-specific" bending magnet beamlines for micro-XPS and X-PEEM. We first describe aberration theory of grazing-incidence mirrors and gratings and apply this in detail to the design of an ALS undulator beamline for ultra-high-resolution spectroscopy. We then present a guide for optimizing diffraction efficiency, discuss the special requirements of the various forms of x-ray microscopy, and illustrate the foregoing principles with examples from the ALS.

2.2 Grating Theory

The first type of focusing grating to be analyzed theoretically was that formed by the intersection of a substrate surface with a set of parallel equi-spaced planes: the so-called Rowland grating. The full optical theory of such gratings was in place before 1967 and was reviewed in the first edition of this book. The theory of spherical-grating systems was established first [23, 45, 46], and was described comprehensively in the 1945 paper of Beutler [2]. Treatments of toroidal [9] and ellipsoidal [27] gratings came later, and the field was reviewed in 1965 by Welford [55] and in 1985 by Hunter [17].

The major developments since 1967 have been in the use of nonuniformly spaced grooves. The application of holography to spectroscopic gratings was first reported by Rudolph and Schmahl [47, 48] and by Labeyrie and Flamand [21]. Its unique opportunities for optical design were developed initially by Jobin-Yvon [42] and by Namioka and coworkers [29, 30]. A different approach was followed by Harada [11] and others, who developed the capability to produce gratings with variable-line spacing through the use of a computer-controlled ruling engine. The application of this class of gratings to spectroscopy has been developed still more recently, principally by Hettrick [13].

In this section, we will give a treatment of grating theory up to the sixth order in the optical path, which is applicable to any substrate shape and any groove pattern that can be produced by holography or by ruling straight grooves with (possibly) variable spacing.

2.2.1 Calculation of the Path Function for a Rowland Grating

Following normal practice, we will analyze the imaging properties of gratings by means of the path function F. The most comprehensive account of this method is given in the paper by Noda and coauthors [30]. We begin, without knowing anything about where the rays will go, by making a purely geometrical calculation of the path length ⟨AP⟩ + ⟨PB⟩ from any point A(x, y, z) to any point B(x', y', z') via a variable point P(ξ, w, l) on the grating surface. We suppose that the zeroth groove (of width d_0) passes through O (the grating pole) while the nth groove passes through P. The overall notation, which is roughly that of Noda and colleagues [30], is explained in Fig. 1. Since we are interested in a *diffracted* beam in mth order, we include the term $mn\lambda$ in the path function so that F changes by an additional m waves for each groove moved by P when the position of P is allowed to vary [1]. The sign conventions we use are similar to those used in the first edition of this book. That is, λ, d_0, and α are positive, and α and β are of opposite sign if they are on opposite sides of the normal. Inside order is considered positive, and the directions of the ingoing ray and of increasing n are both toward $+y$.

We consider first the case of a Rowland grating since it will be simple to extend the treatment of this case to cover the other interesting ones. For a Rowland grating, F is given by

$$F = \langle AP \rangle + \langle PB \rangle + mn\lambda = \langle AP \rangle + \langle PB \rangle + \frac{m\lambda}{d_0} w \qquad (1)$$

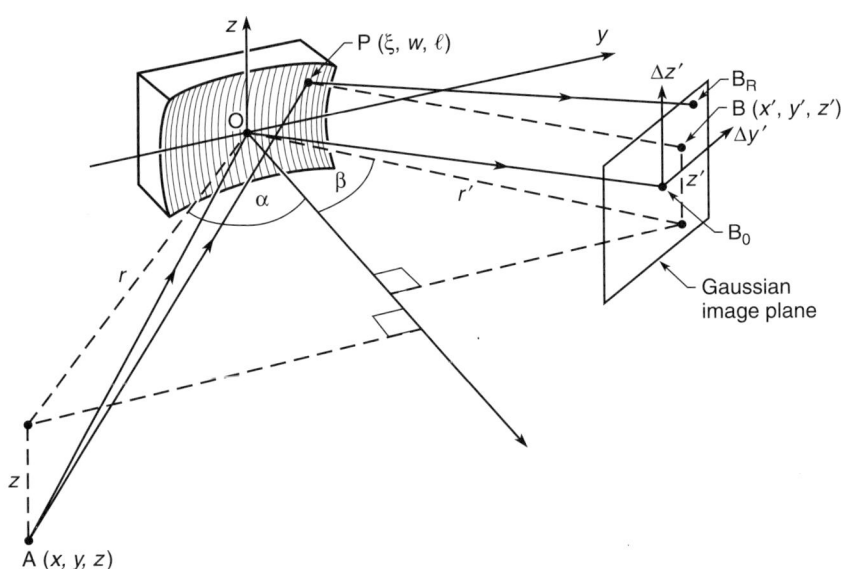

FIG. 1. Geometry and notation for grating theory. The axes and the rays are represented by solid lines and other distances by dashed lines. B is a general point, B_0 the Gaussian image point, and B_R the arrival point of the ray.

where

$$\langle AP \rangle = \sqrt{(x - \xi)^2 + (y - w)^2 + (z - l)^2} \qquad (2)$$

and $\langle PB \rangle$ equals a similar expression with x', y', and z'. We now substitute for x, y, and ξ (see Fig. 1) in Eq. (2) according to

$$x = r \cos \alpha, \qquad y = r \sin \alpha, \qquad \xi = \sum_{ij} a_{ij} w^i l^j \qquad (3)$$

where r and α are constants and the summation replacing ξ expresses the shape of the grating surface. We then expand $\langle AP \rangle$ as a Maclaurin series in w, l, and z,

$$\langle AP \rangle = \sum_{n=0}^{\infty} \frac{1}{n!} \left\{ w \frac{\partial}{\partial w} + l \frac{\partial}{\partial l} + z \frac{\partial}{\partial z} \right\}^n \langle AP \rangle \bigg|_{0,0,0} \equiv \sum_{ijk} C_{ijk} w^i l^j z^k \qquad (4)$$

and use algebraic software (Mathematica™) [58] to compute the coefficients C_{ijk}, which are functions of α, r, and the a_{ij}'s. Applying the same method to $\langle PB \rangle$ as well, Eq. (1) now becomes

$$F = \sum_{ijk} C_{ijk}(\alpha, r) w^i l^j z^k + \sum_{ijk} C_{ijk}(\beta, r') w^i l^j z'^k + \frac{m\lambda}{d_0} w \qquad (5)$$

which can be written

$$F = \sum_{ijk} F_{ijk} w^i l^j \qquad (6)$$

where

$$F_{ijk} = z^k C_{ijk}(\alpha, r) + z'^k C_{ijk}(\beta, r')$$

except

$$F_{100} = C_{100}(\alpha, r) + C_{100}(\beta, r') + \frac{m\lambda}{d_0}.$$

The coefficient F_{ijk} is related to the strength of the i,j aberration of the wavefront diffracted by the grating. The coefficients C_{ijk} are given up to the sixth order in Table I where the following notation is used:

$$T = T(r, \alpha) = \frac{\cos^2 \alpha}{r} - 2a_{20} \cos \alpha \qquad S = S(r, \alpha) = \frac{1}{r} - 2a_{02} \cos \alpha \qquad (7)$$

$$T' = T(r', \beta) \qquad\qquad S' = S(r', \beta)$$

To use the C_{ijk}'s, one must also have the a_{ij}'s, and these are given to the sixth order for ellipses and toroids [43] in Tables II and III, respectively. The a_{ij}'s for spheres, circular cylinders, paraboloids, and hyperboloids can also be obtained from Tables II and III by suitable choices of the parameters r, r', and θ.

TABLE I. Coefficients C_{ijk} of the expansion of F [Eq. (4) et seq.][a]

$$C_{011} = -\frac{1}{r}$$

$$C_{020} = \frac{S}{2}$$

$$C_{022} = -\frac{S}{4r^2} - \frac{1}{2r^3}$$

$$C_{031} = \frac{S}{2r^2}$$

$$C_{040} = \frac{4a_{02}^2 - S^2}{8r} - a_{04}\cos\alpha$$

$$C_{042} = \frac{a_{04}\cos\alpha}{2r^2} + \frac{3S^2 - 4a_{02}^2}{16r^3} + \frac{3S}{4r^4}$$

$$C_{100} = -\sin\alpha$$

$$C_{102} = \frac{\sin\alpha}{2r^2}$$

$$C_{111} = -\frac{\sin\alpha}{r^2}$$

$$C_{120} = \frac{S\sin\alpha}{2r} - a_{12}\cos\alpha$$

$$C_{131} = -\frac{a_{12}\cos\alpha}{r^2} + \frac{3S\sin\alpha}{2r^3}$$

$$C_{122} = \frac{a_{12}\cos\alpha}{2r^2} - \frac{3S\cos\alpha}{4r^3} - \frac{3\sin\alpha}{2r^4}$$

$$C_{013} = \frac{1}{2r^3}$$

$$C_{200} = \frac{T}{2}$$

$$C_{202} = -\frac{T}{4r^2} + \frac{\sin^2\alpha}{2r^3}$$

$$C_{211} = \frac{T}{2r^2} - \frac{\sin^2\alpha}{r^3}$$

$$C_{300} = -a_{30}\cos\alpha + \frac{T\sin\alpha}{2r}$$

$$C_{140} = -a_{14}\cos\alpha + \frac{1}{2r}(2a_{02}a_{12} + a_{12}S\cos\alpha - a_{04}\sin 2\alpha) + \frac{\sin\alpha}{8r^2}(4a_{02}^2 - 3S^2)$$

$$C_{220} = -a_{22}\cos\alpha + \frac{1}{4r}(4a_{20}a_{02} - TS - 2a_{12}\sin 2\alpha) + \frac{S\sin^2\alpha}{2r^2}$$

$$C_{222} = \frac{1}{2r^2}a_{22}\cos\alpha + \frac{1}{8r^3}(3ST - 4a_{02}a_{20} + 6a_{12}\sin 2\alpha) + \frac{3}{4r^4}(T - 2S\sin^2\alpha)$$
$$- \frac{3\sin^2\alpha}{r^5}$$

$$C_{231} = -\frac{1}{r^2}a_{22}\cos\alpha + \frac{1}{4r^3}(-3ST + 4a_{02}a_{20} - 6a_{12}\sin 2\alpha) + \frac{3S\sin^2\alpha}{r^4}$$

$$C_{240} = -a_{24}\cos\alpha$$
$$+ \frac{1}{2r}(a_{12}^2\sin^2\alpha + 2a_{04}a_{20} + a_{22}S\cos\alpha + a_{04}T\cos\alpha - a_{14}\sin 2\alpha + 2a_{02}a_{22})$$
$$+ \frac{1}{16r^2}(-4a_{02}^2T - 8a_{02}a_{20}S + 12a_{12}S\sin 2\alpha + 3TS^2 + 16a_{02}a_{12}\sin\alpha$$
$$- 8a_{04}\sin 2\alpha) + \frac{\sin^2\alpha}{4r^3}(2a_{02}^2 - 3S^2)$$

continues

TABLE I. Continued

$$C_{302} = \frac{a_{30}\cos\alpha}{2r^2} - \frac{3T\sin\alpha}{4r^3} + \frac{\sin^3\alpha}{2r^4}$$

$$C_{311} = -\frac{a_{30}\cos\alpha}{r^2} + \frac{3T\sin\alpha}{2r^3} - \frac{\sin^3\alpha}{r^4}$$

$$C_{320} = -a_{32}\cos\alpha + \frac{1}{2r}(2a_{20}a_{12} + 2a_{30}a_{02} + a_{30}S\cos\alpha + a_{12}T\cos\alpha - a_{22}\sin 2\alpha)$$

$$+ \frac{1}{4r^2}(4a_{20}a_{02}\sin\alpha - 3ST\sin\alpha - 4a_{12}\cos\alpha\sin^2\alpha) + \frac{S\sin^3\alpha}{2r^3}$$

$$C_{400} = -a_{40}\cos\alpha + \frac{1}{8r}(4a_{20}^2 - T^2 - 4a_{30}\sin 2\alpha) + \frac{T\sin^2\alpha}{2r^2}$$

$$C_{402} = -\frac{1}{16r^3}(4a_{20}^2 + 3T^2 + 12a_{30}\sin 2\alpha) + \frac{a_{40}\cos\alpha}{2r^2} - \frac{3T\sin^2\alpha}{2r^4} + \frac{\sin^4\alpha}{2r^5}$$

$$C_{411} = -\frac{a_{40}\cos\alpha}{r^2} + \frac{1}{8r^3}(4a_{20}^2 - 3T^2 - 12a_{30}\sin 2\alpha) + \frac{3T\sin^2\alpha}{r^4} - \frac{\sin^4\alpha}{r^5}$$

$$C_{420} = -a_{42}\cos\alpha + \frac{1}{2r}(2a_{20}a_{22} + 2a_{12}a_{30}\sin^2\alpha + 2a_{02}a_{40} - a_{32}\sin 2\alpha + a_{40}S\cos\alpha$$

$$+ a_{22}T\cos\alpha) + \frac{1}{16r^2}(-4a_{20}^2S - 8a_{02}a_{20}T + 3ST^2 + 12\sin 2\alpha[a_{30}S + a_{12}T]$$

$$+ 8\sin\alpha[2a_{02}a_{30} - 2a_{22}\sin 2\alpha + 2a_{12}a_{20}])$$

$$+ \frac{1}{2r^3}(2a_{02}a_{20}\sin^2\alpha - 3ST\sin^2\alpha - 2a_{12}\cos\alpha\sin^3\alpha) + \frac{S\sin^4\alpha}{2r^4}$$

$$C_{500} = -a_{50}\cos\alpha + \frac{1}{2r}(2a_{20}a_{30} + a_{30}T\cos\alpha - a_{40}\sin 2\alpha) + \frac{\sin\alpha}{2r^2}(a_{20}^2 - a_{30}\sin 2\alpha)$$

$$- \frac{3T^2\sin\alpha}{8r^2} + \frac{T\sin^3\alpha}{2r^3}$$

$$C_{600} = -a_{60}\cos\alpha + \frac{1}{2r}(a_{30}^2\sin^2\alpha + 2a_{20}a_{40} + a_{40}T\cos\alpha - a_{50}\sin 2\alpha)$$

$$+ \frac{1}{16r^2}(-4a_{20}^2T + T^3 + 16a_{20}a_{30}\sin\alpha + 12a_{30}T\sin 2\alpha - 16a_{40}\cos\alpha\sin^2\alpha)$$

$$+ \frac{1}{4r^3}(2a_{20}^2\sin^2\alpha - 3T^2\sin^2\alpha - 4a_{30}\cos\alpha\sin^3\alpha) + \frac{T\sin^4\alpha}{2r^4}$$

[a] The coefficients for which $i \leq 6$, $j \leq 4$, $k \leq 2$, $i + j + k \leq 6$, $j + k =$ even are included in this table. The only addition to those is C_{013}, which has some interest, because when the system is specialized to be symmetrical around the x axis, it represents a Seidel aberration, namely distortion.

TABLE II. Ellipse coefficients Q_{ij} from which the a_{ij}'s of Eq. (3) are obtained[a] [43]

i \ j	0	1	2	3	4	5	6
0	0	0	1	0	$C/4$	0	$C^2/8$
1	0	0	A	0	$3AC/4$	0	—
2	1	0	$(2A^2 + C)/2$	0	$3C(4A^2 + C)/8$	0	—
3	A	0	$A(2A^2 + 3C)/2$	0	—	0	—
4	$(4A^2 + C)/4$	0	$(8A^4 + 24A^2C + 3C^2)/8$	0	—	0	—
5	$A(4A^2 + 3C)/4$	0	—	0	—	0	—
6	$(8A^4 + 12A^2C + C^2)/8$	0	—	0	—	0	—

[a] If r, r', and θ are the object distance, image distance, and incidence angle to the normal, respectively, and

$$a_{20} = \frac{\cos\theta}{4}\left(\frac{1}{r} + \frac{1}{r'}\right), \quad A = \frac{\sin\theta}{2}\left(\frac{1}{r} - \frac{1}{r'}\right), \quad C = A^2 + \frac{1}{rr'},$$

then

$$a_{ij} = a_{20}\frac{Q_{ij}}{\cos^j\theta}.$$

2.2.2 Calculation of the Path Function for Non-Rowland Gratings

Next we calculate F for a grating with the groove pattern formed by holography using two coherent point sources, $C(r_C, \gamma, z_C)$ and $D(r_D, \delta, z_D)$. If the sources C and D are both real or both virtual, their equi-phase surfaces are a family of confocal hyperboloids of revolution around CD. If one is real and the other virtual, then the equi-phase surfaces are the corresponding family of ellipsoids. This follows from the fact that interference fringes always bisect the angle between the *forward directions* of the rays forming them. Given that the

TABLE III. Toroid a_{ij}'s of Eq. (3)[a] [43]

i \ j	0	1	2	3	4	5	6
0	0	0	$1/(2\rho)$	0	$1/(8R^3)$	0	$1/(16\rho^5)$
1	0	0	0	0	0	0	—
2	$1/(2R)$	0	$1/(4\rho R^2)$	0	$(2\rho + R)/(16\rho^3 R^3)$	0	—
3	0	0	0	0	—	0	—
4	$1/(8R^3)$	0	$3/(16\rho R^4)$	0	—	0	—
5	0	0	—	0	—	0	—
6	$1/(16R^5)$	0	—	0	—	0	—

[a] R and ρ are the major and minor radii of the toroid.

TABLE IV. Coefficients n_{ijk} for a grating with variable line spacing [Eq. (12)]

$n_{100} = 1/d_0$	$n_{400} = (-v_1^3 + 2v_1 v_2 - v_3)/4d_0$
$n_{200} = -v_1/2d_0$	$n_{500} = (v_1^4 - 3v_1^2 v_2 + v_2^2 + 2v_1 v_3 - v_4)/5d_0$
$n_{300} = (v_1^2 - v_2)/3d_0$	$n_{600} = (-v_1^5 + 4v_1^3 v_2 - 3v_1 v_2^2 - 3v_1^2 v_3 + 2v_2 v_3 + 2v_1 v_4 - v_5)/6d_0$

recording process, at wavelength λ_0, delivers n fringes between O and P, we can write

$$n\lambda_0 = (\langle CP \rangle \pm \langle PD \rangle) - (\langle CO \rangle \pm \langle OD \rangle) \tag{8}$$

where the upper sign is for the ellipsoid case and the lower one for the hyperboloid. We can calculate $\langle CP \rangle$ and $\langle PD \rangle$ in the same way as $\langle AP \rangle$ and express them in the same way. Therefore, inserting n from Eq. (8) into Eq. (1) and dropping the second bracket of Eq. (8), which is just a constant, we obtain

$$F_{ijk} = z^k C_{ijk}(\alpha, r) + z'^k C_{ijk}(\beta, r') + \frac{m\lambda}{\lambda_0} \{z_C^k C_{ijk}(\gamma, r_C) \pm z_D^k C_{ijk}(\delta, r_D)\}. \tag{9}$$

Finally, we calculate F for a ruled grating with straight, parallel grooves that may have a variable spacing [26].

$$d(w) = d_0(1 + v_1 w + v_2 w^2 + \cdots). \tag{10}$$

In this case, the calculation proceeds as for a Rowland grating up to Eq. (5), but now that n and d are functions of w (although not of l and z), the term $n(w)m\lambda$ must be expanded as a power series and will contribute, in principle, to all the F_{ijk}'s for which $j = k = 0$. Therefore, recognizing that the local groove frequency is $1/d(w) = \partial n/\partial w$ and that $n(0) = 0$, we have

$$n(w) = \sum_{i=1}^{\infty} n_{i00} w^i = \sum_{i=1}^{\infty} \frac{1}{i!} \left.\frac{\partial^i n}{\partial w^i}\right|_0 w^i$$

$$= \sum_{i=1}^{\infty} \frac{1}{i!} \left.\frac{\partial^{i-1}}{\partial w^{i-1}} \left(\frac{1}{d_0(1 + v_1 w + v_2 w^2 + \cdots)}\right)\right|_0 w^i. \tag{11}$$

Evaluating the derivatives, we find

$$F_{ijk} = z^k C_{ijk}(\alpha, r) + z'^k C_{ijk}(\beta, r') + n_{ijk} m\lambda \tag{12}$$

where $n_{ijk} = 0$ ($j, k \neq 0$) and the n_{i00} are given up to sixth order in Table IV.

2.2.3 Location of the Gaussian Image Point

So far we have been making a purely geometrical calculation of the path F for two *arbitrary* points A and B. We now proceed to apply Fermat's principle to determine the actual direction of the outgoing beam. First consider the incoming principal ray AO. If this ray is to follow the path AOB_0 to the Gaussian image point $B_0 = (r'_0, \beta_0, z'_0)$, then Fermat's principle requires

$$\left.\frac{\partial F}{\partial w}\right|_{\substack{w=0\\l=0}} = 0, \qquad \left.\frac{\partial F}{\partial l}\right|_{\substack{w=0\\l=0}} = 0. \tag{13}$$

This effectively sets the coefficients of the linear terms (F_{100} and F_{011}) equal to zero, which implies

$$\frac{m\lambda}{d_0} = \sin\alpha + \sin\beta_0, \qquad \frac{z}{r} + \frac{z'_0}{r'_0} = 0 \tag{14}$$

giving the grating equation and the law of magnification in the sagittal direction and thence the direction of the outgoing principal ray. To find the tangential focal distance r'_0, we set the focusing term F_{200} equal to zero.

$$T(r, \alpha) + T(r'_0, \beta_0) = 0 \qquad \text{(Rowland)} \tag{15}$$

$$T(r, \alpha) + T(r'_0, \beta_0) + \frac{m\lambda}{\lambda_0}\{T(r_C, \gamma) \pm T(r_D, \delta)\} = 0 \qquad \text{(holographic)} \tag{16}$$

$$T(r, \alpha) + T(r'_0, \beta_0) - \frac{v_1 m\lambda}{d_0} = 0 \qquad \text{(variable line spacing)} \tag{17}$$

Equations (14)–(17) determine the Gaussian image point $B_0 = (r'_0, \beta_0, z'_0)$, and in combination with the sagittal focusing condition ($F_{020} = 0$), describe the focusing properties of grating systems under the paraxial approximation.

2.2.4 Calculation of Ray Aberrations

In an aberrated system, the outgoing ray will arrive at the Gaussian image plane at a point B_R displaced from B_0 by the ray aberrations $\Delta y'$ and $\Delta z'$ (Fig. 1) which we would like to calculate [28]. To do so, consider a Gaussian reference sphere with center B_0 and radius $\mathcal{R}$ ($< r'_0$), chosen so that the rays OB_0 and PB_R and the line PB_0 intersect the sphere at real points O', Q, and Q', respectively (Fig. 2). Let the tangent plane to the sphere at O' be the exit pupil plane with pupil coordinates (Y, Z) parallel and perpendicular to the principal plane, respectively. The ray aberrations may then be found from Eq. (10) of paragraph 5.1 of the book by Born and Wolf [3].

$$\Delta y' = \mathcal{R}\frac{\partial \Phi}{\partial Y}, \qquad \Delta z' = \mathcal{R}\frac{\partial \Phi}{\partial Z} \tag{18}$$

In this equation, Φ is the "wave aberration function" [3], which is defined as the ray path length between the actual wave front and the Gaussian reference sphere

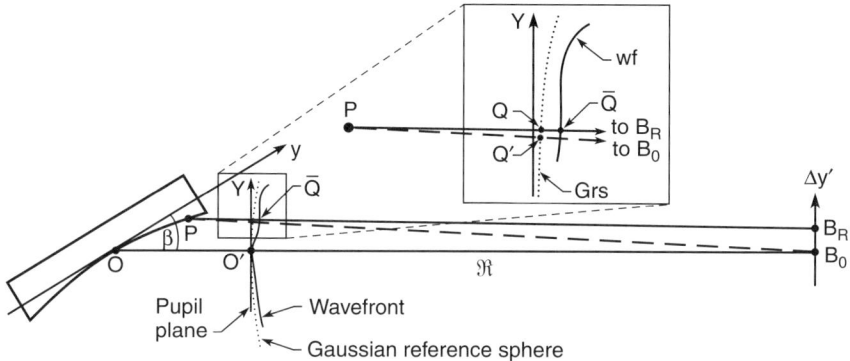

FIG. 2. Geometry for calculation of the wave front aberration function and the definition of the points Q, Q', and $\bar{Q}$. PB_R represents the ray and B_0 the Gaussian image point.

when they instantaneously coincide at O'. We can calculate Φ as follows:

$$\Phi = APQ - AP\bar{Q}$$
$$\cong APQ' - AP\bar{Q}$$
$$= APB_0 - \mathcal{R} - AP\bar{Q} \qquad (19)$$

where $\bar{Q}$ is the point where the ray PB_R intersects the wave front. Now $AP\bar{Q}$ is a constant (by the definition of a wave front), $\mathcal{R}$ is a constant, and we have approximated the ray length PQ with the known length PQ' measured along the line PB_0 (Eq. 7.8 of Welford 1974 [56]). We also know that when the position of $P(\xi, w, l)$ is allowed to vary, the increments in the coordinates (Y, Z) of Q' are given by $dY = dw \cos\beta_0(\mathcal{R}/r'_0)$ and $dZ = dl(\mathcal{R}/r'_0)$. Therefore, the derivatives of Φ with respect to the coordinates of Q' in Eq. (18) can be replaced by derivatives of F with respect to the coordinates of P as follows:

$$\Delta y' = \frac{r'_0}{\cos\beta_0} \frac{\partial F}{\partial w}, \qquad \Delta z' = r'_0 \frac{\partial F}{\partial l} \qquad (20)$$

where F is to be evaluated for $A = (r, \alpha, z)$, $B = (r'_0, \beta_0, z'_0)$. From this development, we can see that the function F is related to the characteristic function (V) of Hamilton [3, 4, 56] but is not identical to it because specification of V requires a knowledge of the ray path whereas specification of F does not.

The importance of Eq. (20) is that, by means of the expansion of F, it allows the ray aberrations to be calculated separately for each aberration type. This is very useful for optical design and for understanding the results of exact ray tracing. Thus we have the equations

$$\Delta y'_{ijk} = \frac{r'_0}{\cos\beta_0} \frac{\partial}{\partial w} \{F_{ijk} w^i l^j\}, \qquad \Delta z'_{ijk} = r'_0 \frac{\partial}{\partial l} \{F_{ijk} w^i l^j\} \qquad (21)$$

Since the coefficients F_{ijk} are independent of w and l, the only way that an aberration can vanish when P is not at the origin is for its F_{ijk} to vanish. This is the justification for our use of $F_{200} = 0$ to determine the focal distance in Eqs. (16)–(18). Moreover, provided that the aberrations are not too large, they are additive,

$$\Delta y' = \sum_{ijk} \Delta y'_{ijk}, \qquad \Delta z' = \sum_{ijk} \Delta z'_{ijk}, \qquad (22)$$

implying that aberrations can either add or cancel.

2.2.5 The Astigmatic Curvature of Focal Lines

As an illustration of some of the principles described in the preceding paragraphs, we calculate the largest resolution-determining aberration of the grazing-incidence toroid. Such a surface has steep sagittal curvature, which typically leads to a strong curvature $\Delta y' = k\Delta z'^2$ of the tangential focal line. Such line curvature, termed "astigmatic curvature" by Beutler [2], is an important consideration in the design of toroidal condensing mirrors and in determining the resolution of toroidal grating monochromators (TGMs). To understand it, we note first that from Eq. (21) the astigmatism is given by

$$\Delta z'_{020} = r' \frac{\partial}{\partial l} \{F_{020} l^2\} = r' l (S + S'), \qquad (23)$$

so we need those terms that give a $\Delta y'$ proportional to $\Delta z'^2$ after we have taken the derivative with respect to w and substituted for l from Eq. (23). There are three such terms [55]

$$\Delta y'_{lc} = \frac{r'}{\cos\alpha} \frac{\partial}{\partial w} \left(\frac{1}{2} w l^2 F_{120} + w l F_{111} + w F_{102} \right) \qquad (24)$$

all of which are nonzero even though $z = 0$. Making the substitutions for l and the F's and noting that, for the case at hand, $a_{12} = 0$ and $\Delta z' = z'$, we get

$$\Delta y'_{lc} = \frac{\Delta z'^2}{2r' \cos\beta (S + S')^2}$$

$$\times \left[\frac{S \sin\alpha}{r} + \frac{S' \sin\beta}{r'} - \frac{2(S + S') \sin\beta}{r'} + (S + S')^2 \sin\beta \right] \qquad (25)$$

This equation can be applied to mirrors ($\alpha = -\beta$), and it shows [16] that the line curvature vanishes if the imaging is stigmatic with unity magnification or if the tangential and sagittal magnifications are related by $M_s = 2M_t/(1 + M_t)$. For a grating system, Eq. (25) allows an estimate of the blurring of the spectral resolution resulting from curvature of the focal line. If the toroid is close to stigmatic, that is, z and $z' \approx 0$, the F_{111} and F_{102} terms in Eq. (24) vanish and

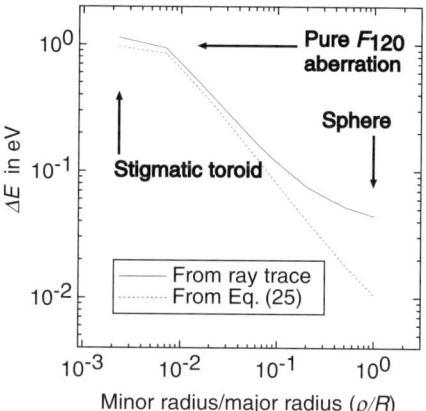

FIG. 3. Relation of the energy resolution of a soft x-ray toroidal grating monochromator to the minor radius of the grating and thus to the curvature of the focal line. Parameters: included angle = 174°, d_0^{-1} = 1100/mm, λ = 20 Å, R = 55 m, arm lengths: set to satisfy the Rowland condition.

$\Delta y'_{lc}$ is determined mostly by the F_{120} aberration, which is large and dominates the spread function. If we now evaluate $\Delta y'_{lc}$ [14] as the sagittal curvature of the toroid is gradually reduced so as to progress from a stigmatic toroid toward a sphere, we see that the resolution broadening caused by line curvature diminishes because of increasing cancellation among the three terms. By the time the toroid becomes a sphere, the line curvature contribution to the resolution becomes negligible for long-radius soft-x-ray systems and much reduced for shorter-radius VUV systems. This process is illustrated for a soft x-ray TGM/SGM in Fig. 3. As shown in the figure, these conclusions are confirmed by ray tracing, and they provide an explanation for the superiority of the resolution of monochromators with long-radius spherical gratings [50] compared with their toroidal predecessors.

2.3 Application of Aberration Theory to the Design of an Undulator-Based SGM

2.3.1 General Design Principles

To illustrate the design concepts described in the previous section, we have chosen to examine in detail the design adopted for an existing undulator beamline at the ALS. This beamline, 9.0.1, has a 4.5-m-long, 10-cm-period undulator source; a Kirkpatrick–Baez condenser system, focused on the entrance slit in the vertical plane and the sample in the horizontal plane; and a spherical-grating monochromator. This monochromator system [5, 6, 14, 31] was first demonstrated by Chen and Sette in 1986 and has since become one of the standard types. The basic system has a fixed entrance slit, a spherical grating with a fixed rotation axis, and an exit slit. The grating-to-exit-slit distance can be varied to achieve focus, but the included angle (2θ) at the grating is fixed. The beamline 9.0.1 instrument uses three interchangeable gratings to cover the energy range from 20–300 eV [12]. It is used for atomic and molecular spectroscopy and was designed so that high resolution could be achieved while still maintaining good throughput. The system has the following basic parameters:

Entrance arm length, r	1.45 m
Exit arm length, r'	4.02–4.78 m
Included angle, 2θ	165°
Grating groove density, $1/d_0$	380, 925, 2100/mm
Diffraction order, m	+1
Grating radius, R	21 m
Vertical demagnification, source-to-slit, M_V	8.04 : 1
Horizontal demagnification, source-to-sample, M_H	1 : 1

We will now examine how these parameters were derived, using the value of the minimum wavelength ($\lambda_{\min}$) as a starting point. The first choice to be made is that of the overall size of the monochromator. This should be as large as space allows for maximum phase-space acceptance of the instrument. In the case at hand, the length from entrance to exit slit was set at 6 m.

The included angle of the monochromator is determined by requiring adequate reflectivity at angle θ (not α) at the highest intended photon energy. In the present case, using a nickel coating for the lowest-wavelength grating, reasonable reflectivity (~45% at 300 eV) is obtained at an included angle of 165°.

Given the slit-to-slit length (AB_0) of the instrument (see Fig. 1) and the complement θ_G of θ, we can make a good estimate of the grating radius R, which plays the role of an overall scale factor for the system. From a diagram of the Rowland circle and the principal ray AOB_0, we find $AB_0 \approx 2\theta_G R$. To make further progress we use the grating equation [Eq. (14)] and by taking derivatives we can obtain the following useful quantities: the reciprocal linear dispersion $d\lambda/d(\Delta y')$, the entrance- and exit-slit-width-limited resolutions, and the horizon wavelength λ_H (at which either α or β equals 90°):

$$\frac{d\lambda}{d(\Delta y')} = \frac{d_0 \cos \beta}{mr'} \qquad \Delta\lambda|_{\text{ent}} = \frac{s d_0 \cos \alpha}{mr} \qquad (26)$$

$$\Delta\lambda|_{\text{exit}} = \frac{s' d_0 \cos \beta}{mr'} \qquad \lambda_H = 2d_0 \cos^2 \theta,$$

where s and s' are the entrance and exit slit widths.

To analyze the focusing conditions, we first consider the Rowland solution, in which the two terms of Eq. (15) are set separately equal to zero, leading to $r = R \cos \alpha$, $r' = R \cos \beta$. We will not generally be using this solution in the case of an SGM because r' will be constantly changing to achieve focus. However, it can be shown that the value of r' needed for focus has a stationary point that occurs approximately at the Rowland wavelength (λ_R), and therefore the amount of slit travel can be minimized by setting λ_R at the center of the wavelength range. Adopting this condition then allows us to determine values for r and r' at λ_R. However, it turns out that, even with the help of the stationary point, we can still only build the r' motion to cover about a factor of 2.5 in wavelength. This fact completes our definition of the wavelength range and the entire monochromator geometry for the lowest-wavelength grating except for the groove density to which we now turn.

The choice of groove density depends on the resolution requirement, and there are two main possibilities:

(i) the best possible resolution capability at the minimum practical slit widths
(ii) the maximum flux at some particular value of the resolution

The latter requirement applies especially to monochromators dedicated to a single application and is similar to the needs of microscopy beamlines, which are discussed in a later section. For the best resolution, the groove density should be maximized, which means that it will be limited by the horizon effect according to

$$3.5 \lambda_{min} \approx 0.8 \lambda_H = 1.6 d_0 \cos^2 \theta \qquad (27)$$

The maximum flux (i.e., the maximum grating efficiency) requires the *minimum* groove density as discussed in Section 2.5. The phase-space acceptance of an SGM is given by $N \Delta \lambda |_{ent}$ [14] where N is the number of illuminated grooves. Thus, given a grating of the largest practical size, reducing the groove density at fixed resolution amounts to a reduction of the phase-space acceptance. This will still be advantageous (for flux) until the usable beam emittance is matched. Thus, the groove density should be chosen to achieve a phase-space match or to approach a match as nearly as possible with an available grating.

Having thus obtained a candidate design for the lowest wavelength grating we can obtain a design for the next one by scaling up the groove width by a factor 2.5 so that $m\lambda/d_0$ will be unchanged. All the design parameters will then be the same as for the lowest-wavelength grating as will also the resolving power. According to the 2.5× multiplication rule, our present example would then have three wavelength ranges 40–100, 100–250, and 250–625 Å.

2.3.2 Calculation of the Effect of Grating Aberrations on Energy Resolution

We have shown in Section 2.2.4 how individual terms of the optical path function can be associated with aberrations that scale with powers of the ray coordinates on the grating, w and l. In addition, Eq. (21) shows how these terms can be related directly to the ray aberrations at an image plane. Combining this with the dispersion given by Eq. (26), we can calculate the effect in terms of a deviation (ΔE) from the nominal photon energy registered by the principal ray. The most important wavelength aberrations in SGMs are F_{200} (defocus), F_{300} (aperture defect or coma) and F_{400} (spherical aberration). As previously shown, the astigmatic curvature [Eq. (25)], although dominant in TGMs, is usually small in SGMs.

Taking a grating of length 150 mm and width 44 mm, Fig. 4 shows the contributions from all the above aberrations for $w = w_{max}$, $l = l_{max}$, and their sum at photon energies from 40–120 eV, with a *fixed* exit arm length r' of 4.4 m. The same thing is shown in Fig. 5 except that r' has been set for focus. This also brings about consequential changes in aberrations other than defocus, including a shift of the coma zero. Note that, in both these plots, the spherical aberration and line curvature terms are scaled by 10× and 1000×, respectively. The defocus curve goes through two zeroes, as expected, and the coma curve goes through one. The spherical aberration is small and always positive, and the line curvature, as expected, is negligibly small (see Section 2.2.5). The individual aberrations are shown in their signed form, meaning that the signs are those of

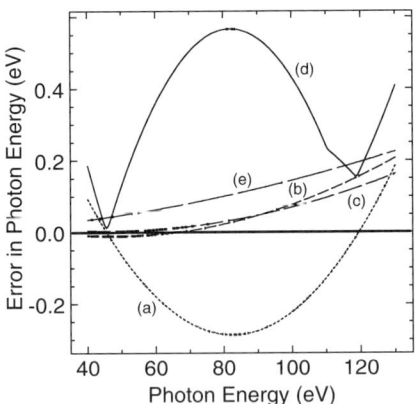

FIG. 4. Aberration components and their sum are shown for the 9.0.1 SGM with a fixed exit arm length r' of 4.4 m, and a ruled length and width of 150 mm and 44 mm, respectively. The curves show (a) defocus, (b) coma, (c) spherical aberration ×10, (d) the sum, and (e) line curvature ×1000.

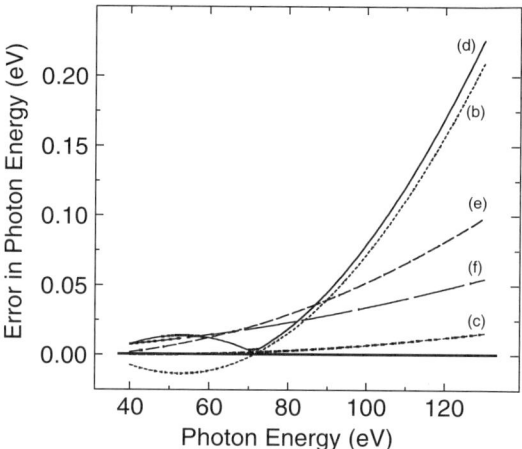

FIG. 5. Aberration components and their sum are shown similarly to Fig. 4 but with the exit arm length set to be in focus. The ruled length and width are again 150 mm and 44 mm, respectively. The curves show (b) coma, (c) spherical aberration, (d) the sum, (e) entrance slit, and (f) exit slit.

the aberration coefficients, w_{max} and l_{max} being positive. In this way we can see how the individual components may be summed. The ray aberrations resulting from defocus and spherical aberration scale as odd powers of w, so they change sign when w goes negative. Those from coma, on the other hand, scale as w^2 and therefore always have the sign of the coefficient. Representing defocus, coma, and spherical aberration by d, c, and s respectively, we obtain the full-width-at-zero-height value of ΔE by looking for the maximum algebraic difference among $(d + c + s)$, (0), and $(-d + c - s)$. An interesting effect of this algebraic differencing can be seen in the region greater than 83 eV, where defocus is negative but becoming smaller and coma is positive and becoming larger (see Fig. 4). At 110 eV, the coma equals the sum of defocus and spherical aberration. For photon energies less than this, the extreme rays are given by $(d + c + s)$ and $(-d + c - s)$, and for those greater, they are given by $(d + c + s)$ and (0). The result is the jump in the rate of change of aberration at 110 eV as one half of the defocus is eliminated. This effect is typical of the analysis based only on the principal ray and the two marginal rays. A more complete analysis would include all the rays, and we examine this in more detail in the next section.

2.3.3 Calculation of Spectral Lineshapes

The calculation of aberrations in terms of the principal and marginal rays gives an approximate understanding of the magnitude of the error and is useful in performing fast optimizations. However, for a detailed understanding of the resolution of a system, we must calculate the spectral lineshape. This can be done by applying the method previously outlined to a distribution of points across the grating surface. Figure 6 shows the photon energy errors resulting from individual aberration terms as a function of the w coordinate (Fig. 1) for a fixed exit arm length of 4.133 m at a photon energy of 100 eV. This shows that, for the negative side of the grating, defocus and spherical aberration subtract, and on the positive side they add. Combined with the coma term, the net effect (dashed curve) is to produce a small negative aberration on the negative side and a large positive one on the positive side.

The information plotted in Fig. 6 allows us to predict spectral lineshapes. If we draw a horizontal line at a particular ΔE value in Fig. 6, it could cross the summed curve at up to two locations. Each of these will correspond to a position, w, on the grating, illuminated with a local photon flux per unit width, $I(w)$. The flux per unit energy range $di/d(\Delta E)$ in the image plane therefore depends on $I(w)$ at the w values of the crossing points and also on the gradient $\partial w/\partial(\Delta E)$ in Fig. 6. For example, in the case of pure coma, the calculation proceeds as follows:

$$\Delta y'_{300} = \frac{r'_0}{\cos \beta_0} 3w^2 F_{300} \quad \text{[by Eq. (21)]}$$

$$\frac{d\lambda}{d(\Delta y')} = \frac{d_0 \cos \beta_0}{mr'_0} \quad \text{[by Eq. (26)]}$$

$$|\Delta E| = \frac{hc\,\Delta\lambda}{\lambda^2}, \quad \text{so that } |\Delta E_{300}| = \frac{3hcd_0 F_{300}}{m\lambda^2} w^2 \quad (28)$$

$$\frac{di}{d(\Delta E_{300})} = \frac{di}{dw}\frac{dw}{d(\Delta E_{300})} = \frac{I(w)}{2}\sqrt{\frac{m\lambda^2}{3hcd_0 F_{300}}}\frac{1}{\sqrt{|\Delta E_{300}|}}$$

For a uniformly illuminated grating ($I(w)$ = constant), the intensity in the image plane caused by coma will thus be proportional to $1/\sqrt{\Delta E}$, and the intensity will be infinite on axis and have long tails. Also from Fig. 6, it is clear that we should expect steps in intensity, corresponding to the ends of the summed curve, for example at around -3 meV, where the right-hand part of the summed curve reaches the edge of the grating.

These effects can be seen in Fig. 7, where the same case of 100 eV is used, but the lineshapes have been assessed for exit-arm lengths of 4.113–4.193 m in steps of 0.02 m. Curve (e) (4.193 m) corresponds to the in-focus condition, and we can see its expected single-sided distribution (because the dominant aberration, which is coma, scales with an even power of w), and the asymptotic rise in the gradient $\partial w/\partial(\Delta E)$ as ΔE approaches its minimum value. We can also see

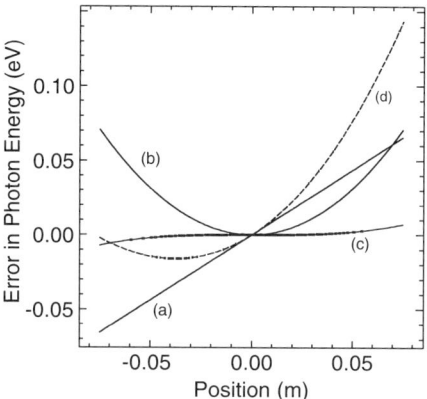

FIG. 6. Aberration components are shown as a function of position in the w direction on the grating, with $l = 0$, an exit arm length r' of 4.133 m, and a photon energy of 100 eV. The curves show (a) defocus, (b) coma, (c) spherical aberration, and (d) the sum.

steps in intensity at +65 and +79 meV. From Fig. 6, we can see that the maximum coma (with a slightly different exit-arm length) is 71 meV, and the maximum spherical aberration is ±7 meV. The negative-w end of the grating therefore will have a maximum error of 64 meV, and the positive-w end, 78 meV. We can therefore associate the two steps with the total aberration for the extreme rays on the grating. If we look at curve (b) in Fig. 7, which is the

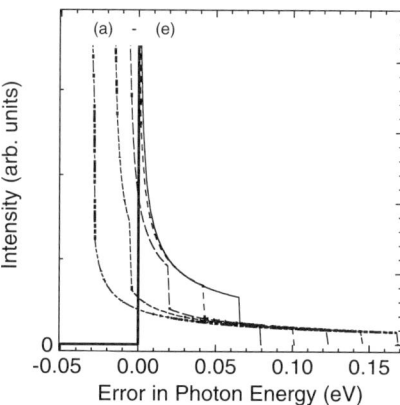

FIG. 7. The spectral distribution is shown for a photon energy of 100 eV, with a uniformly illuminated grating with a length of 150 mm and a width of 44 mm. Distributions are shown for exit arm lengths r' of (a) 4.113 m, (b) 4.133 m, (c) 4.153 m, (d) 4.173 m, and (e) 4.193 m. The latter case is in focus.

$r' = 4.133$ m curve (previously described in Fig. 6), we see that the features can be easily correlated. The asymptotic limit is at -17 meV, corresponding to the minimum in the sum curve of Fig. 6, and there are sharp breaks at -3 meV and $+144$ meV corresponding to the endpoints of the sum curve at the edges of the grating.

The case of Fig. 7 would correspond to an overfilled grating, a situation that is common on bending magnet beamlines. To simulate an undulator source, we have to take into account the narrow angular divergence of the beam and diffraction from the entrance slits. In the case shown in Fig. 5, we have taken a 4.5-m undulator, radiating at 100 eV and a demagnification of 8:1 onto an entrance slit of width 10 μm. We have approximated the two angular distributions delivered by the undulator and the slit by their Gaussian equivalents.

$$\sigma_u = \sqrt{\frac{\lambda}{L}}, \quad \sigma_s = \frac{\lambda}{2s}, \quad \Sigma^2 = M_v^2 \sigma_u^2 + \sigma_s^2, \quad I(w) = Are^{-w^2 \cos^2\alpha/(2r'^2 \Sigma^2)} \tag{29}$$

where L is the length of the undulator and we have neglected the effect of the vertical electron beam divergence. In this case, Fig. 8 shows the resulting intensity distribution for $r' = 4.133$–4.193-m exit-arm lengths in 0.02-m steps, this time for undulator radiation. We can see some remarkable differences between the two cases of uniform illumination and radiation from an undulator. The most striking is that the steps evident in Fig. 7 have disappeared and that, when significantly out of focus (4.133 m), the asymptote disappears. The reason

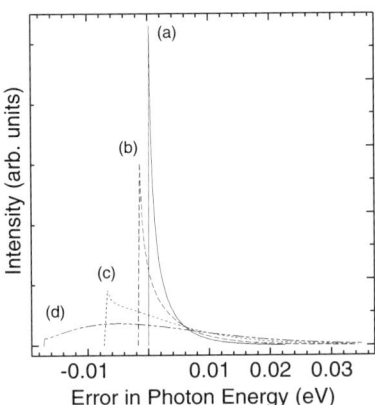

FIG. 8. The spectral distribution is shown for a photon energy of 100 eV, with a grating length of 150 mm and width of 44 mm illuminated by an undulator source. The undulator is 4.5 m in length, demagnification from the source to slit is 8:1 and diffraction is included for 10 μm slits. Distributions are shown for exit arm lengths r' equal to (a) 4.193 m, (b) 4.173 m, (c) 4.153 m, and (d) 4.113 m. The case (a) is in focus.

for both is that the undulator beam is illuminating a region near the center of the grating, and so the termination of the $di/d(\Delta E)$-versus-ΔE curve happens smoothly because the intensity falls to zero smoothly. In addition, near the center the dominant aberration is defocus, which is seen as a slow intensity variation. The absence of the asymptotic cusp is due to the fact that there is no intensity at the minimum of the ΔE curve at $w = -37$ mm [curve (d), Fig. 6].

The inclusion of slit diffraction is critical if an accurate estimate of the spectral line shape is to be made. The same effect guarantees (via the van Cittert–Zernike theorem) to coherently illuminate the appropriate number of grooves N_0 on the grating to produce the calculated slit-width-limited resolution and the diffraction-limited resolution mN_0, which can be seen to be the same. We show the effect of slit width on line shape in Fig. 9 for slit sizes of 2 µm, 6 µm, 10 µm, 25 µm, and fully open for a photon energy of 64 eV and for the in-focus condition. We have chosen this energy because extensive studies of the double-ionization series of helium have been conducted in this region, and in particular, recent work by Kaindl and collaborators [50], using the SGM on ALS beam-line 9.0.1 that we have been studying, has demonstrated a resolution of 65000 at this energy. The helium spectrum in question is given in Fig. 10, which shows the extraordinary resolving power achieved. The $2p3d$ peak at 64.117 eV is particularly interesting as it has a natural linewidth of a few microelectron volts, so that what we see is the instrumental line-shape function.

It can be seen from Fig. 9 that the calculated aberration-limited line width broadens as the slits are closed. This occurs because the divergence of the light

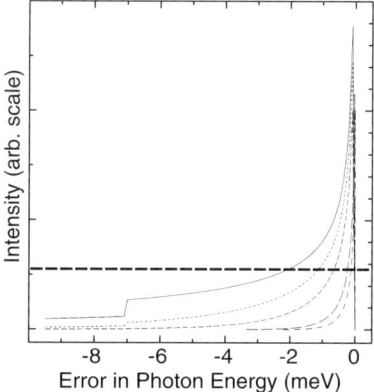

FIG. 9. The spectral distribution is shown for 64 eV, undulator illumination, and for entrance slit sizes of 2, 6, 10, 25 µm and fully open. 2 µm corresponds to the widest distribution. The horizontal line gives the 50% integral point for each curve, yielding widths of 2.1, 1.1, 0.67, 0.23, and 0.13 meV, respectively, for the above slit sizes.

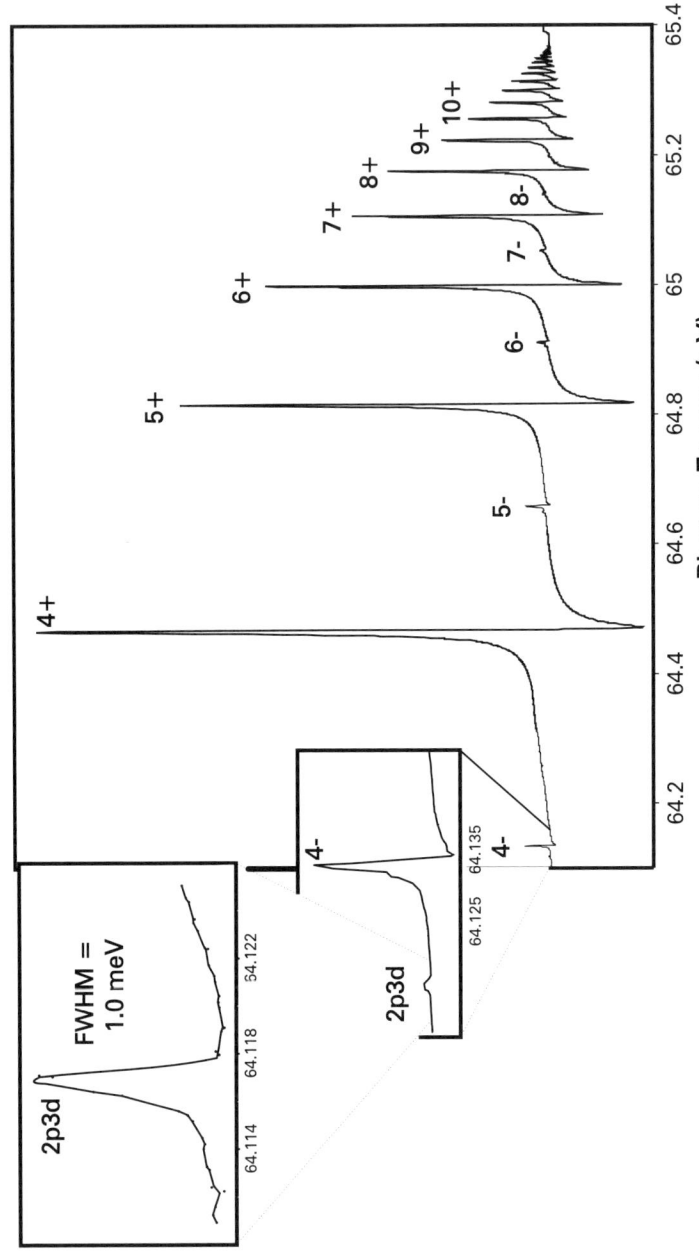

FIG. 10. The double-ionization series of helium is shown near the ionization limit of the + series. The 2p3d feature has a natural linewidth of a few μeV, and so the feature here should directly map the line shape function of the monochromator.

increases as the slit is closed, filling more of the grating and increasing all the aberrations. In order to make a fair comparison of line shapes, we clearly cannot take the peak height into account, because the distribution on axis would go to infinity, which is only prevented here by the finite sampling. However, the integral is finite, and so we take the 50% integral point from the asymptotic cusp as a measure of the width, and the data are normalized to that value. We therefore arrive at aberration-limited widths of 2.1, 1.1, 0.67, 0.23, and 0.13 meV for the 2-μm, 6-μm, 10-μm, 25-μm, and open slit cases. (The demagnified undulator source geometry is used for the latter). The corresponding slit-width-limited resolutions are 0.3, 0.9, 1.5, and 3.8 meV for the 2-μm, 6-μm, 10-μm, and 25-μm slit openings. Combining the slit and aberration limits indicates an optimum around 6 μm slit width, which is in reasonable agreement with the measured resolution of 1 meV. Moreover, it is clear from the shape of the experimental curve that the broadening is caused by residual coma. Note also that the sign of the coma aberration is the same in the experimental and theoretical curves at 64 eV but reversed from the curve at 100 eV. This is because the sign of the coma changes at the coma zero, which is at 71 eV (Fig. 5). In the theoretical curves, also note that for the smallest slit size, sufficient intensity is diffracted to the edge of the grating so that the sharp truncations of the coma and spherical aberration again become apparent.

Geometrical aberration theory can therefore be used at several different levels. On the one hand, it can be rapidly used to assess gross performance by prediction of the extreme ray aberrations. On the other hand, it can also be used to predict line shapes in a more precise way. In this sense, grating aberration theory is a powerful tool for understanding design possibilities and arriving at candidate designs. Once a design is established, it is useful to run a ray trace from a point source with only a few rays positioned on a regular grid. (The SHADOW code [22] maintained by the Center for X-ray Lithography at the University of Wisconsin is the one most suited for synchrotron-radiation applications.) One should be able to explain quantitatively why every ray goes where it does on the basis of the aberration calculations from Eq. (21) and observations such as the relative positions of the rays from w_{max} and $w_{max}/2$, etc. Once a candidate design that meets the requirements has been established, it then becomes useful to run SHADOW again, this time with a realistic source and a larger number of rays to verify the correctness of the design and to study other aspects of it.

2.4 Focusing in Variable-Included-Angle Monochromators

In the previous section, we showed how grating aberration theory could be applied to a fixed-deviation-angle SGM to extract information on focusing, aberrations, and line shapes. The same approach can also be applied to other monochromators, where there is a single focusing element, including the case where a movable plane mirror is used to vary the included angle at the grating. Two such cases are the variable-included-angle SGM [18, 22, 32, 36] and the SX700 [38, 39, 41]. In these monochromators, the included angle may become a user-controlled variable and not only determines focusing but also modifies the efficiency behavior of the system.

We have seen how it is possible in a fixed-deviation-angle SGM to keep the defocus term zero by moving the exit slits. Sometimes, particularly in microscopy where the microscope optics are fixed, the exit slit has to be in a fixed location. The variable-angle SGM achieves focus at fixed r and r', that is, with fixed slits, by varying the included angle. According to Eq. (15) with $a_{20} = 1/2R$, the condition for focus is

$$\frac{\cos^2 \alpha}{r} - \frac{\cos \alpha}{R} + \frac{\cos^2 \beta}{r'} - \frac{\cos \beta}{R} = 0 \qquad (30)$$

If α is regarded as the independent variable, then this equation can be solved for β as a function of α. Then, using the sign convention defined in Section 2.2.1, the included angle, 2θ, is obtained as $\alpha - \beta$. For each α, β pair, the grating equation yields the wavelength, and we can plot included angle against photon energy (Fig. 11). By selecting differing grating radii, we can then select different solutions with different angles while keeping r and r' constant. Three different solutions are illustrated in Fig. 11. The action of changing the grating radius causes the focusing curve to be moved on both the energy scale and the included-angle scale. The figure shows curves for both solutions to the quadratic, which join at the low-energy limit, although only the lower solution has practical value. Changing the radius has the desirable effect that the mean

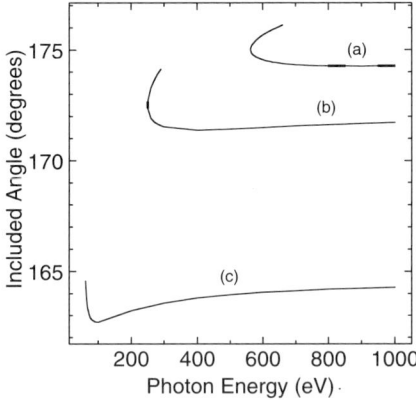

FIG. 11. Three different solutions for a variable-angle SGM with a 1200/mm grating in positive first order: the values of r/R and r'/R are (a) 0.1 and 0.2, (b) 0.05 and 0.1, and (c) 0.033 and 0.066.

included angle can be set to a value appropriate for the energy ranges of each separate grating. This is advantageous for collection aperture, order sorting, and, as we will see in Section 2.5, for diffraction efficiency. However, Fig. 11 shows that the amount of change in the angle within the scan range of any one grating is always small (about a degree or less). On the other hand, if we want to optimize diffraction efficiency, say for a rectangular phase grating in mth order, then the path length difference between rays diffracted at the top and bottom of the grooves has to be m wavelengths. Therefore, the included half-angle θ should vary roughly as $\theta = \cos^{-1}(m\lambda/2h)$, where h is the groove depth, whereas in fact, as Fig. 11 shows, it must be kept roughly constant to hold focus. The variable-angle SGM thus allows us to have a fixed exit slit and optimum efficiency at one wavelength for each grating, but it still has the problem of the standard SGM; it has a limited tuning range per grating and still cannot track the diffraction efficiency maximum.

These disadvantages are overcome to a considerable degree in the SX700 system originally proposed by Petersen in 1980 [38, 39, 41]. It was the first instrument to use included-angle control by means of a plane mirror and this was combined with a plane grating and fixed focusing mirror. The latter was originally an ellipsoid of revolution, but in later instruments that was superseded by a more-easily-manufactured spherical mirror [32, 44]. The astigmatism of the spherical mirror at grazing incidence was overcome using a separate mirror outside the monochromator focusing in the nondispersive direction, and the coma of the sphere was reduced sufficiently by selecting an appropriate magnification.

The focusing condition of the monochromator is given by Eq. (30) with R equal to infinity.

$$r' = r \frac{\cos^2 \beta}{\cos^2 \alpha} = -rC^2 \qquad (31)$$

We can see that a virtual image is formed at a distance rC^2 behind the grating. The function of the plane premirror in this design is again to select an included angle that will maintain focus. Other operational modes of the monochromator are possible with different included angles, at the expense of no longer being in focus, for example, to reduce harmonics or to exactly track the efficiency maximum. In a later version of the system, moveable exit slits were used so that, for an entrance-slitless version of the monochromator, a larger value of α could be used to improve the source-size-limited resolution at the expense of flux. If α is eliminated between Eq. (31) and the grating equation, we obtain

$$1 - \left(\frac{m\lambda}{d_0} - \sin \beta\right)^2 = \frac{\cos^2 \beta}{C^2} \quad \text{whence}$$

$$\sin \beta = \frac{(m\lambda/d_0) - \sqrt{(1 - C^{-2})^2 + (m\lambda/Cd_0)^2}}{1 - C^{-2}} \qquad (32)$$

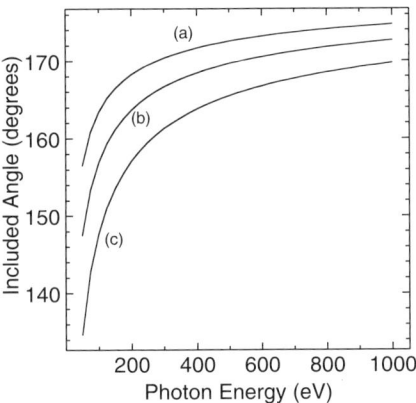

FIG. 12. Three solutions of Eq. (32) for a 1200/mm grating in +1 order with C values of (a) 4.4, (b) 2.1, and (c) 1.45.

Three solutions are shown in Fig. 12. One can see that, unlike the SGM and variable-angle SGM, wide ranges of both wavelength and included angle are spanned for a single grating. In addition, the curve falls strongly as the photon energy decreases, as required for maximizing diffraction efficiency.

2.5 Diffraction Efficiency

One starting point for a monochromator design is to know the maximum diffraction efficiency that can be obtained at a particular photon energy under optimum conditions. In principle, this would need many time-consuming calculations, but a range of these have been done by Padmore and colleagues [34] for gold- and nickel-coated rectangular phase gratings from 300 to 1200/mm and from 100 to 2000 eV. These provide values of the maximum possible efficiency and the deviation angle, groove depth, and groove width necessary to get it for each line density and energy. (Note that the groove width is not the same as the period.) The calculated results were shown to agree with both experiment and reciprocity. Maximum efficiency can vary significantly, depending on line density. For example, a 300/mm grating at 1500 eV has a diffraction efficiency maximum of 33% for a deviation angle of 2°. For a 1200/mm grating, the maximum efficiency is 8% for a deviation angle of 4°. The tendency is always for the diffraction efficiency and the optimum deviation angle to decrease at high line density. For the present purpose, the diffraction process is best described in terms of the *deviation* angle because this is directly related to the important physical quantity, which is the momentum transfer.

To illustrate the characterization of grating efficiency for particular geometries, we have chosen the SX700 case. Figure 13 shows the "nomogram" for a 1200/mm gold-coated grating. The upper panel gives the maximum diffraction efficiency; the middle panel, the required deviation angle; and the lower panel, the required groove depth and width. The figure shows two cases, one (dotted) in which the deviation angle, groove width, and groove depth are variable parameters, and one in which the deviation angle is constrained to the SX700 fixed-focus condition for the same groove parameters. It can be seen that the two curves are virtually identical. It can also be seen that the diffraction efficiency is above 10% from 100 to 1400 eV, dropping only to 6% at 2 keV. Of course, having continuously variable groove parameters is not realizable in practice, and Fig. 14 shows what happens when they are fixed. The groove depth and width have been optimized at 500 eV, and it can be seen that only at energies less than 300 eV is there significant variance between the fully optimized case and the case with fixed groove parameters.

The SX700 geometry therefore offers the enormous advantage of staying always close to the blaze-maximum condition. In addition, if we allow ourselves the possibility to have more than one groove depth and width, then we can have optimized performance in the region below 300 eV. This can be done in practice by having either more than one grating or more than one ruling on a single grating.

We have presented the preceding case where a monochromator has to tune over a very wide energy region. There are many cases where this is unnecessary, and we are increasingly designing for highly specific experiments where simpler monochromator systems can be used. However, the starting point is still to understand the relationship between deviation angle and efficiency, using the tabulations described in reference [34].

2.6 An Optimized Beam Line for Microscopy by Photoelectron Emission Microscopy and Micro-X-Ray Photoelectron Spectroscopy

To illustrate some of the issues of monochromator design and to show some of the special considerations necessary when designing systems for microscopes, we will describe beam line 7.3.1 at the ALS. The layout of the beam line is given in Fig. 15 [34, 35]. The system was originally designed for photoelectron emission microscopy (X-PEEM), applied to the study of magnetic materials. The materials of interest, the upper $3d$ transition metals ($2p$–$3d$ edges) and up to the middle of the rare earths ($3d$–$4f$ edges) have a relatively narrow band of energies from around 650–1300 eV. The PEEM itself requires an illuminated field of view of about 30 μm with maximum flux density. As the PEEM selects

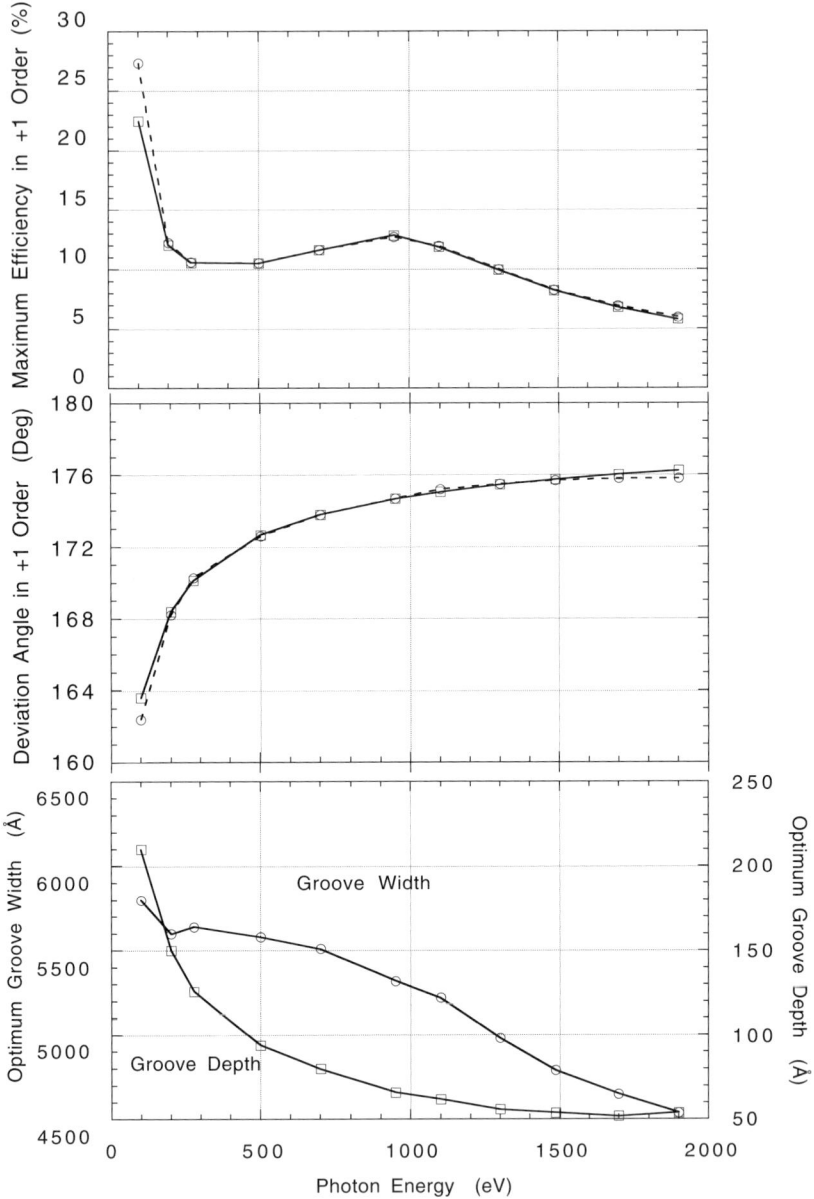

FIG. 13. Optimum efficiency of a 1200 /mm gold-coated grating and the deviation angle, groove depth, and groove width needed to get it. For the dashed curve (best case), the deviation angle and groove depth and width are all varied in the search for an optimum, whereas for the continuous curve the groove depth and width are the same but the deviation angle is chosen to focus an SX700 with $C = 2.166$.

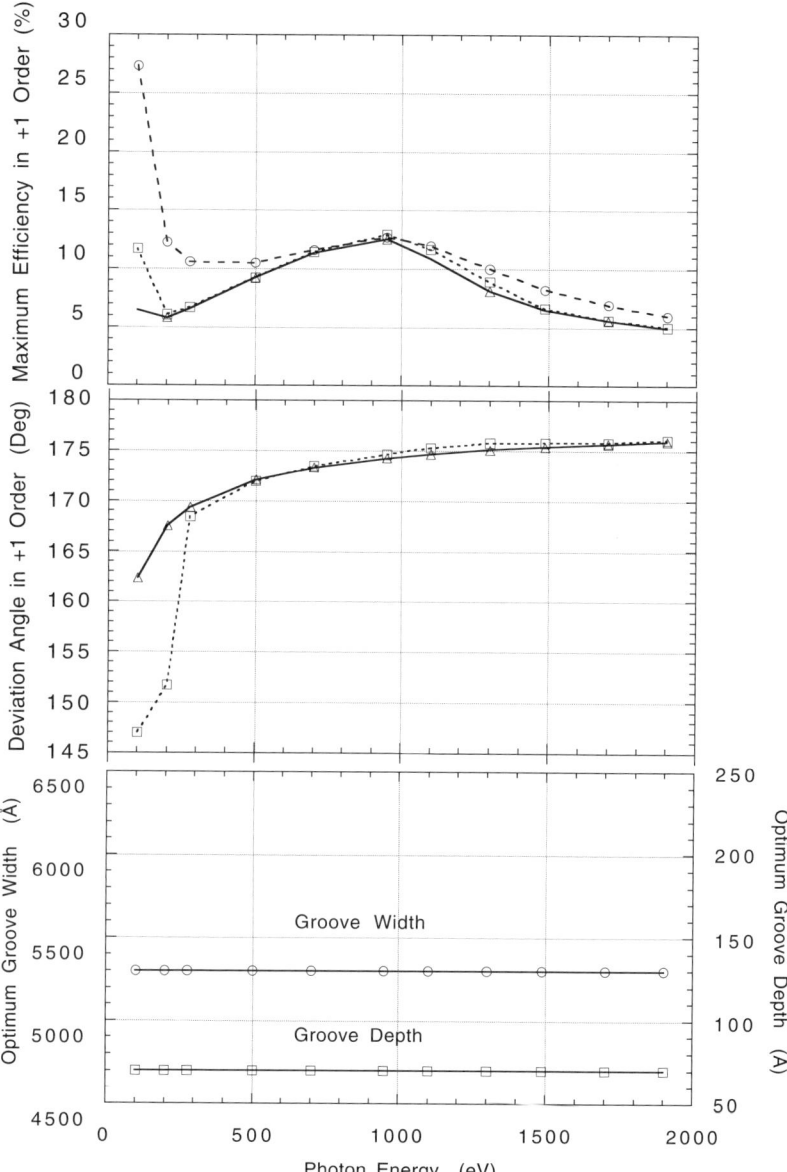

FIG. 14. Similar to Fig. 13 except that the groove width and depth are fixed at the values needed for an optimum at 500 eV. The short-dash curve is for the best deviation angle while the continuous curve is for the angle that would focus an SX700 with $C = 2.094$. The "best-case" curve of Fig. 13 is shown (long dash) for comparison.

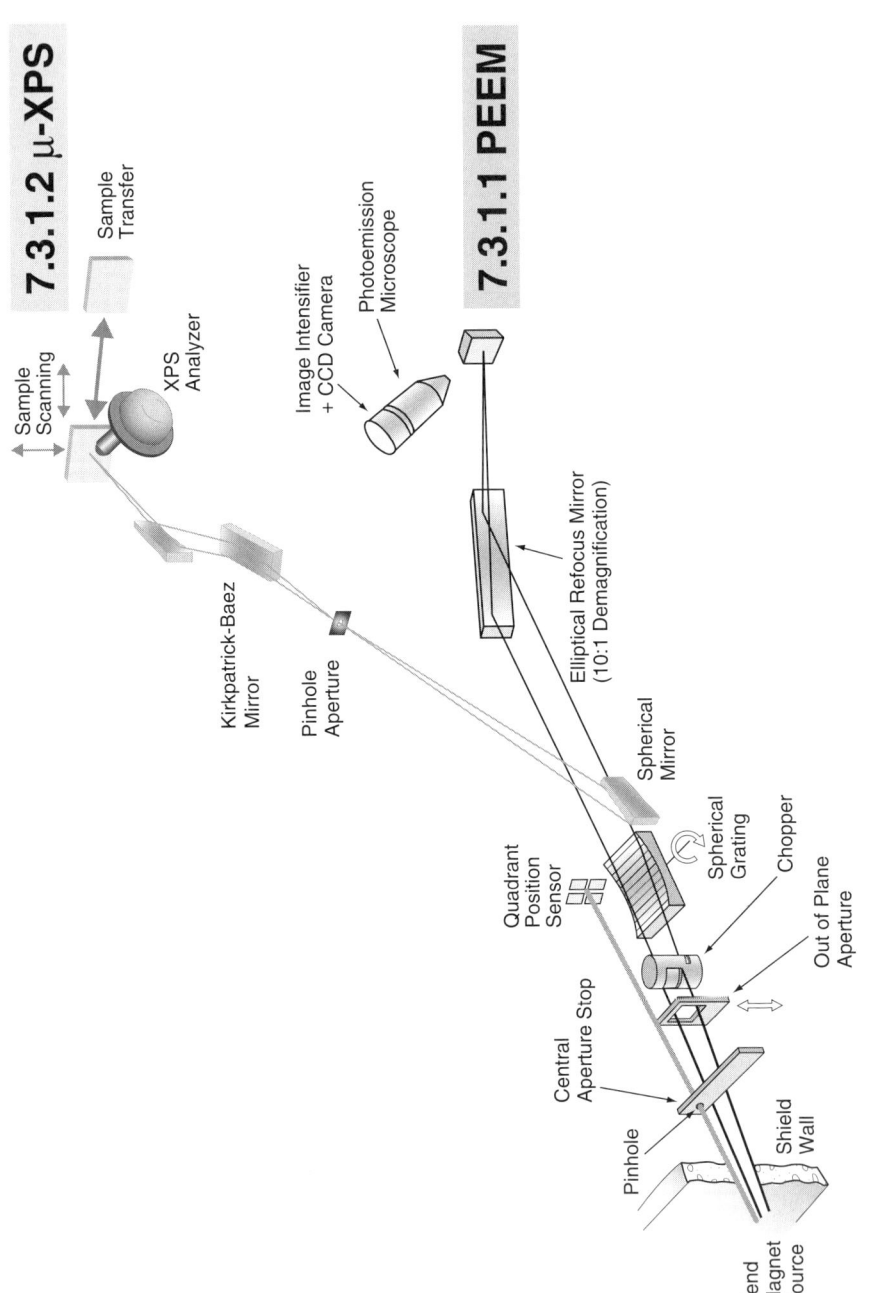

FIG. 15. Optical systems of ALS beam line 7.3.1 as discussed in the text.

its own field of view, it can also act as an effective exit slit. Thus, considering the vertical source size of around 30 µm full width at half maximum (FWHM), we chose to build a single-grating fixed-angle SGM in entrance- and exit-slitless mode at 1:1 monochromatic magnification. Because the core-level widths of the edges we wished to study were rather large, we also designed the system to have a deliberately moderate resolution. By making the instrument sufficiently long, a very low line density grating (200/mm) could be used. This combined with the relatively small energy range and the resulting small change of focal length with energy meant that we could operate with fixed exit slits, or in this case, with a fixed sample. The use of such a low line density also has the added advantage of a very high diffraction efficiency, typically 30% at 1 keV. However, it is noteworthy that the low line density places a greater burden on the optical quality of the grating.

In the horizontal direction, the object was to collect as much light as possible and condense it to the field size of 30 µm. We have a horizontal source size of around 300 µm FWHM that requires a demagnification of 10. The maximum angular spread that can be passed by the horizontal condenser mirror is about one-half the critical angle at the maximum photon energy. At 1.3 keV, this means a maximum of 25 mrads convergence, or 2.5 mrads collection from the source. The resulting elliptical-cylinder condenser mirror is particularly challenging. It is 1 m in length and requires an rms slope error tolerance of 3 µrads to preserve the source brightness as well as a superfine finish. A rigid mirror of this specification is beyond the state of the art, and we have adopted a method based on the bending of a flat plate made of nickel-plated steel with specially calculated width variation. This beam line indeed delivers the expected flux of 3×10^{12} ph/sec in a 0.5 eV bandpass at 1 keV in a 30-µm-diameter spot size.

The application-specific nature of the beam line, together with the need to have the maximum flux density in the imaging field of the PEEM, has led us to a unique design that is both highly efficient and much simpler than that of a traditional beam line. Because of the special needs of x-ray microscopy and of the materials issues that are being addressed with the new generation of microscopes, there are likely to be more systems of this type—highly specific to one task, highly optimized, but simplified by removing the demand for universal capabilities. The exact matching of the system to a particular function also tends to imply exclusive use of the beam line by a single user group, and such an arrangement can normally be considered only on a bending magnet.

A further example of this approach is given by a branch of the same beam line. This branch splits off 0.2 mrads of horizontal aperture from the X-PEEM beam line using a horizontally deflecting mirror. This produces a 2:1 demagnified image of the horizontal source at the monochromatic focal plane of the monochromator. At this location, we have a pair of bilaterally adjustable slits that define the source for the microscope that follows. The nominal setting is

20 μm vertical by 40 μm horizontal. The horizontal image of the ALS is 150 μm wide at this location, so the slits are significantly overfilled. We can now apply the same phase-space arguments as for X-PEEM, but in this case we wish to demagnify to a 1-μm spot size, and we use a slightly smaller grazing angle of 1.6° on the probe-forming optics. These are a Kirkpatrick–Baez pair, again elliptical and produced by the bending of width-profiled flats by the application of unequal couples [33]. The vertical demagnification is 20:1, and using our half-critical-angle rule, we derive a convergence angle of 12 mrads. This means that we can have an acceptance from the source of 0.6 mrads in the vertical direction. This value is approximately equal to the divergence of the light from the source at 1 keV. We can therefore say that, in the vertical direction, light from an ALS bending magnet can be focused by a mirror operating at 1 keV to an image size of 1 μm without geometric loss. In the horizontal direction, the convergence onto the sample is the same, 12 mrads, but the demagnification is 40:1. The acceptance from the source is therefore 0.15 mrads, with a loss of a factor of 4 in throughput at the horizontally defining slits. Even so, the predicted (and measured) flux in the focused spot is 3×10^{10} ph/sec in the so-far-achieved spot size of 1.0×1.3 μm^2, with a 0.5 eV bandpass. This is sufficient for x-ray photoelectron spectroscopy (XPS) using modern high-aperture analyzers with multichannel detection. In the present case, the system was developed for the microelectronics industry, specifically to perform scanning micro-XPS on large-area (50-mm-diameter) samples. It was important to design a system that was economical and had sufficient flux at its design resolution to enable a high throughput of samples.

It can be seen from the preceding examples that the design of monochromators to be used in conjunction with microscopes is significantly different from that for high-resolution spectroscopy. Wherever possible, these optical systems should be separated, as they clearly have very different design constraints and lead to very different optimized solutions.

Acknowledgments

This work was supported by the Director, Office of Energy Research, Office of Basic Energy Sciences, Materials Sciences Division of the U.S. Department of Energy, under Contract no. DE-AC03-76SF00098.

References

1. Abramson, N., *J. Opt. Soc. Am. A*, **6**, 627–629 (1989).
2. Beutler, H. G., *J. Opt. Soc. Am.*, **35**, 311–350 (1945).
3. Born, M., E. Wolf, *Principles of Optics*, Pergamon, Oxford, 1980.

4. Buchdahl, H. A., *An Introduction to Hamiltonian Optics*, Cambridge University Press, Cambridge, 1970.
5. Chen, C. T., *Nucl. Instr. Meth.*, **A 256**, 595–604 (1987).
6. Chen, C. T., F. Sette, *Rev. Sci. Instrum.*, **60**, 1616–1621 (1989).
7. Ederer, D. L., J. B. West (Eds.), *Proc. Nat. Conf. Synchrotron Radiation Instrumentation*, *Nucl. Instrum. Meth.* 172 (1980).
8. Gudat, W., C. Kunz, in Kunz, C. (Ed.), Vol. **10**, *Springer Series on Topics in Current Physics*, Berlin, 1979.
9. Haber, H., *J. Opt. Soc. Am.*, **40**, 153–165 (1950).
10. Haensel, R., C. Kunz, *Z. Angew. Phys.*, **23**, 276–295 (1967).
11. Harada, T., T. Kita, *Appl. Opt.*, **19**, 3987–3993 (1980).
12. Heimann, P., T. Warwick, M. Howells, W. McKinney, D. DiGennaro, B. Gee, D. Yee, B. M. Kincaid, *Nucl. Instrum. Meth.*, **A319**, 106–109 (1992).
13. Hettrick, M. C., *Appl. Opt.*, **23**, 3221–3235 (1984).
14. Hogrefe, H., M. R. Howells, E. Hoyer, in *Soft X-ray Optics and Technology*, Koch, E.-E., G. Schmahl (Eds.), Proc. SPIE, Vol. **733**, SPIE, Bellingham, 1986.
15. Howells, M., in *Synchrotron Radiation Sources and Applications; Proceedings of the Thirtieth Scottish Universities Summer School in Physics (SUSSP), Aberdeen, Sept. 1985*, Greaves, G. N., I. H. Munro (Eds.), SUSSP, IOP Publishing (Adam Hilger), Edinburgh, 1989.
16. Howells, M. R., in *New Directions in Research with Third-Generation Soft X-ray Sources*, Schlachter, A. S., F. J. Wuilleumier (Eds.), Kluwer, London, 1994.
17. Hunter, W. R., in *Spectrometric Techniques*, Vol. **IV**, Academic Press, Orlando, 1985.
18. Jark, W., P. Melpignano, *Nucl. Instrum. Meth.*, **A349**, 263–268 (1994).
19. Johnson, R. L., *Handbook of Synchrotron Radiation*, Vol. *1A*, E. E. Koch (Ed.), North Holland, Amsterdam, 1983.
20. Koch, E. E., R. Haensel, C. Kunz (Eds.), *Proc. 4th Int. Conf. on Vacuum Ultraviolet Radiation Physics*, Hamburg, 1974.
21. Labeyrie, A., J. Flammand, *Opt. Comm.*, **1**, 5–8 (1969).
22. Lai, B., F. Cerrina, *Nucl. Instrum. Meth.*, **A246**, 337–341 (1986).
23. Mack, J. E., J. R. Stehn, B. Edlen, *J. Opt. Soc. Am.*, **22**, 245–264 (1932).
24. Madden, R. P., in *X-ray Spectroscopy*, Azaroff, L. V. (Ed.), McGraw-Hill, New York, 1974.
25. Marr, G. V., I. H. Munro (Eds.), *International Symposium for Synchrotron Radiation Users*, Vol. DNPL/R26, Daresbury Nuclear Physics Laboratory, Daresbury, 1973.
26. McKinney, W. R., *Rev. Sci. Inst.*, **63**, 1410–1414 (1992).
27. Namioka, T., *J. Opt. Soc. Am.*, **51**, 4–12 (1961).
28. Namioka, T., M. Koike, *Nucl. Instrum. Meth.*, **A319**, 219–227 (1992).
29. Namioka, T., H. Noda, M. Seya, *Sci. Light*, **22**, 77–99 (1973).
30. Noda, H., T. Namioka, M. Seya, *J. Opt. Soc. Am.*, **64**, 1031–1036 (1974).
31. Padmore, H. A., in *Soft X-ray Optics and Technology*, Koch, E.-E., G. Schmahl (Eds.), Proc. SPIE, Vol. **733**, Bellingham, 1986.
32. Padmore, H. A., *Rev. Sci. Instrum.*, **60**, 1608-1616 (1989).
33. Padmore, H. A., M. R. Howells, S. Irick, T. Renner, R. Sandler, Y.-M. Koo, in *Optics for High-Brightness Synchrotron Radiation Beam Lines II*, Arthur, J. (Ed.), Proc. SPIE, Vol. **2856**, SPIE, Bellingham, 1996.
34. Padmore, H. A., V. Martynov, and K. Holis, *Nucl. Instrum. Meth.*, **A347**, 206–215 (1994).
35. Padmore, H. A., T. Warwick, *J. Electron. Spect.*, **75**, 9–22 (1995).

36. Peatman, W. B., J. Bahrdt, F. Eggenstein, G. Reichardt, F. Senf, *Rev. Sci. Inst.*, **66**, 2801–2806 (1995).
37. Peatman, W. B., F. Senf, in *Vacuum Ultraviolet Radiation Physics (VUV11)*, Wuilleumier, F. J., Y. Petrof, I. Nenner (Eds.), World Scientific, Singapore, 1992.
38. Petersen, H., *Opt. Comm.*, **40**, 402–406 (1982).
39. Petersen, H., *Nucl. Inst. Meth.*, **A246**, 260–263 (1986).
40. Petersen, H., in *Soft X-ray Optics and Technology*, Koch, E.-E., G. Schmahl (Eds.), Proc. SPIE., Vol. **733**, SPIE, Bellingham, 1986.
41. Petersen, H., *Rev. Sci. Instrum.*, **66**, 1777–1779 (1995).
42. Pieuchard, G., J. Flamand, Final report on NASA contract number NASW-2146, GSFC 283-56,777, Jobin-Yvon, 1972.
43. Rah, S. Y. The authors are grateful to Dr. S. Y. Rah of the Pohang Accelerator Laboratory for calculating the a_{ij} expressions (1997).
44. Reininger, R., V. Saile, *Nucl. Instrum. Meth.*, **A288**, 343–348 (1990).
45. Rowland, H. A., *Phil. Mag.*, Supplement to Vol. **13** (5th series), 469–474 (1882).
46. Rowland, H. A., *Phil. Mag.*, **16** (5th series), 197–210 (1883).
47. Rudolph, D., G. Schmahl, *Umsch. Wiss. Tech.*, **67**, 225 (1967).
48. Rudolph, D., G. Schmahl, in *Progress in Optics*, Vol. **XIV**, Wolf, E. (Ed.), North-Holland, Amsterdam, 1977.
49. Saile, V., P. Gurtler, E. E. Koch, A. Kozevnikov, M. Skibowski, W. Steinmann, *Appl. Opt.*, **15**, 2559–2564 (1976).
50. Schultz, K., G. Kaindl, M. Domke, J. D. Bozek, P. A. Heimann, A. S. Schlachter, J. M. Rost, *Phys. Rev. Lett.*, **77**, 3086–3089 (1996).
51. SRI82, in *Proc. Int. Conf. on X-ray and VUV Synchrotron Radiation Instrumentation*, Hamburg, 1982, *Nucl. Instrum. Meth.* **208** (1983).
52. SRI85, *Proc. 2nd Int. Conf. on Synchrotron Radiation Instrumentation*, Stanford, 1985, *Nucl. Instrum. Meth.* **246** (1986).
53. SRI88, *Proc. 3rd Int. Conf. on Synchrotron Radiation Instrumentation*, Tsukuba, 1988, *Rev. Sci. Instrum.* **60** (1989).
54. SRI91, *Proc. 4th Int. Conf. on Synchrotron Radiation Instrumentation*, Chester, UK, 1991, *Rev. Sci. Instrum.* **63** (1992).
55. Welford, W., in *Progress in Optics*, Wolf, E. (Ed.), Vol. **IV**, 1965.
56. Welford, W. T., *Aberrations of the Symmetrical Optical System*, Academic Press, London, 1974.
57. West, J. B., H. A. Padmore, in *Handbook on Synchrotron Radiation*, Marr, G. V., I. H. Munro (Eds.), Vol. **2**, North-Holland, Amsterdam, 1987.
58. Wolfram Research, Inc., 100, Trade Center Drive, Champaign, IL 61820-7237.
59. Wuilleumier, F., Y. Farge (Eds.), *Int. Conf. on Synchrotron Radiation Instrumentation and New Developments*, *Nucl. Instrum. Meth.* **152** (1978), North-Holland, Orsay, France, 1977.

3. SPECTROGRAPHS AND MONOCHROMATORS USING VARIED LINE SPACING GRATINGS

James H. Underwood
Center for X-ray Optics, Lawrence Berkeley National Laboratory
Berkeley, California

3.1 Limitations of Uniformly Spaced Grating Instruments

Spherical concave gratings, with equally spaced grooves ruled along a chord, form the basis of a whole family of glancing incidence spectrographs, spectrometers, and monochromators for extreme ultraviolet (EUV) and soft x-ray spectroscopy (Chapter 2). Their usefulness arises from their ability to both disperse and focus the radiation. However, the optical power of a spherical grating varies as the wavelength is tuned, so that for a fixed entrance slit distance, the distance of the dispersed image varies with wavelength [Eq. (3)]. For example, in the simplest form of spectrograph/spectrometer using a single optical element—the grating—the source is fixed on the Rowland circle, the spectrum is dispersed around the Rowland circle, and the detector must conform to it or be scanned along it. Plane gratings have similar properties: Although the substrate has no optical power, the grating itself has optical power, which also varies as the instrument is tuned. Hence, the primary geometrical optics design consideration for any spectrometer or monochromator is the maintenance of focus while the grating (plane or spherical) is changing its focal properties as the wavelength is tuned over its range. There are additional restrictions on monochromators, which must deliver a single narrow band of wavelengths to an experiment, usually through a slit. Most commonly, it is required that the slit, and the direction of the dispersed radiation, be fixed. Until recently, most monochromators and spectrometers developed for use in synchrotron radiation research and other applications have used plane or spherical gratings with uniformly spaced rulings, and tuning/focusing schemes using a combination of rotation and translation of the grating and/or other optical elements have been developed, as described in Chapter 2. The present chapter describes the advantages to be gained by using varied line spacing (VLS) gratings, in which the groove spacing varies across the ruled width. These include, but are not limited to, simplification of the focusing and scanning conditions.

An aspect of performance common to all spectroscopic instruments using glancing incidence gratings is the correction of astigmatism. In the Rowland circle spectrograph, the spectral features are dispersed as vertical lines on the focal plane, because of the weak focusing of the grating in the sagittal direction (perpendicular to the plane of dispersion). This may be acceptable if a film or area detector is used, but it results in a reduction of the signal-to-noise ratio. A similar situation occurs in monochromators. The weak sagittal focusing by the grating leads to a line image at the exit slit, and a beam that continues to diverge after the slit.

Reduction of astigmatism can be achieved by methods analogous to those used for the astigmatism correction of a spherical mirror (Volume 31, Chapter 9). The spherical grating can be replaced by a toroid, with a radius of curvature in the equatorial direction smaller than that in the meridional direction. This is essentially the principle of the toroidal grating monochromator mentioned briefly in Chapter 2. Astigmatism can also be corrected with an auxiliary sagitally-focusing mirror; a cylindrical or elliptical mirror set orthogonally to the grating as in Fig. 27 of Volume 31, Chapter 9. This scheme allows the positions of the two orthogonal foci to be chosen independently, an advantage in many experimental situations. Alternatively, a toroidal mirror can be used, as in the spherical grating monochromator (SGM) configuration discussed in the previous chapter. Whichever scheme is chosen, if the auxiliary mirror is considered to operate only in the sagittal direction (the direction perpendicular to the plane of dispersion) then we can confine the treatment that follows to the meridional plane of the grating.

3.2 Paraxial Focusing Equations for VLS Gratings

Rowland noted that, for a given wavelength, a theoretically perfect grating for one position of the entrance and exit slits could be ruled on *any* surface, plane or otherwise [1]. This result holds in general for only one wavelength, but it clearly applies to virtual as well as to real sources. This construction clearly requires a grating with nonuniform spacing. Interferometrically and numerically controlled ruling engines have made it possible to introduce such programmed variations in the spacing. This concept, pioneered by Harada and Kita [2] of the Hitachi Corporation, has led to a new family of designs for EUV/soft x-ray spectroscopic instrumentation. Most designs use VLS gratings that have been mechanically ruled by such a programmable engine; this is the most general case, since the ruling variation can be almost any arbitrary function. It is also possible to make VLS gratings by the holographic recording method, and recording geometries have been devised which allow duplication of some groove patterns produced by mechanical ruling. The discussions of this chapter apply equally to either type of grating.

We first derive paraxial formulae for the grating variation required for point-to-point focusing with a plane VLS grating. "Paraxial" means that the results are valid for an infinitesimal pencil of rays near the optical axis, which we take to mean the principal ray through the center C of the grating. We use the grating equation in the form:

$$m\lambda = \sigma(\sin \alpha + \sin \beta) \tag{1}$$

Consider a grating whose groove period varies with the ruled width w, with period σ_0 at its center C. If we express the period σ as a function of the ruled width:

$$\sigma(w) = \varepsilon_0 + \varepsilon_1 w + \varepsilon_2 w^2 + \varepsilon_3 w^3 + \cdots \tag{2}$$

where $\varepsilon_0 = \sigma_0$, $\varepsilon_1 = d\sigma/dw$, $\varepsilon_2 = d^2\sigma/dw^2$ ...

We take as a starting point the equation for the defocus term F_{20} of the expansion of the light path function F for a concave VLS grating of radius R_G:

$$2F_{20} = \frac{\cos^2 \alpha}{r} + \frac{\cos^2 \beta}{r'} - \frac{\cos \alpha + \cos \beta}{R_G} + \frac{m\lambda}{\sigma_0^2} \frac{d\sigma}{dw} \tag{3}$$

The focal condition requires that $F_{20} = 0$, so that for a plane grating ($R_G = \infty$) we obtain:

$$\frac{m\lambda}{\sigma_0^2} \frac{d\sigma}{dw} + \frac{\cos^2 \alpha}{r} + \frac{\cos^2 \beta}{r'} = 0 \tag{4}$$

This equation provides the linear variation $d\sigma/dw$ of the grating period required to focus from the arbitrary point O to the arbitrary point I at wavelength λ. It is the equivalent of the optical "thin lens formula" for a plane VLS grating, and correspondingly requires sign conventions. The focusing grating produces a multiplicity of images corresponding to the values of m. For $r = \infty$ Eq. (4) becomes:

$$\frac{\cos^2 \beta}{r'} = -\frac{m\lambda}{\sigma^2} \frac{d\sigma}{dw} \tag{5}$$

The possibilities are illustrated in Fig. 1. If $d\sigma/dw$ is negative (upper panels), that is, the groove period decreases toward the right of the figure, the positive orders produce real images whereas the negative orders produce virtual ones. When $d\sigma/dw$ is positive (lower panels); positive orders produce virtual images and negative orders real images. The case for real and virtual objects can be found by reversing the rays. The sign convention is that r and r' are positive for real, and negative for virtual, objects or images. A VLS grating is thus converging if $m(d\sigma/dw) < 0$, diverging if $m(d\sigma/dw) > 0$. This equation also implies that for a finite object distance, there exists a "collimating wavelength" where the image goes to infinity.

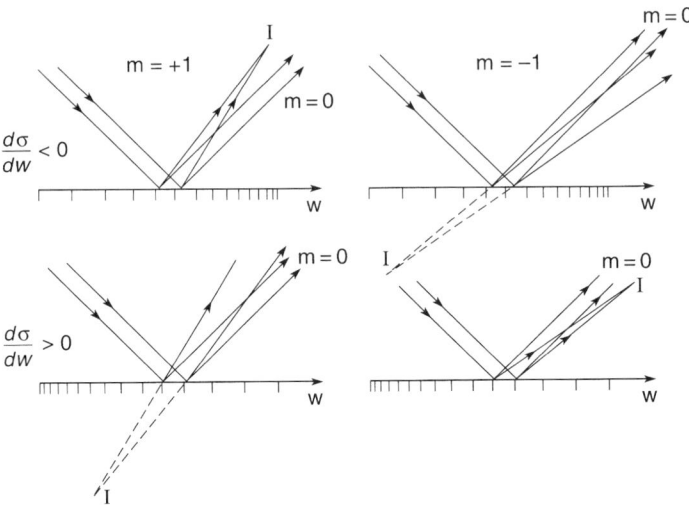

FIG. 1. Focusing by a plane VLS grating with source at infinity. If $d\sigma/dw$ is negative (upper panels), that is, the groove period decreases toward the right of the figure, the positive orders produce real images whereas the negative orders produce virtual ones; when $d\sigma/dw$ is positive (lower panels) positive orders produce virtual images and negative orders produce real images.

3.3 Harada-Style Focusing

The simplest focusing system consists of a single plane converging VLS grating forming a real image of a real object (Fig. 2a). The choice of positive or negative order depends on other factors such as the required dispersion. Using this principle, Harada et al. [3] designed the first soft x-ray VLS monochromator for synchrotron radiation for installation at the Photon Factory (Fig. 2b). The grating operates in the weakly divergent, almost parallel light, from the distant undulator source with object distance $OC = 20$ m, image distance $CI = 1$ m, $m = +1$, and angle of incidence $\alpha = 89°$. With a 2400 l/mm grating ($\sigma_0 = 417$ nm) we see from Eq. (4) that focusing at a wavelength of 5 Å ($\beta = 87.02°$), requires a linear variation of grating spacing $d\sigma/dw = 9.4375 \times 10^{-7}$. This results in a groove spacing variation of 511–322 nm across the 200 mm grating width, a variation easily achieved with a numerically controlled ruling engine. The range of the monochromator is 5 to 100 Å. This requires variable included angle (the horizon wavelength for fixed included angle $\alpha - \beta = 176.02°$ is 10.05 Å). A scanning curve (a functional relationship between α and β) is thus required. Harada and colleagues chose α and β to maintain focus, that is, so that the monochromator remains in focus for fixed r and r'. In a later paper, Itou et al. [4] developed

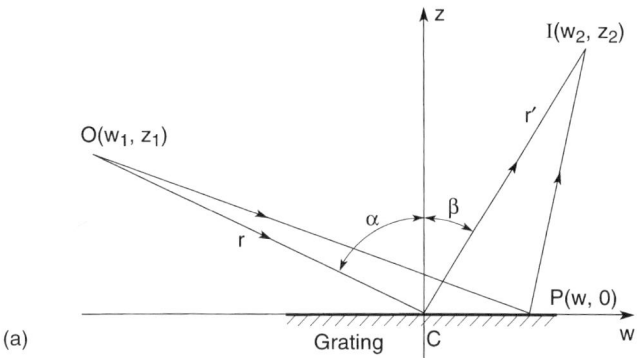

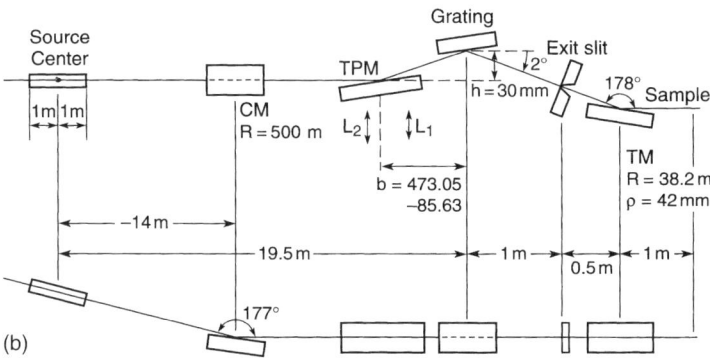

FIG. 2. (a) The first soft x-ray monochromator for synchrotron radiation [2] used a VLS grating that formed a real image (I) of a real object (O). (b) Side view (above) and top view (below) of the VLM19 beamline at the Photon Factory, where the source is a 2-m long undulator. The cylindrical mirror (CM) provides horizontal focusing direction. The tilting plane mirror (TPM) is positioned by two precise linear slides, L_1 and L_2, to maintain the focusing condition of Eq. (6). Toroidal mirror (TM) focuses the light on the sample.

an exact formula for this optimization; this may be obtained from Eq. (4). Substituting for $\cos^2 \beta$ using the grating Eq. (1), we obtain:

$$-\left(\frac{1}{r} + \frac{1}{r'}\right)\sin^2 \alpha + \frac{2M}{r'}\sin \alpha + \frac{1}{r} + \frac{1 - M^2}{r'} + \frac{M}{\sigma_0}\frac{d\sigma}{dw} = 0 \quad (6)$$

where $M = m\lambda/\sigma_0$. This quadratic in $\sin \alpha$ can be solved for any value of λ to yield the appropriate pair of (α, β) values.

A monochromator of this type is installed on an undulator beamline at the Photon Factory [5]. Horizontal focusing is accomplished with an orthogonal cylindrical mirror. The resolving power $E/\Delta E$ was computed by ray tracing to be 4600 at a photon energy of 1239.85 eV and 10,000 at 91.2 eV. The measured performance was similar. Advantages of this style of monochromator are the large (20:1) demagnification of the source (which results in a short and compact instrument); efficient tracking of the efficiency curve of the blazed grating; and plane, easily manufactured, optical surfaces. A disadvantage is the relatively complicated motion required to maintain the (α, β) relationship required for focusing. This is implemented by translating and rotating an auxiliary mirror whose motion is synchronized with the grating rotation by an arm running on a cam. A further disadvantage is that zero order light is not focused, because only plane elements are used. Focused zero order light is valuable for calibration and alignment as we shall see.

3.3.1 Higher Order Corrections

Equation (6) gives only the linear variation of the d-spacing. The other terms in Eq. (2), which are analogous to the spherical aberration terms of Eq. (25) of Volume 31, Chapter 9, are required to specify the groove spacing for maximum correction. Itou *et al.* [4] developed these up to the fourth order. Harada *et al.* [3], originally calculated the groove spacing variation using an analytical ray trace formula; from Fig. 1, we see that the local groove spacing $\sigma(w)$ is:

$$\sigma(w) = m\lambda/[\sin \alpha(w) + \sin \beta(w)] \qquad (7)$$

where

$$\alpha(w) = \tan^{-1}[(x_1 - w)/y_1]; \qquad \beta(w) = \tan^{-1}[(x_2 - w)/y_2]$$

Evidently, this procedure provides correction of all orders for one pair of conjugates (r, r') and value of λ. The small aperture ensures that the correction will be relatively good at other wavelengths.

3.4 Hettrick-Style Focusing

With Harada-style focusing both source image points are real; it is also possible to have a real object and a virtual image (or vice versa). Hettrick [6] pointed out the advantages of operating a plane VLS grating in the converging light produced by a concave mirror. In this case, the object is virtual and a real image is formed. In this arrangement, the image distance can be held constant for any kind of scanning motion, which can be seen as follows. In Eq. (4), r and r' may be chosen independently; having chosen either, the other can be chosen to satisfy the focal condition [Eq. (4)] at *two* wavelengths λ_1 and λ_2. Let the corresponding incidence and diffraction angles be (α_1, β_1) and (α_2, β_2). Then from the grating Eq. (1) we obtain:

$$\frac{-1}{\sigma}\frac{d\sigma}{dw} = \frac{(\cos^2\alpha_1)/r + (\cos^2\beta_1)/r'}{\sin\alpha_1 + \sin\beta_1} = \frac{(\cos^2\alpha_2)/r + (\cos^2\beta_2)/r'}{\sin\alpha_2 + \sin\beta_2} \qquad (8)$$

or:

$$\frac{-r'}{r} = \frac{\cos^2\beta_2(\sin\alpha_1 + \sin\beta_1) - \cos^2\beta_1(\sin\alpha_2 + \sin\beta_2)}{\cos^2\alpha_2(\sin\alpha_1 + \sin\beta_1) - \cos^2\alpha_1(\sin\alpha_2 + \sin\beta_2)}$$

$$= \frac{\lambda_1\cos^2\beta_2 - \lambda_2\cos^2\beta_1}{\lambda_1\cos^2\alpha_2 - \lambda_2\cos^2\alpha_1} \qquad (9)$$

After the optimum value of r'/r has been found for Eq. (9), Eq. (4) is used to find the corresponding optimized value of the linear term $d\sigma/dw$ in the grating period variation.

By introducing a large grating radius of curvature R_G as another free parameter, it is possible to optimize at three wavelengths to widen the tuning range. Returning to Eq. (3) for F_{20}, for a given value of r, we seek the values of r', R_G, and $d\sigma/dw$ that make $F_{20} = 0$ at three separate wavelengths λ_1, λ_2, and λ_3 or, equivalently, three incidence and diffraction angles pairs (α_1, β_1), (α_2, β_2), and (α_3, β_3). If the *two-wavelength* optimizations of (r'/r) are denoted as:

$$-\left(\frac{r'}{r}\right)_{12} = \frac{\lambda_1\cos^2\beta_2 - \lambda_2\cos^2\beta_1}{\lambda_1\cos^2\alpha_2 - \lambda_2\cos^2\alpha_1}, \text{ etc.,} \qquad (10)$$

then the *three-wavelength* optimization $(r'/r)_{123}$ is given by:

$$(r'/r)_{123} = [(r'/r)_{12} - S(r'/r)_{13}]/(1-S) \qquad (11)$$

where

$$S = \frac{(\lambda_1 a_3^2 - \lambda_3 a_1^2)[\lambda_1(a_2 + b_2) - \lambda_2(a_1 + b_1)]}{(\lambda_1 a_2^2 - \lambda_2 a_1^2)[\lambda_1(a_3 + b_3) - \lambda_3(a_1 + b_1)]} \qquad (12)$$

and $a_1 = \cos(\alpha_1)$; $b_1 = \cos(\beta_1)$, etc.

The grating radius R_G and $d\sigma/dw$ can be found by back substitution, for example:

$$\frac{R_G}{r} = \frac{\lambda_1(a_2 + b_2) - \lambda_2(a_1 + b_1)}{(\lambda_1 a_2^2 - \lambda_2 a_1^2)[1 - (r'/r)_{12}/(r'/r)_{123}]} \qquad (13)$$

Application of this optimization leads to values of R_G in the kilometer range; the curvature can be concave or convex. Eqs. (8)–(13) are completely general and can be applied to *any* instrument with a real and a virtual focus. The particular values of α_i and β_i will be determined by the kind of instrument, but the optimization is valid independently of this choice. Thus, the mechanical motion for scanning wavelength in the spectrometer or monochromator can be freed from the requirement that it maintain focus and can be designed to achieve other desirable features. In the mechanical area, these may include simplicity of motion and ease of fabrication, stability of geometry, and retrofitting of an existing design. In the optical area, these may include high (or low) resolving power, operation in either positive or negative order, high throughput, good tracking of the grating efficiency curve, optimum suppression of higher orders, and zero order in focus. Some specific functional relationships between α and β are:

a. $\alpha = $ constant ("spectrograph")
b. $\alpha - \beta = 2\phi = $ constant ("constant deviation angle 2ϕ")
c. $\alpha + \beta = 2\psi = $ constant ("strictly on-blaze, blaze angle $= \psi$"). This last condition gives optimum tracking of the efficiency of a blazed grating.

3.5 Flat Field Spectrographs and Spectrometers

3.5.1 High Resolution Tunable Plane Grating Spectrograph

If α is held constant in Eq. (9) (condition a in the previous section), we can bring two wavelengths λ_1, λ_2 to the same focal distance, that is, to form a spectrum on a focal plane which is essentially flat and perpendicular to the incident rays (an "erect" focal plane). This can be an advantage over the Rowland circle style of spectrograph if the detector cannot be curved to conform to the Rowland circle; this applies, for example, to the streak cameras used for the spectroscopy of fast phenomena. If the grating is now rotated maintaining a constant deviation angle between the entrance slit-center of the grating-center of the detector (condition b of the previous section), a third wavelength λ_3 is brought into focus (although the focal plane will not necessarily be flat at this wavelength). Thus, the instrument can be tuned over a considerable range while maintaining a substantially fixed and erect focal plane.

Using these principles, Hettrick et al. [7] designed and built a high resolution spectrograph for the wavelength region 80–200 Å and measured a resolving power $\lambda/\Delta\lambda = 35{,}000$ using a Penning source. This instrument was later used to measure the gain-dependent line width of selenium x-ray lasers [8]. The optical design (Fig. 3), was described in detail by Underwood and Koch [9]. The grating was mechanically ruled with a central groove density of 1800 mm^{-1} and a spacing variation optimized for $\lambda_1 = 103.09$ Å, $\lambda_2 = 204.88$ Å. The groove density varied from 1838 to 1763, or $\pm 2.1\%$ across the ruled width of the grating. Thus, this style of focusing requires a less extreme spacing variation than the Harada style. Figure 4 shows the variation of r' with wavelength under the condition of constant deviation angle $2\phi = 164°$. It can be seen that the focal length remains essentially constant until the "collimating wavelength"

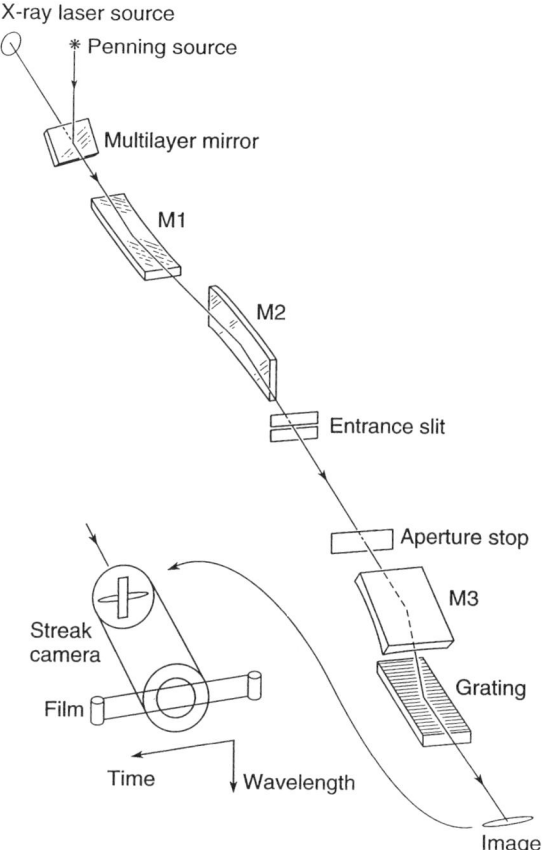

FIG. 3. The tunable spectrograph of Hettrick *et al.* (1988) as used to measure the gain-dependent line width of selenium x-ray lasers [7]. M1 images the source on the entrance slit, M2 images the source in the horizontal direction on the streak camera cathode, M3 provides the converging beam to the plane VLS grating. The retractable multilayer mirror allows calibration and alignment with EUV lines from the Penning source.

[cf. Eq. (5)] is approached. Figure 5 shows the shape of the fields with the spectrograph tuned to various central wavelengths. It can be seen that the true "erect field" occurs at a wavelength of 217.75 Å.

3.5.2 Spherical Grating Spectrometers

Osborn and Callcott [10] showed how a VLS *spherical* grating could be used to modify the focal curve of a Rowland circle style of spectrograph/spectrometer and to achieve a similar effect, a focal plane better optimized to the required flat

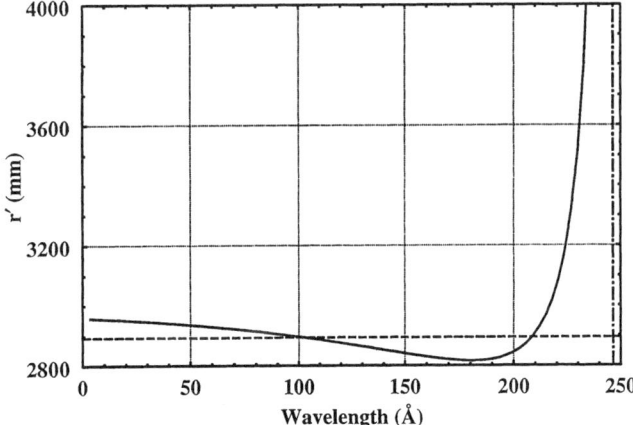

FIG. 4. The variation of r' with wavelength under the condition of constant deviation angle $2\phi = 164°$ for the tunable spectrograph of Fig. 3. The focal length remains essentially constant up to the "collimating wavelength" of 246.6 Å.

detector configuration. This is a simplification of the spherical mirror/plane VLS grating (SM/PVLSG) configuration in that the focusing and dispersive elements have been combined into one. However, this configuration gives up the "tunability" aspect of the SM/PVLSG design, since rotation of the grating results in strong defocusing.

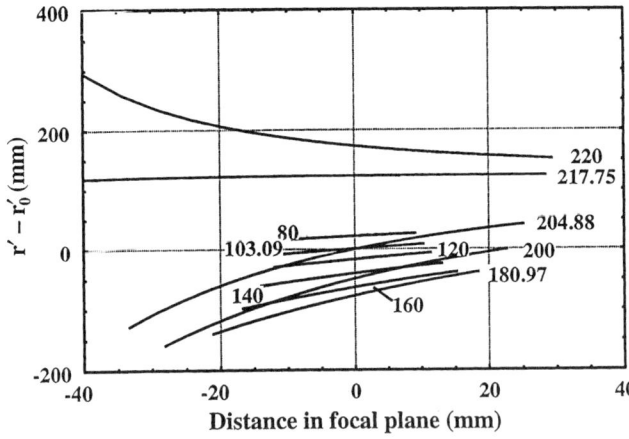

FIG. 5. The shape of the tunable spectrograph fields at a variety of wavelength tunings. Each curve is labeled with the corresponding wavelength in Å at the field center. While most fields are essentially flat, the true "erect field" occurs at a wavelength of 217.75 Å.

With either of these configurations, it is clear that if the focal plane detector is replaced by an exit slit situated at the center of the field and wavelength is scanned in some manner, this instrument becomes a monochromator. Before proceeding to monochromators, we discuss methods for obtaining the higher terms ε_2, ε_3, ... in the groove spacing variation, and the optimum correction of aberrations over a range of wavelengths.

3.6 Light Path Function for a System of a Mirror and a VLS Grating

In the design of the tunable spectrograph (Section 3.5.1), the grating spacing variation was calculated via the algorithm of Eq. (7). This clearly provides correction of all orders for a single wavelength λ. In fact, because of the small aperture, terms higher than ε_2 were negligible and this method provided good correction over the wavelength range of the instrument. It is possible to compute ε_2 and higher terms from the expansion of the light path function for a plane grating and a point source, but this is of little utility for this type of system, because:

a. The focusing mirror introduces aberrations, especially if used at large demagnification, and the virtual source for the grating is not a point. The light path function for a plane grating illuminated by a perfectly converging light beam is not appropriate. We require the appropriate expression for a combination of a spherical mirror and a grating.
b. Determination of the coefficients from the terms in the expansion of the light path function provides correction at a single wavelength only. A procedure for optimizing over an extended spectral range is needed.

Concerning point (a), it is possible to use an elliptical mirror to eliminate spherical aberration. However, aspherical optics are best avoided in monochromator design. Also, it is not necessary; the variation of grating spacing can be chosen for perfect correction of the mirror aberrations. This can be seen from the following discussion. Figure 6 is a reproduction of Fig. 19 of Volume 31, Chapter 9; it depicts the spherical aberration caustic of a mirror operating at $M = -0.5$ with a plane VLS grating interposed between the mirror and the Gaussian focus I_g. The paraxial bundle of rays from the center C of the mirror strikes the grating at an incidence angle $\alpha(0)$ at the center C' ($w = 0$). $C'I_g$ is the distance $r(0)$ of the virtual source for this bundle, which is brought to a focus at I by the grating, where $r'(0) = C'I$ and $\beta(0)$ are determined by Eq. (4). A ray bundle through another point A' at a finite value of w is incident at $\alpha(w) \neq \alpha(0)$, is focused at I_A, where $A'I_A = r(w) \neq r(0)$, and $\beta(w) \neq \beta(0)$. This bundle can

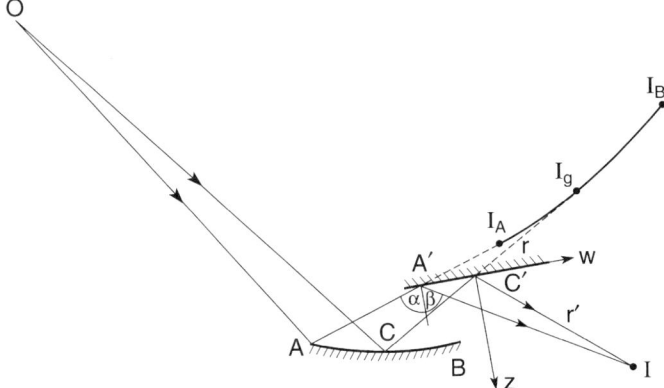

FIG. 6. The spherical aberration caustic of a mirror operating at M = −0.5 with a plane VLS grating interposed between the mirror and the Gaussian focus I_g. I_A and I_B are the images of O produced by the points A and B on the mirror. Proper choice of the line spacing variation collapses the caustic to the single point I.

also be focused at I by the proper choice of the local grating spacing $\sigma(w)$ via Eq. (4).

It is clear that this approach can be developed into an analytical expression for the higher order corrections using the equations for the caustic of a spherical mirror. The same result can be obtained by developing the expansion for the light path function of a spherical mirror-plane VLS grating system. Amemiya et al. [11] have given explicit expressions for the terms F_{j0} in the light path function up to $j = 4$. Koike et al. [12] adopted a different approach. Using a simple ray tracing procedure, they developed an analytical merit function that closely represents the variance of an infinite number of ray traced spots in the focal plane. This method allows optimization over a finite wavelength range. It was used to determine the ruling parameters for a series of gratings used in three monochromators having strongly demagnifying (M = −0.1) converging mirrors.

3.7 Monochromators for Synchrotron Radiation

3.7.1 Scanning by Grating Rotation $\alpha - \beta$ = Constant ("Constant Deviation Angle $2\phi''$")

Since the image distance remains constant as the grating is rotated, the tunable spectrograph design described in Section 3.5 can be converted into a fixed deviation monochromator by rotating the grating about its pole [13, 14].

For synchrotron radiation beamlines, considerable advantages ensue from making the converging mirror demagnifying. One of these is a compact instrument. On a synchrotron radiation beamline, the source is 10–20 m behind the shield wall, and a unit magnification mirror places the exit slit at a further, approximately equal distance, resulting in a long and cumbersome beamline. Demagnification by a factor ~10 yields a 1–2 m exit arm length, and a monochromator that can be mounted on a single relatively small stand even if the monochromator is operated in the inside (m > 0) orders. Furthermore, many experiments using third generation synchrotron light sources *require* a demagnifying monochromator to reduce the linear dispersion at the exit slit plane and avoid loss of flux as a result of excessive spectral resolution.

To illustrate: Suppose that to perform micro-x-ray photoelectron spectroscopy (μ-XPS) or some analogous technique on a solid specimen, a spot of x-rays (or EUV) of 1 μm diameter is formed on the sample. In the vertical (or dispersion) direction, this spot will either be formed by demagnifying the exit slit or by demagnifying an intermediate aperture placed at the position of a real image. In the second case this aperture becomes the effective exit slit, so we need only consider the first case. Suppose the total demagnification factor of the post-slit optics is 10×, then the effective exit slit size is 10 μm. Setting the actual exit slit larger than this value has no effect on either the photon flux or the spectral resolution. For a typical monochromator (175° deviation, -1 order, 1000 l/mm grating, 10 meter exit arm length) the resolving power in the middle of the spectral range is $\sim 5 \times 10^4$ for a 10 μm exit slit. This is more than an order of magnitude greater than almost any solid state experiment requires, and flux will be wasted as a result of this "empty" resolving power. For microscopy experiments that require even smaller focal spots (~0.1 μm) the mismatch will be even more extreme. Better matching requires a combination of (a) post-slit optics with higher demagnification and (b) lower linear dispersion in the monochromator. It is in fact quite difficult to construct a monochromator with sufficiently low linear dispersion, and thus the advantage offered by the demagnifying plane grating VLS design is very useful.

Three monochromators have been constructed using these principles: Two are installed on bend magnet sources (at the ALS and BESSY), and the third on an undulator source at the ALS. The first of these was built for a calibration and standards beamline (Fig. 7) equipped with a reflectometer (Volume 31, Chapter 10) for measuring the optical properties of mirrors, gratings, multilayers, etc. [15]. The stability and small size of the ALS source allows entrance-slitless operation, with a consequent large gain in intensity over a design equipped with an entrance slit. The horizontally collecting and focusing mirror M1 reimages the bending magnet source in the horizontal direction at the center of the reflectometer. M2 demagnifies the ALS source by a factor of 10 and provides the virtual source for the three interchangeable gratings. These are

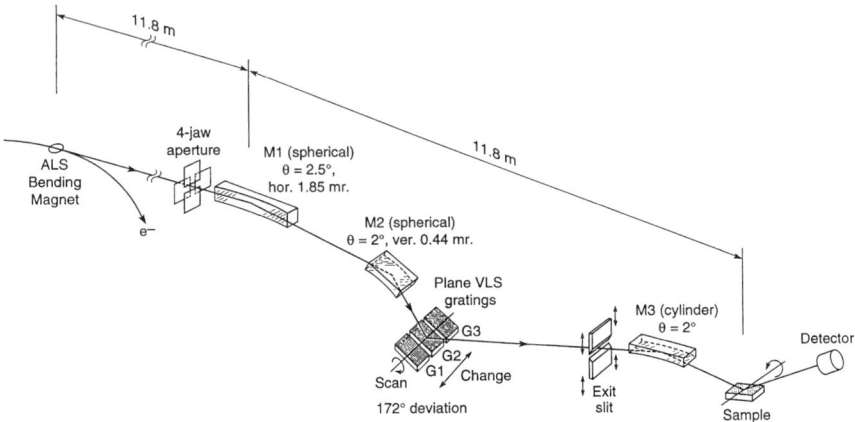

FIG. 7. Calibration and standards beamline at the ALS [14]. M1 images the source on the sample in the horizontal direction. M2 provides the converging beam for the three interchangeable gratings with 300, 600, and 1200 l/mm central groove density. M3 is a bendable mirror that reimages the slit on the sample. Alternatively, it can be adjusted to focus at other locations, including infinity.

mechanically ruled with central groove densities of 300, 600, and 1200 l/mm. With a fixed deviation angle of 172° an energy range of 50–1000 eV is covered, with a source size limited resolution $\lambda/\Delta\lambda \approx 7000$ at 400 eV, and a maximum throughput of 10^{12} photons/sec/0.1% BW at 100 eV.

The compactness of the monochromator and its stable, well-determined geometry allow absolute determination of the wavelength. This is aided by the in-focus zero-order image, which is used as a beam position monitor to track residual source motions. This feature also allows visible synchrotron light to be focused in the experimental chamber as an aid to alignment.

An almost identical monochromator is installed on a bending magnet at BESSY. This monochromator is also slitless, but as the BESSY source size is relatively large, it is designed to operate in +1 order to maximize dispersion and resolving power. A third monochromator, designed as part of a beamline for interferometric testing of EUV lithography projection optics at 100 eV [16], is installed on an undulator beamline at the ALS. In this application, the beamline/monochromator combination was required to illuminate a sub-micron pinhole with sufficient flux for point diffraction interferometry. It thus provides a very good illustration of the requirement for high flux with moderate spectral resolution ($\lambda/\Delta\lambda \sim 1000$). This goal was achieved by lowering the linear dispersion of the monochromator by (a) using a demagnification of 10:1 in the converging mirror, (b) setting the grating halfway between the converging mirror and the exit slit, and (c) using a coarsely ruled grating (200 lines/mm).

Amemiya et al. [11] have described a monochromator of this type using a holographically recorded grating. The recording parameters were optimized using the terms F_{j0} in the light path function (see Section 3.6), although in this case the converging mirror was used at magnification close to unity, so that minimum correction of spherical aberration was required from terms higher than F_{20}. Amemiya et al. showed also that, with a single grating, a variation of the included angle could be achieved through the use of a plurality of fixed spherical mirrors.

3.7.2 Scanning with an Auxiliary Mirror (α = Constant)

The spectrograph configuration of Section 3.5 can also be converted into a monochromator by the addition of an auxiliary mirror to sweep the spectrum over the exit slit while maintaining fixed incidence angle. The mirror M2 in this figure simultaneously translates and rotates while the grating remains fixed. Callcott et al. [17] (see also [18]) used this concept to retrofit a monochromator at NSLS, originally equipped with a transmission grating, with a VLS plane grating. The instrument was later moved to CAMD, where it is operating over the energy range 100–800 eV with a resolving power of ~750 and a throughput of ~10^{12} photons/sec at 310 eV. Thus, this design provides a high throughput, moderate resolution design for exciting soft x-ray emission in compounds for analysis by a soft x-ray spectrometer or spectrograph.

3.7.3 Variable Included Angle

Equation (9) [or alternatively, Eq. (11)] can be used to optimize *any* form of PGM, including variable-included angle designs, which have the advantages of optimizing grating efficiency over a wide energy range with a single grating, and efficient higher order suppression. However, many such designs, such as the SX-700 style of PGM developed by Petersen [19] are already optimized for focus by maintaining $\cos^2 \beta / \cos^2 \alpha$ = constant, which with $d\sigma/dw = 0$ ensures that r'/r = constant [Eq. (4)]. In this case, the VLS grating can be used to improve resolution by correcting the higher order aberrations of the mirror (if spherical and not elliptical) and the grating. An example of this approach is given by Lu [20]. Alternatively, the monochromator can be designed for fixed focus while maintaining an α-β relationship that maximizes grating efficiency and/or higher order suppression. Koike and Namioka [21] have described a scheme in which the required incidence and diffraction angles are achieved through the use of an auxiliary plane mirror that simultaneously rotates and translates.

3.7.4 Spherical Grating—VLS Monochromators

Schemes for maintaining focus using spherical VLS gratings have also been devised. The "in-focus monochromator" (IFM) described by Hettrick [22], is a fixed deviation angle spherical grating monochromator using a VLS grating. To scan wavelength, the grating is rotated about an axis fixed in space and is simultaneously translated laterally to bring the appropriate set of groove parameters into the illuminated region. The optimum set of groove parameters is found by numerically iterating to eliminate defocusing at all wavelengths. This scheme requires a grating that is only 50% larger than a conventional (nontranslating) spherical grating. A monochromator ("HERMON") designed on this principle has been installed at the Aladdin storage ring [23] and a performance approaching the theoretical resolving power of 10,000 has been reported.

3.8 Holographically Recorded Gratings

All the spectrometers and monochromators described in this chapter used gratings ruled mechanically on an interferometrically controlled ruling engine. It is now possible to make VLS gratings by holographic recording techniques. The monochromator of Amemiya *et al.* [11], which uses a holographic grating, was described in the previous section. Koike and Namioka [24, 25] have shown how VLS plane gratings, which are interchangeable with the conventionally ruled, blazed gratings ruled for the monochromators described in Section 3.7.1, can be made holographically by the interference of a spherical wavefront and an aspheric wavefront. A prototype grating has been recorded in this manner for the monochromator in the calibration and standards beamline at the ALS (Section 3.7.1). As expected, the holographic grating exhibited somewhat lower scattered light than the conventional ruled grating [26].

3.9 Conclusions

The VLS principle provides a powerful tool for the design of EUV and soft x-ray spectrographs, spectrometers, and monochromators, providing an extra degree of freedom to escape the restrictive focusing conditions of the constant-spacing plane or spherical grating. Additionally, the spherical aberration of a concave mirror, and other forms of aberration, can be corrected with the appropriate line-spacing variation. This allows the construction of essentially aberration-free instruments. The number of instruments constructed using mechanically ruled VLS gratings is growing, and it is expected that the development of holographic VLS recording methods will lead to wider acceptance of such designs.

References

1. Stroke, G. W., Diffraction gratings. *Handbuch der Physik* **XXIX**, p. 473 (1967).
2. Harada, T. and T. Kita, Mechanically ruled aberration corrected concave gratings. *Appl. Opt.* **19**, 3987 (1980).

3. Harada, T., M. Itou, and T. Kita, A grazing incidence monochromator with a varied-space plane grating for synchrotron radiation. Proc. SPIE **503**, *Application, Theory and Fabrication of Periodic Structures*, J. Lerner (Ed.), 114–118 (1984).
4. Itou, M., T. Harada, and T. Kita, Soft x-ray monochromator with a varied-space plane grating for synchrotron radiation: design and evaluation. *Appl. Opt.* **28**, 146–153 (1989).
5. Fujisawa, M., A. Harasawa, A. Agui, M. Watanabe, A. Kakizaki, S. Shin, T. Ishii, T. Kita, T. Harada, Y. Saitoh, and S. Suga, Varied line-spacing plane grating monochromator for undulator beamline. *Rev. Sci. Instrum.* **67**, 345–349 (1996).
6. Hettrick, M. C., and S. Bowyer, Variable line-space gratings: new designs for use in grazing incidence spectrometers. *Appl. Opt.* **22**, 3921–3924 (1983).
7. Hettrick, M. C., J. H. Underwood, P. J. Batson, and M. J. Eckart, Resolving power of 35,000 (5 mÅ) in the extreme ultraviolet employing a grazing incidence spectrometer. *Appl. Opt.* **27**, 200–202 (1988).
8. Koch, J. A., B. J. McGowan, L. B. Da Silva, D. L. Matthews, J. H. Underwood, P. J. Batson, R. W. Lee, R. A. London, and S. Mrowka. Experimental and theoretical investigation of neonlike selenium x-ray laser spectral linewidths and their variation with amplification. *Phys. Rev.* **A50**, 1877–1898 (1994).
9. Underwood, J. H., and J. A. Koch, High resolution tunable spectrograph for x-ray laser line width measurements with a plane varied line spacing grating. *Appl. Opt.* **36**, 4913–4921 (1997).
10. Osborn, K. D., and T. A. Calcott, Two new optical designs for soft x-ray spectrometers using variable-line-space gratings. *Rev. Sci. Instrum.* **66**, 1–6 (1995).
11. Amemiya, K., Y. Kitajima, T. Ohta, and K. Ito, Design of a holographically recorded plane grating with a varied line spacing for a soft x-ray grazing incidence monochromator. *J. Synchrotron Rad.* **3**, 282–288 (1996).
12. Koike, M., R. Beguristain, J. H. Underwood, and T. Namioka, A new optical design method and its application to an extreme ultraviolet varied line spacing plane grating monochromator. *Nucl. Instrum. Methods* **A347**, 273–277 (1994).
13. Hettrick, M. C., and J. H. Underwood, Stigmatic high throughput monochromator for soft x-rays. *Appl. Opt.* **25**, 4228–4231 (1986b).
14. Hettrick, M. C., and J. H. Underwood, Varied-space grazing incidence gratings in high resolution scanning spectrometers, in *Short Wavelength Coherent Radiation, Generation and Applications*, J. B. Bokor and D. T. Attwood (Eds.). AIP Conf. Proc. **147**, 237–245 (1986a). See also: Hettrick, M. C., and J. H. Underwood, Optical system for high resolution spectrometer/monochromator. U.S. Patent 4,776,696 (Oct. 11, 1988).
15. Underwood, J. H., E. M. Gullikson, M. Koike, P. C. Batson, P. E. Denham, R. Steele, and R. Tackaberry, Calibration and standards beamline 6.3.2 at the ALS. *Rev. Sci. Instrum.* **67**(9), 1–5, (1996) (available on CD-ROM only).
16. Tejnil, E., K. A. Goldberg, S. Lee, H. Medecki, P. J. Batson, P. E. Denham, A. A. MacDowell, J. Bokor, and D. T. Attwood, At-wavelength interferometry for EUV lithography. *J. Vac. Sci. Tech.* **B15**, 2455–2461 (1997).
17. Callcott, T. A., W. L. O'Brien, J. J. Jia, Q. Y. Dong, D. L. Ederer, R. Watts, and D. R. Mueller, A simple variable line space grating monochromator for synchrotron radiation light source beamlines. *Nucl. Instrum. Methods* **A319**, 128–134 (1992).
18. Haass, M., J. J. Jia, T. A. Callcott, D. L. Ederer, K. E. Miyano, R. N. Watts, D. R. Mueller, C. Tarrio, and E. Morikawa, Variable groove spaced grating monochromators for synchrotron light sources. *Nucl. Instrum. Methods* **A347**, 258–263 (1994).
19. Petersen, H., The plane grating and elliptical mirror: a new optical configuration for monochromators. *Opt. Comm.* **40**, 402–406 (1982).

20. Lu, L.-J., Coma correction and extension of the focusing geometry of a soft x-ray monochromator. *Appl. Opt.* **34**, 5780–5786 (1995).
21. Koike, M., and T. Namioka, High-resolution grazing incidence plane grating monochromator for undulator radiation. *Rev. Sci. Instrum.* **66**, 2144–2146 (1995b).
22. Hettrick, M. C., In-focus monochromator: theory and experiment of a new grazing incidence mounting. *Appl. Opt.* **29**, 4531–4535 (1990). See also M. C. Hettrick, U.S. Patent 4,991,934 (1991).
23. Fisher, M.V., N. Steinhauser, D. Eisert, B. Winter, B. Mason, F. Middleton, and H. Hoechst, Combining rotation and translation in a variable line space high resolution soft x-ray monochromator: design requirements and performance evaluation of a novel grating mount. *Nucl. Instrum. Methods* **A347**, 264–268 (1994).
24. Koike, M., and T. Namioka, Aspheric wave-front recording optics for holographic gratings. *Appl. Opt.* **34**, 2180–2186 (1995a).
25. Koike, M. and T. Namioka, Plane gratings for high resolution grazing incidence monochromators: holographic grating versus mechanically ruled varied-line-spacing grating. *Appl. Opt.*, in press.
26. Underwood, J. H., E. M. Gullikson, M. Koike, and S. Mrowka, Experimental comparison of a holographically ruled and a mechanically ruled VLS grating, in *Gratings and Grating Monochromators for Synchrotron Radiation*, Proc. SPIE **3150**, 40–46 (1997).

4. INTERFEROMETRIC SPECTROMETERS

Anne P. Thorne
Blackett Laboratory
Imperial College
London, England

Malcolm R. Howells
Advanced Light Source
Lawrence Berkeley National Laboratory
Berkeley, California

4.1 Introduction

In contrast to the well established field of grating spectrometry, interferometric spectrometry is a new technique for the vacuum ultraviolet (VUV). There is no mention whatever of it in Samson's 1967 *Techniques of Vacuum Ultraviolet Spectroscopy*, although there is a precursor in the form of a short section on broadband interference filters. Of the two types of interferometric spectrometer described here, the Fabry–Perot was then widely used in the near UV, but Fourier transform spectrometry (FTS) was considered to be an infrared technique only. The situation now is quite different. Whereas the range of Fabry–Perot (F–P) interferometers has been somewhat extended, leading to limited use in the VUV, there has been a dramatic increase in both the wavelength range and the use of FTS. Operation in the visible and the near UV is now routine, and recent developments have shown it to be a viable technique for the VUV, with a current limit of about 140 nm. Moreover, instruments designed to work at still shorter wavelengths are under construction.

The fundamental differences between dispersive and interferometric spectrometry arise from the ways in which the spectral information is accessed. All spectrometers work by superposing a number of rays of varying phase, the number being 10^5–10^3 for a grating, a few tens for a F–P interferometer, and just two for the Michelson interferometer that is used for FTS. However, prisms and gratings distinguish different wavelengths by spreading them out spatially, using slits with their narrow dimension in the plane of dispersion, whereas interferometers work by imposing a wavelength-dependent spatial or temporal modulation on the signal and have axial symmetry. The well-known multiplex and throughput advantages of interferometric spectrometry, relevant in the IR but not necessarily in the VUV, stem from this difference. Because interferometric methods are somewhat unfamiliar to most VUV spectroscopists, we start with an initial overview of the instrumentation. More detail can be found in Refs. [1, 2].

An F–P interferometer consists of a pair of transmitting plates mounted parallel to one another with highly reflecting coatings on their facing surfaces. The incident light undergoes multiple reflections between these surfaces, with constructive interference giving a maximum of intensity for wavelengths λ satisfying

$$n\lambda = x = 2\mu t \cos\theta$$

where n is an integer, t is the plate separation, μ is the refractive index of the medium between the plates, and θ is the angle of incidence. The interference pattern consists of fringes of equal inclination (constant θ) localized at infinity. The traditional method of use was to fix μt and record the ring pattern photographically. Modern practice is to limit the angular spread ($\theta \sim 0$—see the discussion on the throughput advantage following) and vary either μ (by changing gas pressure) or t, recording the interference signal photoelectrically. Whichever method is used, the narrow free spectral range of the F–P requires an auxiliary monochromator for order-sorting, as discussed in Section 4.2.1.

FTS is usually based on a Michelson interferometer in which one (or sometimes both) of the mirrors is scanned to change the optical path difference x from zero up to some large positive or negative value L. As with the scanning F–P, the field of view is limited to the center of the ring pattern, and the signal is recorded as a function of x. This signal, the "interferogram," is the Fourier transform of the spectrum, and the latter is recovered, potentially with no ambiguity or "order overlap," by performing the inverse FT. Variants of this basic instrument are described in Sections 4.5–4.7.

FTS has largely superseded grating spectrometry in the infrared because of the high gains in signal-to-noise ratio arising from the multiplex and throughput advantages. In the visible and UV, where the dominant noise is photon noise or light source fluctuations rather than detector noise, the multiplex advantage does not exist; indeed, there can actually be a multiplex disadvantage, as shown in Section 4.3. The throughput advantage, however, applies to all interferometers in which the interfering beams are split by division of amplitude so that the interference pattern has circular symmetry. This is true for the F–P and for most of the FT interferometers in the following discussion. It is not true for division-of-wavefront interferometers, which require a slit geometry.

However, there are two other strong reasons for pursuing FTS into the VUV, one of which applies also to F–P spectrometry: the high spectral resolution that can be achieved. The resolution limit is determined by the maximum optical path difference between the interfering beams. In a grating spectrometer, this is limited by the physical size of the grating, whereas in an interferometer it is determined by the length of the scan (FTS) or the plate separation and the effective number of multiple reflections (F–P), both of which can be made much larger than the width of a diffraction grating. The second reason, which applies only to FTS, is the wavelength advantage that arises from the strictly linear wavenumber scale derived from the sampling intervals of the interferogram. This makes accurate wavelength calibration possible from a very small number of reference lines (in principle, only one is needed).

One might add as further advantages of interferometric spectrometry the smaller size and greater flexibility of the instruments: High resolution can be achieved without the long focal length required in a dispersive instrument, and, whereas the performance of the latter is essentially fixed during its manufacture, that of F–P and FT spectrometers can be to a large extent determined by the method of use.

Before discussing these two types of interferometer in more detail, we return to the throughput advantage, which they hold in common. The optical path difference x between the beams reflected at two parallel surfaces separated by a distance t in a medium of refractive index μ is given by

$$x = 2\mu t \cos \theta \simeq 2\mu t(1 - \theta^2/2) \tag{1}$$

where θ is the angle of incidence (Fig. 1), which is always small enough to justify the expansion. As t is increased, the ring spacing for any one wavelength gets finer, and the maximum value of θ must be correspondingly reduced if it is to include only the "central ring." For the Michelson interferometer, the criterion is that x should not vary by more than half a wavelength over the field of view at the maximum value of t (i.e., a phase change of π from center to edge) [1, 2]. Since L, the maximum path difference, is $2t_{\max}$, we have (for $\mu \sim 1$)

$$2t_{\max}\theta^2/2 \leq \lambda/2 \text{ giving } \theta_m^2 = \lambda/L$$

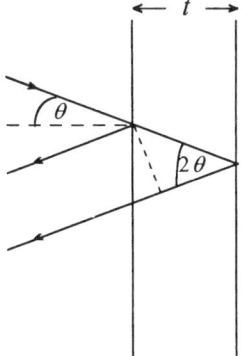

FIG. 1. The path difference x between the two beams reflected from parallel surfaces separated by t is given by $x = t \sec \theta(1 + \cos 2\theta) = 2t \cos \theta$, where θ is the angle of incidence.

By Eq. (16) the resolution limit in wave numbers ($\sigma = 1/\lambda$) is related to L by $\delta\sigma = \delta\lambda/\lambda^2 = 1/2L$, so in terms of the resolving power, $\mathscr{R} = \lambda/\delta\lambda$, we have

$$\theta_m = \sqrt{2/\mathscr{R}} \qquad (2)$$

This last relation holds also for the F–P interferometer. The angle $2\theta_m$ defines the field of view of the interferometer—that is, the angle subtended by the (circular) entrance aperture at the collimating lens or mirror. The throughput is proportional to the corresponding solid angle Ω. Thus, for an interferometric spectrometer the maximum allowable solid angle Ω_i is

$$\Omega_i = \pi\theta_m^2 = 2\pi/\mathscr{R} \qquad (3)$$

To quantify the throughput advantage, we compare this with the field of view for a grating spectrometer. The resolution of the latter is assumed to be slit-limited rather than diffraction-limited, usually a valid assumption for high-resolution spectrometry in the VUV. For a Littrow mounting with the grating at approximately 45°, the angle subtended by the slit width is given by $\theta_w \simeq 2/\mathscr{R}$. If β is the angle subtended by the slit height, the solid angle subtended by the slit area at the grating is given by $\Omega_g = 2\beta/\mathscr{R}$. The throughput advantage is then conventionally described by the ratio of the two solid angles for the same resolving power:

$$\Omega_i/\Omega_g = \pi/\beta \qquad (4)$$

At moderate to high resolution slit curvature and other aberrations effectively limit β to less than 0.01 radian (a 1 cm slit for a 1 m spectrometer), so there is an apparent throughput advantage of over two orders of magnitude. In the VUV, this gain is reduced for two reasons: The area of the interferometer beamsplitter is likely to be significantly smaller than that of the grating, and the number of reflections in the interferometer is inevitably greater than one. The real throughput ratio is likely to be below 100. Nevertheless, the gain in signal-to-noise ratio is still on the order of 10 for a photon-noise limited system. If the dominant noise is from source fluctuations, the noise simply scales with the signal, and a larger throughput confers no advantage.

In summary, therefore, the unique advantages offered by interferometric over grating spectrometry in the VUV are the higher resolution and (in the case of FTS) the more accurate wavenumbers attainable. For resolving powers of order 100,000 that can be reached with either technique in the region down to 100 nm or so, interferometry offers a significant gain in signal-to-noise for photon-limited systems and also more compact and flexible instrumentation. Because the attainable resolving power of grating spectrometers falls off almost linearly with wavelength (60,000 is the best value achieved at 20 nm), the resolution advantage is the primary reason for extending interferometric techniques into the grazing-incidence region (Section 4.7).

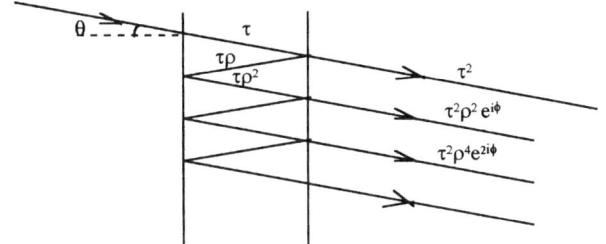

FIG. 2. Complex amplitudes of successive reflections in F–P interferometer. τ and ρ are the coefficients of transmission and reflection, respectively, and ϕ is the phase increment, given by $\phi = 2\pi x/\lambda = 4\pi \sigma t \cos\theta$.

4.2 Fabry–Perot Interferometry

4.2.1 Instrument Function and Resolution

The F–P interferometer consists of a pair of transmitting plates mounted with their facing surfaces accurately parallel, polished to a high degree of flatness, and coated with partially reflecting films. Figure 2 shows schematically the complex amplitudes of the reflected rays: Each double pass introduces an additional factor $\rho^2 e^{i\phi}$, where ρ is the amplitude coefficient of reflection and ϕ is the phase increment, given by $\phi = 2\pi\sigma x = 4\pi\sigma t \cos\theta$. Summing these complex amplitudes and multiplying by the complex conjugate gives the well known Airy intensity distribution

$$I = I_0 \left(\frac{T}{1-R}\right)^2 \frac{1}{1 + f\sin^2(\phi/2)} \tag{5}$$

where I_0 is the incident intensity, T and R are the transmissivity and reflectivity, respectively, of the coatings ($R = \rho^2$) and $f = 4R/(1-R)^2$ [1, 2, 3]. This function is illustrated for two different values of R in Fig. 3. It has maxima whenever $\phi/2 = n\pi$—that is,

$$I_{max} = I_0 T^2/(1-R)^2 \quad \text{for} \quad \sigma x = n \quad \text{or} \quad 2t = n\lambda \tag{6}$$

The minima, $I_{min} = I_0 T^2/(1+R)^2$, lie halfway between the maxima.

The characteristics of the Airy distribution are as follows. First, for any given value of x the pattern is repeated at wavenumber intervals defined by unit change in the order n—that is, $\Delta\sigma = 1/x$. The free spectral range is therefore given by

$$\text{FSR} = \Delta\sigma = \Delta\lambda/\lambda^2 = 1/2t \tag{7}$$

Second, the intensity does not fall to zero in the minima between the peaks. The contrast is given by

$$\text{Contrast} = I_{max}/I_{min} = (1+R)^2/(1-R)^2 \tag{8}$$

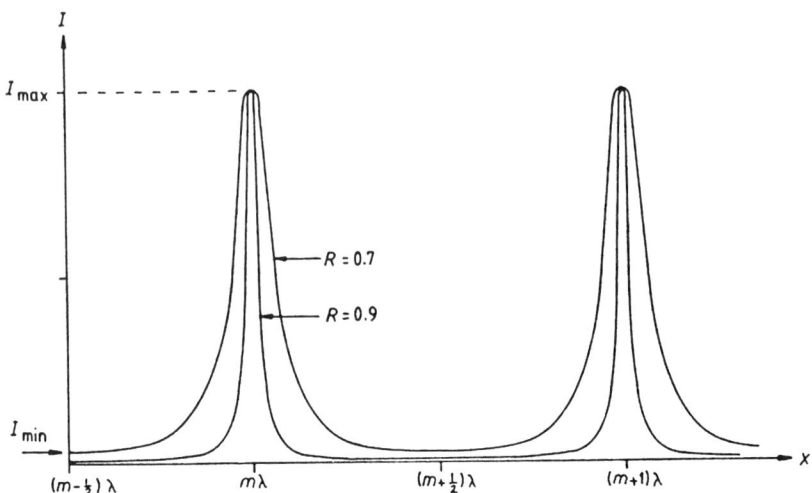

FIG. 3. Airy distribution as a function of optical path difference x for two different reflectivities, $R = 0.7$ and $R = 0.9$.

Third, the intensity transmitted by the interferometer at the peaks of the distribution is a fraction $T^2/(1 - R)^2$ of the incident intensity [Eq. (6)], and if the absorption in the coatings is negligible this is 100%. On the other hand, if the absorption A is not negligible, then

$$I_{max} = \left(\frac{1 - R - A}{1 - R}\right)^2 = \left(1 - \frac{A}{1 - R}\right)^2 \tag{9}$$

which drops rapidly with increasing A at the high values of R normally used. Finally, the resolving power of the interferometer is also strongly dependent on R. The Rayleigh criterion for resolution in its original form applies to a sinc² and not to an Airy instrumental function, but the criterion in a more generally applicable form states that two lines of equal intensity are just resolved if the minimum in the dip between them is 81% ($8/\pi^2$) of the peak intensity. Applying this to the Airy distribution, making allowance for the finite contribution from the wing of each line to the peak intensity of the other, gives for the resolution limit $\delta\sigma$ in wavenumbers

$$\delta\sigma = \frac{2}{\pi\sqrt{f}} \Delta\sigma = \frac{1 - R}{\pi\sqrt{R}} \Delta\sigma \tag{10}$$

The ratio of free spectral range to resolution limit measures the fine-ness, or "finesse," of the fringes. In the ideal case represented by Eq. (10) this is given by the reflection finesse N_r, where

$$N_r = \pi\sqrt{R}/(1 - R) \tag{11}$$

By rewriting the free spectral range $\Delta\sigma = 1/2t$ [Eq. (7)] in terms of the order number $n(= 2\sigma t)$ from Eq. (6), we obtain the resolving power as

$$\mathcal{R} = nN_r \tag{12}$$

By analogy with the similar expression for a grating, N_r can be interpreted as the effective number of interfering beams.

The resolution of the F–P does not match that expected from the reflectivity unless the reflecting surfaces are sufficiently flat. A bump of $\lambda/10$ changes the local path difference by $\lambda/5$, or a fifth of an order, and tends to spread the line correspondingly. This is taken into account by defining a plate finesse, N_p. Increasing N_r much above N_p does little for the resolution and loses light. A detailed treatment [2, 3] shows that the two should be approximately matched, and the effective finesse is then 0.6 of either. For moderate to high resolution, even with the high finesse achievable in the visible region (typically 50 or so), the free spectral range is so small that an auxiliary prism or grating monochromator, rather than an optical filter, is required to avoid order overlap.

4.2.2 Fabry–Perot Spectrometry in the VUV

The problems of F–P interferometry in the VUV are apparent from the preceding discussion. Whereas almost loss-less dielectric coatings can be used in the visible and infrared, with N_r up to 50 or more, the reflectivities achieved with overcoated metal films in the VUV are unlikely to be better than about 75%, for which N_r is about 10. Even then their relatively large absorption makes for poor peak transmission [Eq. (9)]. If N_r is low, high resolution requires a high order number [Eq. (12)] and a correspondingly low free spectral range, leading to more stringent requirements on the auxiliary monochromator. The fringe contrast is also low: For $R \simeq 70\%$ the contrast is less than 6. In addition, the tolerances on surface flatness and parallelism of the plates scale directly with the wavelength. Although it is possible to polish silica plates to reach a plate finesse of 30 at 200 nm, corresponding to 1/200 of a visible wavelength, it is much more difficult to obtain a good surface finish with the VUV-transmitting crystals that have to be used at shorter wavelengths. The problem of maintaining the plates parallel is usually solved by using a fixed spacer and enclosing the interferometer in a gas-tight box in which the pressure can be varied to change the optical path between the plates. An interferometer of this type with a fixed spacer is usually known as an F–P etalon.

The only systematic use of F–P spectrometry in the crystal transmission region of the VUV ($\lambda > 110$ nm) appears to have been by the group at Brest, who used silica F–P plates down to about 170 nm [4] and MgF_2 plates to reach

138 nm [5]. With the silica plates near 200 nm, they were able to achieve reflection and plate finesses of close to 30, giving an overall finesse of about 17. At the shorter wavelengths (~150 nm), the poorer quality of the MgF_2 plates and the absorbance of the coatings combined to limit the overall finesse to about 4 or 5. The resolving powers depended on the choice of spacer and varied from nearly one million at 200 nm to about 100,000 at shorter wavelengths; it must be remembered that opting for high resolution when the finesse is low brings the penalty of a small free spectral range. The problems of realizing better reflectance have been further discussed by this group [6].

At the other end of the VUV, in the soft x-ray region, there has been some work on high-reflection multilayer films that has resulted in the construction of F–P etalons with solid carbon spacers [7–9]. These devices have demonstrated reflection spectra (at grazing incidence) in the region around 0.15 nm and are scanned by changing the angle of incidence rather than the path difference between the "plates." The resolving power achieved is of order 100, with a finesse of about 2 to 3. It is not clear that the technique will have useful spectroscopic applications.

4.3 Fourier Transform Spectrometry: Principal Features

In this section the basic considerations and relations of FTS are presented for those unfamiliar with the technique. More background and detail can be found in the standard works on the subject (e.g., [10–13]), together with several comprehensive review-type articles [14–17]; although none of this material covers FTS in the UV, let alone the VUV, the basic treatment is still valid. Apart from the practical implementation, which is discussed in Section 4.4, the main difference between the IR and UV regions comes in the change from detector-limited noise, independent of the signal, to photon-limited noise, proportional to the square root of the signal. It is this change that loses the multiplex advantage and diminishes the importance of the throughput advantage.

4.3.1 The Interferogram and the Spectrum

The conventional instrument for FTS is a scanning Michelson interferometer in which the path difference x between the two interfering beams is varied up to some maximum value L. The scan can either be symmetric about zero path difference, or one side can be shorter than the other. The angular field of view is restricted as described in Section 4.1 [Eq. (2)] so that the detector records only the center of the ring pattern at maximum path difference. The two-beam interference signal for a given wavenumber σ is then proportional to $(1 + \cos 2\pi\sigma x)$. If the spectral distribution from the source is described by some function $B(\sigma)$, the interference signals from all wavenumbers present are superposed to give $\int B(\sigma)(1 + \cos 2\pi\sigma x)\, d\sigma$.

This signal, recorded as a function of x, is the interferogram

$$I(x) = \int_0^\infty B(\sigma) \cos(2\pi\sigma x + \phi)\, d\sigma \tag{13}$$

In this expression the constant signal $\int B(\sigma)\, d\sigma$ has been subtracted. The phase angle ϕ that has been added allows for the interferogram to be asymmetric around $x = 0$. There are two common causes of asymmetry: the sampling grid (see Section 4.3.2) does not in general coincide exactly with $x = 0$, leading to $\phi = 2\pi\sigma\delta$ where δ is the offset; and the two interfering beams can traverse slightly different thicknesses of the beamsplitter material, leading to a small additional path difference that is itself wavenumber dependent. Other small frequency-dependent contributions to ϕ can come from the electronic processing of the signal. Thus, ϕ is a function of σ but not of x. The asymmetry of the interferogram makes it necessary to use a complex Fourier transform to recover the complete spectral information in the form of the (complex) spectrum:

$$B(\sigma) = \int_{-\infty}^\infty I(x) \exp(-2\pi i\sigma x)\, dx = |B(\sigma)|\, e^{i\phi(\sigma)} \tag{14}$$

Although it is possible just to take the spectrum as being represented by the modulus, a better procedure for several reasons [10, 14, 18] is to "phase-correct": A low resolution phase spectrum $\phi(\sigma)$ is evaluated from a short central section of the interferogram and used to multiply the high resolution spectrum by $e^{-i\phi}$ so as to rotate the complex quantity $B(\sigma)$ into the real plane.

There are two possibilities for varying the path difference x—either step-and-integrate or continuous scan. The latter is more common and is the method used in all the instruments referred to in Section 4.4. If the rate of path change is v (twice the mirror velocity if only one mirror is moved), then $x = vt$ and the frequency f of the signal recorded by the detector for a wavenumber σ is given by $f = v\sigma$. Equation (13) can therefore be written as

$$I(t) = \int_0^\infty B(f) \cos(2\pi ft + \phi)\, df \tag{15}$$

The frequency f is just the optical frequency $v(= c/\lambda = c\sigma)$ downshifted by the factor v/c. Normally, v is chosen to put f in the audio range (say 1–20 kHz), avoiding any frequency ranges particularly sensitive to local noise.

4.3.2 Resolution and Free Spectral Range

There are two important departures from the mathematically simple formulation of Eq. (14): The scan does not go out to $x = \infty$, and the signal is sampled

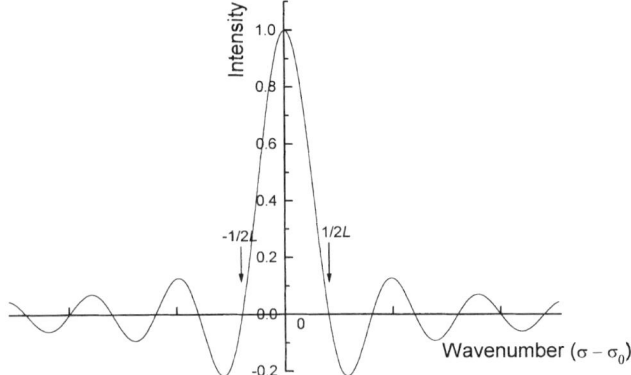

FIG. 4. The intensity distribution sinc $2\sigma L$ for a monochromatic line at σ_0. The first zeroes are at $\sigma - \sigma_0 = \pm 1/2L$.

and digitized at finite intervals rather than recorded in analog form, so that the preceding integrals are actually sums. The consequence of the finite scan length L is to introduce an instrument function

$$I = I_0 \frac{\sin 2\pi\sigma L}{2\pi\sigma L} \equiv \text{sinc } 2\sigma L$$

The sinc function is illustrated in Fig. 4. The resolution limit is taken to be the distance from the central maximum to the first zero:

$$\text{Resolution} \equiv \delta\sigma = 1/2L \text{ cm}^{-1} \qquad (16)$$

This definition does not correspond exactly with the Rayleigh criterion, but it can be justified from the sampling theorem, as shown in the discussion following. It is also close to the full width at half maximum, which is $1.2\delta\sigma$.

The sampling theorem states that, in order to avoid ambiguity in the frequency domain, a signal must be sampled at a frequency at least twice that of the highest frequency present in the spectrum—equivalent to sampling at intervals of not more than one-half the shortest wavelength: $\Delta x \leq \lambda_{\min}/2 = 1/2\sigma_m$. The reason for this minimum sampling frequency (the Nyquist frequency) is apparent from the basic mathematics of the Fourier transform [11]. First, the FT of a single cosine wave of spatial frequency σ_0 is a *pair* of delta functions, one at σ_0 and the other at $-\sigma_0$. The FT always generates both the actual spectrum and a mirror image of it at negative frequencies. Although the negative frequencies are physically unreal, they are important insofar as the mirror image affects the sampling requirements. Second, sampling the interferogram is equivalent to multiplying it by a comb function of spacing Δx; in the spectral domain this represents convolution with a comb function of spacing $1/\Delta x$—that is, the spectrum is replicated at intervals of $1/\Delta x$, and the replication applies to the real spectrum *and* its mirror image, which together occupy a range $2\sigma_m$, as shown in Fig. 5a. It can be seen that the replications will begin to overlap unless the replication interval $1/\Delta x$ is at least equal to $2\sigma_m$. In the UV, however, much of the spectral range from 0 to σ_m is usually empty because of the use of solar-blind detectors or optical filters. When the spectrum is contained within the bandwidth $\sigma_2 - \sigma_1$, as illustrated in Fig. 5b, the sampling interval can be increased up to $\Delta x = 1/2(\sigma_2 - \sigma_1)$ without overlapping. Considered the other way around, the free spectral range is determined by the sampling interval according to

$$\text{Free spectral range} \equiv \Delta\sigma = 1/(2\Delta x) \tag{17}$$

The high-frequency spectral band in Fig. 5b is "aliased" by undersampling into the empty lower frequency regions from 0 to $\Delta\sigma$ and $\Delta\sigma$ to $2\Delta\sigma$. It is often convenient to think of the spectrum as folded backwards and forwards at intervals of $\Delta\sigma = 1/2\Delta x$ as shown in Fig. 5c; each fold represents an alias, and the real spectrum in this case is in the third alias.

Two conditions must be met if undersampling is not to degrade the spectrum. First, some combination of source, detector, and filter characteristics must ensure that the signal really is restricted to the free spectral range $\Delta\sigma$, and,

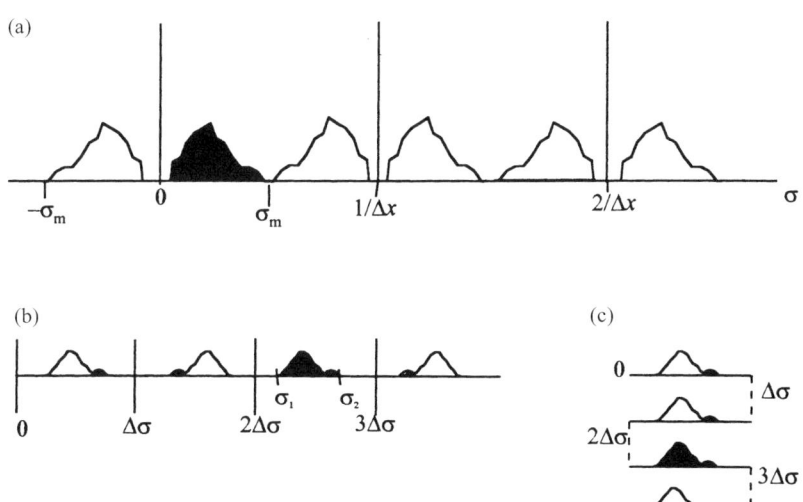

FIG. 5. The effects of the interferogram sampling interval on the recovered spectrum. The spectrum and its mirror image are replicated at intervals of $1/\Delta x$, as shown in (a), where the Nyquist condition $\Delta x \leq 1/2\sigma_m$ is fulfilled. If the spectrum is limited to a finite band, $\sigma_1 \rightarrow \sigma_2$, it is possible to undersample without ambiguity, as shown in (b), provided that $\Delta\sigma \ (\equiv 1/2\Delta x) \geq (\sigma_2 - \sigma_1)$. The spectrum is then "aliased" into the empty lower frequency regions as if folded backwards and forwards, as illustrated in (c).

second, a suitable electrical or digital bandpass filter should be used to prevent noise from other aliases from being folded back into the region of interest (see Section 4.4.1).

It can be seen that the definition of resolution limit in Eq. (16) together with that of free spectral range in Eq. (17) ensures that the number of independent points N is the same in both the interferogram and the spectrum domains:

$$N = \Delta\sigma/\delta\sigma = 2L/(2\Delta x) = L/\Delta x \tag{18}$$

The spectrum sampled at intervals $\delta\sigma$ can be transformed back into the other domain to give the unaliased interferogram. Moreover, for an unaliased spectrum having a maximum frequency of σ_m, N is identical with the resolving power $\mathcal{R}$. If one is working in the nth alias the maximum wavenumber σ_{max} is $n\Delta\sigma$, so in general

$$\mathcal{R} = \sigma_{max}/\delta\sigma = nN \tag{19}$$

equivalent to Eq. (12) for the F–P and to the well-known expression for a diffraction grating.

4.3.3 Wavenumber Accuracy

It was stated in Section 4.1 that high resolution and wavenumber accuracy are usually the most compelling reasons for using FTS rather than grating spectrometry in the VUV. The case for high resolution is obvious, for the resolution is determined only by the length L of the scan, and it should be possible to make this long enough to resolve the true line profiles of most laboratory sources. For atoms over the whole atomic mass range and for temperatures in the range 300–10,000 K, the fractional Doppler widths $\delta\lambda_D/\lambda$, or $\delta\sigma_D/\sigma$, range approximately from 5×10^{-5} to 5×10^{-7}. If the instrumental width is required to be one-half the Doppler width, the corresponding resolving powers are in the range 40,000 to 4×10^6, and only the lower decade of this range is within the reach of even the largest grating spectrometer. With FTS, a mirror displacement of 10 cm, giving $L = 20$ cm, corresponds to a resolving power of two million at 200 nm and nearly three million at 140 nm.

The wavenumber accuracy of FTS comes directly from the fact that a linear x-scale in the interferogram maps into a linear wavenumber scale in the spectrum, and in principle only one reference wavenumber is needed to put this on an absolute scale. Usually, of course, several references are used to improve the accuracy, but it is not necessary to have them well distributed through the spectral range in the manner required for a grating spectrum. Since the x-scale is defined by the sampling intervals, which in practice are obtained from the interference fringes of a helium-neon laser following a path through the interferometer similar to that of the signal beam, it is tempting to use the laser wavenumber as the reference. However, this can lead to errors of the order of

$1/\mathcal{R}$ because the laser beam may not be strictly on axis ($\theta \neq 0$). Having established the absolute wavenumber scale, the next point to consider is the relative accuracy—that is, the uncertainty ε of determining the wavenumber of any given emission or absorption line. It has been shown [17, 19] that this can be expressed by

$$\varepsilon = \frac{W}{\sqrt{n_p}(\text{SNR})} \qquad (20)$$

where W is the width of the line (in the same units as ε), n_p is the number of independent points across W, and SNR is the signal-to-noise ratio at the line peak. Thus, high resolution (allowing Doppler-limited line widths) contributes directly to the wavenumber accuracy, provided it is not at the expense of poor SNR.

4.3.4 Signal-to-Noise Ratios

This last point leads naturally to a consideration of SNR in FT spectra, and in particular to the differences between grating spectrometry and FTS in this respect. The FT, being a linear process, simply adds the noise spectrum to the signal spectrum, and white noise (independent of frequency) therefore appears in the spectrum as a uniform level of noise, independent of the local signal. This is in contrast to a grating spectrum where, in the case of photon noise, the noise is proportional to the square root of the local signal. For FTS it can be shown (e.g., Refs. [12, 15, 17]) directly from the basic FT relations that

$$\left(\frac{\bar{s}}{n}\right)_s = \frac{1}{\sqrt{N}} \left(\frac{s_0}{n}\right)_i \qquad (21)$$

where $\bar{s}$ is the mean signal in the spectrum, s_0 is the signal at zero path difference in the interferogram, N is the number of independent points as given in Eq. (18), and the suffixes s and i refer to the spectrum and the interferogram respectively. For an absorption spectrum, which is a quasi-continuum, the local signal is approximately the same as $\bar{s}$ except in deep absorption minima, and Eq. (21) is a reasonable representation of the SNR. In the case of emission spectra, most of the N spectral points have zero signal, and the signal at the peak of an emission line, s_p, can be much greater than $\bar{s}$. For an emission line we have

$$\left(\frac{s_p}{n}\right)_s = \left(\frac{s_p}{\bar{s}}\right)\left(\frac{\bar{s}}{n}\right)_s = \frac{1}{\alpha}\frac{1}{\sqrt{N}}\left(\frac{s_0}{n}\right)_i \qquad (22)$$

where α is known as the filling factor and can be significantly less than 1% for even a fairly dense emission spectrum. One must be wary of concluding that the

SNR is likely to be several hundred times better in an emission than in an absorption spectrum because the greater flux from a quasi-continuum should lead to a better SNR in the interferogram; if the system is photon-noise-limited, the net gain for an emission spectrum is likely to be about $\sqrt{\alpha}$ rather than α.

4.3.5 Comparison of FT and Grating Spectrometers

It is probably useful to summarize the important differences between these two types of spectrometers for operation in the VUV. Following directly from the last section, we start by comparing SNRs for the same resolving power and the same light source, ignoring initially the throughput advantage of FTS. The direct comparison is most easily made for a quasi-continuous spectrum because the SNR is then the same for each of the N spectral elements for both instruments. In the same total time, a scanning grating records N independent spectral points, and a scanning interferometer records the same number of interferogram points. The signal falling on the interferometer detector [s_0 in Eqs. (21) and (22)] is N times larger because it consists of all spectral elements, so for photon-noise-limits the interferometer has a gain of $\sqrt{N}$ in SNR. This factor is then lost in the Fourier transform, as shown in Eq. (21). Thus, for an absorption spectrum the two instruments are equivalent—*if* the throughput gain of FTS is ignored. For an emission spectrum, the filling factor enters into the equation. As a rough guide, FTS should give better SNRs for all lines down to an intensity of about $\sqrt{\alpha}$ of the stronger lines, whereas the SNR for the very weak lines should be better in a grating spectrum. However, the balance of SNR for both absorption and emission is shifted in favor of FTS by a factor that can realistically be put in the range 5–10 when the throughput advantage is taken into account (Section 4.1).

There are two important exceptions to these conclusions. The first applies to a grating spectrometer equipped with an array detector. If all N spectral elements can be recorded simultaneously, the grating evidently gains a factor $\sqrt{N}$ in SNR for both absorption and emission spectra. Second, the comparison has assumed the noise to be predominantly photon noise; in the case of source noise, proportional to the signal rather than to its square root, the $\sqrt{N}$ (or $\sqrt{\alpha N}$) advantage to FTS from the larger signal falling on the detector is again lost.

Apart from SNRs there are two other obvious differences between FT and grating spectrometry that warrant mention. First, an FT spectrometer cannot be used as a monochromator to produce radiation in a narrow wavelength band for some other experiment. It is true that the exit beam is differentially modulated, the modulation frequency being proportional to the wavenumber for any given scan velocity [Eq. (15)]; in principle this feature could be used to select a given band, but this mode of operation does not appear to have been put to practical use. Second, additional processing is required to obtain the spectrum from the raw interferogram—probably digital filtering and phase correction as well as the actual FT. With modern fast computers, the processing of million-point interferograms is measured in seconds rather than minutes, but this requirement needs to be borne in mind when designing the system.

In summary, the SNR arguments favor an FT spectrometer against a scanning grating for most applications where either could be used. The high resolution and wavenumber accuracy achievable by FTS are not within reach of a grating spectrometer.

4.4 Fourier Transform Spectrometry in Practice in the VUV

There are two obvious reasons why FTS was not considered to be a suitable UV technique for many years after it was firmly established in the infrared. First, as has been shown, the enormous gains in SNR offered by the multiplex and throughput advantages when the noise is independent of the signal (i.e., when it is detector-limited) are either nonexistent or relatively small in the UV. Second, the construction of a scanning Michelson interferometer is perceived to be much more difficult in the UV because all optical and mechanical tolerances scale with wavelength, and the step from the near infrared to 200 nm is an order of magnitude. In this section, we shall discuss design considerations and practical use of a relatively conventional scanning Michelson interferometer. Sections 4.5 and 4.6 will deal respectively with nonscanning and all-reflection interferometers that avoid some of the limitations of VUV operation, at the expense of introducing other constraints. Section 4.7 describes a division-of-wavefront interferometer for the grazing incidence region.

In all cases, the instruments described are state-of-the-art, and most of them are one-off laboratory interferometers. This is in marked contrast to the situation in the infrared; FT–IR is a well-established market in which many manufacturers compete. Of these, only Bomem and Bruker [20, 21] have modified instruments with suitable beamsplitters to take them through the visible into the UV, and the Bruker specifies 220 nm as the end of its range. The Bomem DA8 claims a short wavelength limit of 180 nm, but, as explained in Section 4.4.1, the performance of any interferometer at wavelengths appreciably shorter than those for which it was designed degrades very rapidly. Similarly, the large laboratory IR/visible instruments built at Kitt Peak National Observatory [22] and Los Alamos [23] (the latter now being rebuilt at the National Institute of Standards and Technology) can justifiably claim to be UV spectrometers, but the effective limit of even the well-tried and high quality KPNO instrument is about 220 nm. The only scanning FT spectrometer designed specifically for the UV and actually working into the VUV is that built at Imperial College [24], a commercial version of which has been made by Chelsea Instruments [25].

4.4.1 The Michelson Interferometer: Technical Considerations

The signal from the recombining wavefronts in a Michelson interferometer for a single frequency σ_0 has so far been assumed to have the form of Section 4.3, $I = I_0(1 + \cos 2\pi\sigma_0 x)$, but this is true only for perfectly flat and parallel wavefronts. In practice the signal is

$$I = I_0(1 + m \cos 2\pi\sigma_0 x) \qquad (23)$$

where the modulation m, sometimes called the visibility, is given by

$$m = \frac{I_{\max} - I_{\min}}{I_{\max} + I_{\min}} \qquad (24)$$

for intensities measured at zero path difference. The modulation is governed by the deviations from uniform phase over the whole area of the wavefront that is used. A local deviation ε in optical path difference causes a phase deviation of $2\pi\varepsilon/\lambda$, and the rule of thumb for acceptable modulation is $\varepsilon < \lambda/4$. For example, two perfectly flat wavefronts inclined at an angle such that the extreme path differences are $\pm\lambda/4$ give approximately 70% modulation (the exact value depends on whether the wavefront area is circular or rectangular). If the wavefronts are parallel but have random deviations from flatness, the modulation is 73% when the rms deviation is $\lambda/8$. As each of the recombining wavefronts has undergone at least one transmission and two reflections, the tolerance for each of the transmitting or reflecting optics should be set at $\lambda/8$ or better for the shortest wavelength of interest.

The penalty for low modulation is poor SNR. Whereas the entire dc signal contributes to the noise, it is only the modulated fraction that contributes to the signal. Thus, the SNR is directly proportional to m. For a given value of ε_{rms}, the modulation falls off exponentially with $(\varepsilon_{rms}/\lambda)^2$, and the behavior for other types of wavefront error is similar. For a given interferometer, there will be some wavelength above which the modulation has an acceptable value, say 50%; below this wavelength m falls rapidly while the noise in the transformed spectrum, as we have seen, remains approximately constant. The SNR in the spectrum therefore drops, and the interferometer may have an effective short-wavelength cut-off that is well above the transmission limit of the optics.

Figure 6 shows schematically a conventional Michelson interferometer with plane mirrors, one of which is moved to record the interferogram. Evidently the collimating and focusing lenses can be replaced with mirrors, so that the only transmitting elements are the beamsplitter and compensating plate. For the UV, these are made of fused silica, which can be used into the VUV down to about 180–170 nm, depending on the quality and thickness of the material. This material can be polished very flat and is also very homogeneous. This last point is important, because each of the separated beams is transmitted through one of

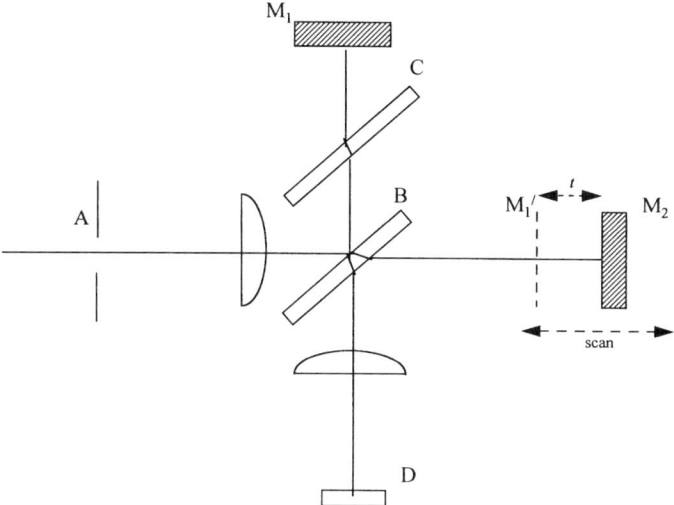

FIG. 6. Conventional scanning Michelson interferometer. A = entrance aperture, B = beamsplitter, C = compensating plate, and D = detector. M_1' is the image of M_1, distant t from the scanning mirror M_2.

the plates—twice in the arrangement of Fig. 6 and once in that of Fig. 7. For a 20 nm error ($\lambda/8$ at 160 nm) and a plate thickness of about 6 mm used at 45°, the refractive index must be uniform to almost one part in 10^6. Below the transmission limit of fused silica it is necessary to use a crystal as beamsplitter. Very pure crystal quartz can transmit to ~160 nm, but its homogeneity is in doubt. Of the other VUV–transmitting crystals, CaF_2 extends to about 130 nm, MgF_2 to

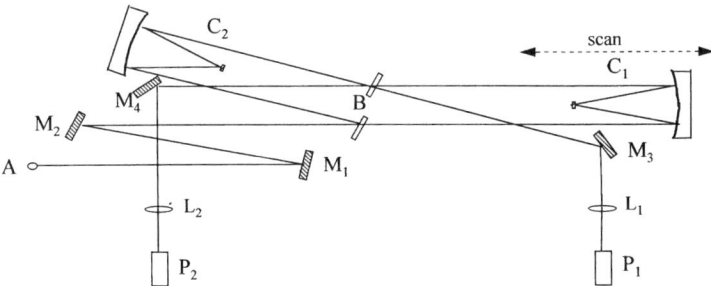

FIG. 7. Schematic diagram of scanning VUV FT spectrometer configuration (see Ref. 24). A = entrance aperture, B = beamsplitter/recombiner (MgF_2), C_1, C_2 = catseye retroreflectors, P_1, P_2 = photomultipliers detecting the two output signals. M_1 is a concave collimating mirror and M_2, M_3, and M_4 are folding plane mirrors. L_1 and L_2 are MgF_2 lenses that also form the windows of the vacuum tank.

about 120 nm, and LiF almost to 100 nm, depending of course on the thicknesses used. Because LiF is very difficult to polish, the material of choice is MgF_2. Although this can be polished adequately, the question of adequate homogeneity is less easy to answer; VUV refractometers do not exist, and measurements at longer wavelengths are of doubtful validity because of the nonlinear behavior of refractive index with wavelength.

The other critical tolerances are those for the surface quality and parallelism of the two mirrors, for the quality (guidance and smoothness) of the drive throughout a scan of several centimeters, and for maintaining the parallelism of the mirrors through this scan. Taking the last point first, the allowed tilt angle for the $\lambda/4$ extreme path difference specified in the preceding discussion and for a beam diameter of 25 mm is 25 microradians, or 5 arcsec, at 120 nm. This cannot be achieved without active adjustment during the scan. The Bomem spectrometer in the preceding discussion uses the sampling He–Ne laser to monitor the parallelism and adjust the fixed mirror to compensate for tilt errors in the moving mirror. The alternative is to replace the fixed mirrors with tilt-invariant retroreflectors, which is the approach adopted in the Bruker, KPNO, Los Alamos, Imperial College, and Chelsea Instruments interferometers. The Bruker has cube-corner retroreflectors, and all the others use catseyes, having a parabolic primary mirror and a small secondary mirror located accurately at its focus. Although the retroreflector ensures that the wavefronts remain accurately parallel, it does introduce an additional tolerance to be met, that of shear, which arises if the catseye does not move accurately along the optic axis of the interferometer. The shear tolerance is governed by the field of view, which in turn is limited by the resolving power required [Eq. (2)]; for high resolution at 120 nm, it is of order 10 µm over the length of the scan. Tilting a retroreflector introduces shear as a second-order effect, but the resulting tilt tolerance is an order of magnitude less stringent than it is for a plane mirror. An incidental advantage of retroreflectors is the ability to offset the return beam from the incident beam and thus gain access to the complementary interferogram traveling back toward the source from the beamsplitter. Figure 7 shows schematically the layout of the Imperial College/Chelsea Instruments UV–FT spectrometers [24] in which both outputs are used.

In most of the instruments mentioned, the guidance and drive requirements are met by using oil bearings and linear motors. The Imperial College/Chelsea Instruments interferometers have successfully exploited a simple rolling bearing in which the moving catseye is carried on three 14-mm balls, two of which are constrained by an accurately machined vee-block rail. One instrument uses a linear motor drive and the other a hydraulic drive; both of these have performed successfully.

The final tolerance to be considered is that of sampling the interferogram. The sampling interval Δx is determined by the free spectral range required [Eq. (17)].

As already stated, in practice the sampling intervals are derived from the interference fringes of a He–Ne laser (633 nm, corresponding to a wavenumber of 15,798 cm^{-1}) following a path through the interferometer similar to that of the signal beam. In the infrared, the samples can be taken at intervals of a whole fringe, or even several fringes, but the free spectral range of about 8000 cm^{-1} obtained from one sample per fringe is only about 20 nm in the VUV. It is therefore almost always necessary to subdivide the laser fringes to some degree: Ten samples per fringe would be required to satisfy the sampling theorem through the region down to 120 nm defined by a MgF$_2$ beamsplitter, but, as discussed in Section 4.3.2, it is common practice to undersample. A fringe division of 3.5, for example, allows the region 180–120 nm to be recorded in the third alias, as illustrated in Fig. 5b, c. Periodic errors in the sampling interval give rise to ghosts, whereas random errors produce noise; in both cases the effect is approximately proportional to the fractional error in the sampling step [17], so SNRs of a few hundred demand sampling accuracy of about one thousandth of a laser fringe. Existing interferometers achieve this sort of accuracy, using in some cases commercial systems adapted for the purpose. However, it is difficult to eliminate entirely "laser ghosts" arising from small systematic errors in the subdivision that repeat with the period of the laser fringes.

In practice, when working in the VUV the bandwidth of the detected radiation is limited as far as possible, partly to allow undersampling but more importantly to avoid the noise that is spread all through the spectrum from radiation outside the spectral region of interest. Unfortunately, VUV interference filters are expensive and not very efficient, and short-pass filters that exclude the near-UV but pass the VUV do not exist. The alternative is to use a detector with a long wavelength cut-off: The conventional solar-blind Cs–Te photocathode is insensitive above about 320 nm, and a more drastic cut-off at about 180 nm is possible by using Cs–I—well matched to the example of undersampling given previously. For absorption experiments conducted with synchrotron continuum radiation, the "filter" problem is solved, in principle, by the provision of a monochromator as part of the standard equipment, but the effects of the dispersion need to be considered, as discussed at the end of Section 4.4.2. It must be remembered when using a high alias that even if a suitable optical filter is in place there will still be additional noise folded in unless a matching electrical bandpass filter is used, either analog or digital. Provided this precaution is taken, the SNR is independent of the free spectral range for the same total integration time.

4.4.2 Applications of VUV FTS

Emission spectrometry in the VUV has been primarily concentrated on the improvement of the database for the interpretation of astrophysical observations from space-borne instruments. This spectral region is rich in lines of the first ionization stage of the transition element spectra, and accurate wavelengths and intensities are required both for reliable line lists and for the spectral analysis that is required to classify the large number of unidentified lines. In some cases, the high resolution yields hyperfine structure and isotope splitting information. The usual light source for this work is a hollow cathode discharge, run at currents up to an ampere or so. Figure 8 shows part (about 15 nm) of the VUV spectrum from a 20 mA platinum–neon hollow cathode lamp at a resolving power of about one million, with a small section expanded to show the resolved line profiles, the asymmetry of which is caused by hyperfine structure. The integration time for the spectrum covering 180–140 nm was about 30 minutes, and the normalization is for unit rms noise, so the ordinates represent the SNR.

The main limitation of FTS for this work is the fall-off of efficiency toward short wavelengths, primarily caused by the decreasing modulation [26]. Although the transmission of a MgF$_2$ beamsplitter is still adequate at 120 nm, in

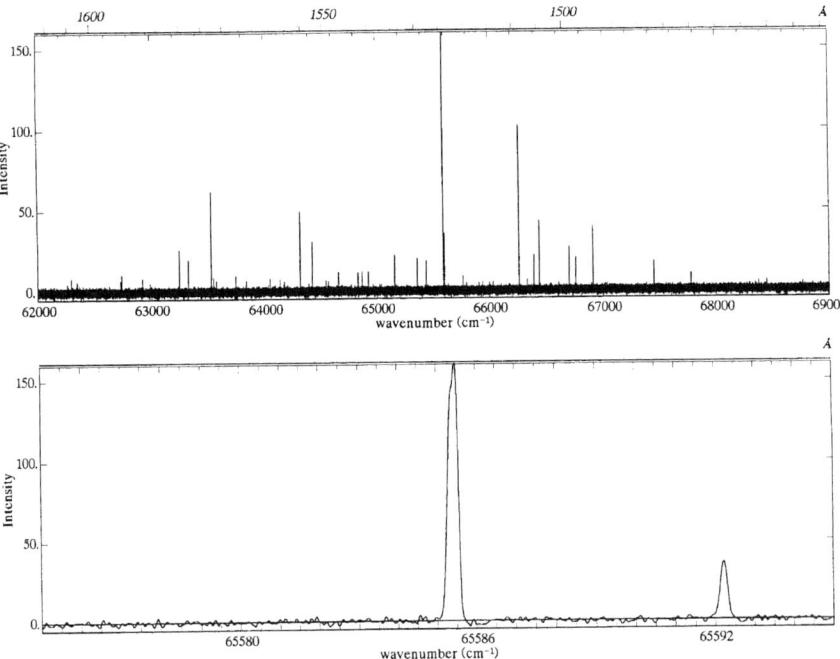

FIG. 8. Part of the VUV spectrum of a 20 mA platinum-neon hollow cathode lamp taken with the Imperial College FT spectrometer in the spectral region 160–145 nm, using a MgF$_2$ beamsplitter. The resolution is 0.06 cm^{-1}, corresponding to a resolving power of about one million. The strong "line" in the top plot is actually two lines separated by 0.015 nm, as shown in the expanded section, and the line asymmetry is caused by unresolved hyperfine structure.

practice it has proved difficult to avoid losing the signal in the noise before this limit is reached. The two other disadvantages that FTS has against a multichannel grating spectrometer (using either a photographic plate or an array detector) are: (1) the SNR is worse for very weak lines (see Section 4.3.5), and (2) the SNR is also likely to be worse for the pulsed sources that may be needed to excite high stages of ionization because of the effects of shot-to-shot irreproducibility. The two techniques should be regarded as complementary in the VUV, with FTS providing the high-quality information for the stronger lines within its wavelength range and grating spectrometry extending the range for wavelength, intensity, and excitation.

Absorption spectrometry in the VUV suffers from the lack of a bright and stable laboratory source of continuum radiation. The high pressure Xe arc falls rapidly in intensity below 200 nm, and deuterium lamps, although stable, are not very bright and are dominated by band structure below 160 nm. High-current positive column hydrogen discharges are brighter than the commercial deuterium lamps, but tend to be less stable and suffer the same breakdown into molecular band structure. The argon mini-arc is the only other laboratory source that can be used into the VUV. The alternative is synchrotron radiation, and this has indeed been successfully tried [26].

The less favorable spectral filling factors from a quasi-continuous source (see Section 4.3.4) result in long integration times for good SNR, even with the high photon flux from a synchrotron. It is strongly advisable whenever possible to limit the bandwidth to the spectral region of interest, thus introducing an effective "filling factor"; the SNR is then proportional to the inverse square root of the bandwidth. A premonochromator has obvious advantages over an interference filter because the center and width of the band can be exactly matched to the requirements and the band edges are more sharply defined. Ideally, this should be a double monochromator with zero net dispersion so that all wavelengths in the bandpass enter the interferometer at the same angle. With a standard single monochromator, it is necessary to correct for the phase and wavelength shifts across the bandpass. It should be noted that there is no net gain in dividing a wide spectral band into several sections by selecting a narrow bandwidth for the monochromator because the gain in SNR for each individual section is cancelled by the necessity for recording several sections.

4.5 Spatially Heterodyned, Nonscanning Interferometers for FTS

Nonscanning interferometers use some form of multichannel detector to record all elements of an interferogram simultaneously, rather than recording them sequentially with a single detector. They offer obvious advantages of ruggedness and stability, together with disadvantages in respect of flexibility and free spectral range. In the form first proposed, the technique was known as holographic FTS; it was achieved by slightly tilting one of the mirrors of a conventional Michelson interferometer, thus producing wedge fringes localized in the plane of the mirrors. These fringes were imaged onto a photographic plate [27]. For a small wedge angle α between the recombining wavefronts, the optical path difference is $x \sin \alpha$ where x is the distance along the plate from zero path difference. The interferogram is thus given by Eq. (13) with x replaced by $x \sin \alpha$, and it can be Fourier transformed in the normal way.

Modern versions of this interferometer replace the photographic plate with an array detector. This immediately focuses attention on the conflicting requirements of free spectral range and resolution. From Eq. (18) we see that the number of independent points N is the same in both the spectrum and the interferogram, and evidently N is now limited by the number of pixels in the detector. Thus, for 1024 pixels the maximum resolving power would be only one thousand, or, taking proper sampling into account, 500. The breakthrough to a useful resolving power came with the introduction of a spatial heterodyning technique [28]. The mirrors are replaced by a pair of diffraction gratings set at the Littrow angle θ for a particular wavenumber σ_0 at one end of the desired bandpass, as shown in Fig. 9. For this wavenumber, the recombining wavefronts are parallel, and the spatial frequency of the interference fringes is zero. The gratings are oppositely tilted, so that for adjacent wavenumbers σ, either larger or smaller than σ_0, the wavefronts are inclined at some angle 2α that depends on $|\sigma_0 - \sigma|$ and the angular dispersion of the gratings. For small values of α, the grating equation gives for the spatial frequency f_x of the fringes along the x-axis

$$f_x = 4|\sigma_0 - \sigma| \tan \theta$$

The equation equivalent to Eq. (13) is then

$$I(x) = \int_0^\infty B(\sigma) \cos\{2\pi(4|\sigma_0 - \sigma|x \tan \theta)\}\, d\sigma \qquad (25)$$

This can be put into the form of Eq. (13) if σ is replaced with $\sigma' \equiv |\sigma_0 - \sigma|$ and x with $x' = 4x \tan \theta$. The spectrum $B(\sigma')$ recovered from Fourier-transforming the interferogram therefore starts from σ_0 instead of zero wavenumber.

The similarity of Eqs. (13) and (25) makes it clear that the instrument function is a sinc function, as in Fig. 4, with L replaced by the maximum value X' of x'. If the interferometer is arranged to give zero path difference in the middle of the interferogram, the maximum value of x is one-half the projected width of the grating, $W \cos \theta/2$, giving $X' = 2W \sin \theta$. The resolution limit [cf. Eq. (16)] is therefore

$$\delta\sigma = 1/(2X') = 1/(4W \sin \theta) \qquad (26)$$

which is just twice the diffraction-limited resolution of each individual grating. Given that the number of independent spectral points is limited to the number N of detector pixels, the free spectral range is given by [cf. Eq. (18)]

$$\Delta\sigma \equiv |\sigma_0 - \sigma_m| = N\delta\sigma = N/(4W \sin \theta) \tag{27}$$

where σ_m is the maximum (or minimum) allowed wavenumber. The same result can be derived from the fringe spatial frequency f_x by setting its maximum value, $4|\sigma_0 - \sigma_m| \tan \theta$, equal to the sampling spatial frequency $N/(W \cos \theta)$ imposed by the detector. For further details, see, for example, [29–31] and references in these papers.

Some arrangements use only one grating, retaining a mirror in the other arm of the interferometer [29]; this leads to some loss of coherence and to drops of factors of two in the relations given previously, but the overall picture is not affected. It is also possible to use two gratings of different spacing, tilted in

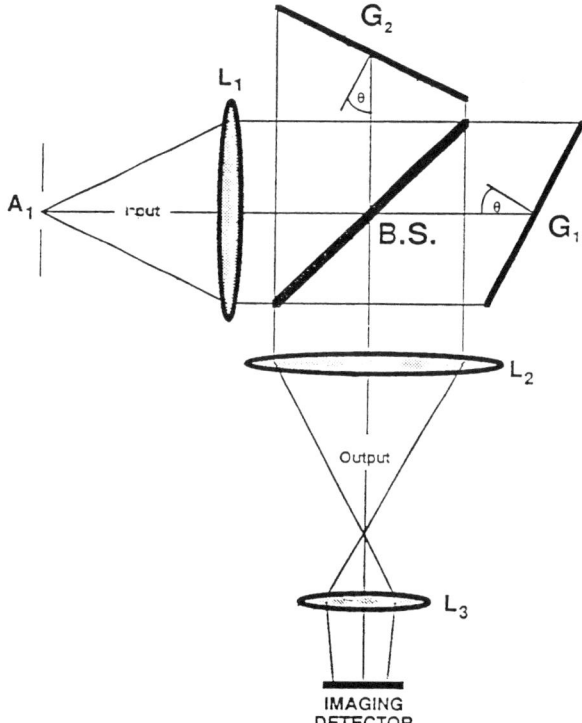

FIG. 9. Schematic diagram of the basic SHS configuration, using a transmitting beamsplitter BS. Wavelength-dependent Fizeau fringes produced by diffraction gratings G_1 and G_2 are focused onto the imaging detector (see Ref. 32).

either the same or opposite directions, to decrease or increase the path difference.

Spatially heterodyned FTS, or SHS, has exactly the same noise characteristics as scanning FTS. Provided the imaging of the fringes is correctly done, the throughput is limited by the resolving power in the same way. If the noise is photon-limited, the SNR is theoretically the same in the two techniques because the spread of the signal over N detectors is exactly compensated by the factor N in integration time per point gained from not scanning. For the source-noise limited case there is actually a gain of $\sqrt{N}$ in SNR.

The advantage or otherwise of SHS over a straightforward dispersion arrangement using the same grating (and hence the same resolving power) and the same detector also needs to be considered. As pointed out in Section 4.3.5, FTS loses a factor of $\sqrt{N}$ in SNR to grating spectrometry when the latter has a multi-channel detector. This is counteracted by the throughput advantage of SHS—particularly if this can be increased by field-widening as described later. The SNR balance between the two techniques is normally in favor of SHS, but it should be assessed for any particular problem. It should also be noted that the use of a dispersive system compromises the high wavelength accuracy of scanning FTS [32].

It is important to remember the band-limiting constraints of SHS. The spectrum is effectively folded over at σ_0, with $\sigma_0 + \Delta\sigma$ indistinguishable from $\sigma_0 - \Delta\sigma$. Wavenumbers outside the range $|\sigma_0 - \sigma_m|$ are aliased in the same way as in a scanning FT spectrometer that is undersampled. It is possible to get around the ambiguity problem by rotating one of the gratings through a small angle β about an axis in its plane, so that the grooves of the two gratings are no longer parallel. This produces an unheterodyned low-resolution interferogram along the y-axis, in which $\sigma_0 \pm \Delta\sigma$ are no longer degenerate, thereby doubling the free spectral range. By an extension of this idea, the gratings can be used in higher orders to produce a two-dimensional interferogram that can be transformed to yield a spectrum in an echelle-like format, with the overlapping orders of the high-resolution spectrum along the x-axis separated by low-resolution cross-dispersion along the y-axis [30].

The SHS technique has been mainly developed in two centers. The Astronomy group at Groningen has been primarily concerned with using it on large ground-based telescopes [29]. Of more relevance to the VUV region is the work of Roesler and his colleagues at Wisconsin [30, 31], who have developed the technique with a view to using it for space-based observations of diffuse interstellar emission lines. In its original form, their interferometer still uses a transmitting beamsplitter; the all-reflection version is discussed in Section 4.6. Elimination of the need to provide an accurate scanning mechanism is not only an advantage in itself, particularly of course for an instrument designed for space-flight applications, but also makes field-widening relatively simple: The addition of two thin prisms to the interferometer in fixed positions, one in front of each of the gratings, increases the throughput by two orders of magnitude [30, 31].

As pointed out by Harlander *et al.* [31], an additional and more subtle advantage of the SHS in the VUV is the relaxation of tolerances in the figure of the various surfaces and the homogeneity of the beamsplitter. This comes about because the interference fringes are localized in the plane of the gratings, so the grating surfaces are imaged on the detector. Each pixel therefore sees only a small element of the grating area. Similarly, the beamsplitter is well enough imaged that only a relatively small part of its area contributes to each pixel. Wavefront errors on a scale larger than the pixel area affect the fringe shape or spacing, but they do not degrade the fringe contrast as they do when the entire wavefront is imaged onto a single detector. Any errors in fringe shape can be detected directly and corrected in software. In view of the uncertainties about optical homogeneity near the transmission limits of the crystals, this is potentially a very important advantage.

4.6 All-Reflection FT Spectrometers

For wavelengths shorter than about 120 nm, FTS is possible only with an all-reflection interferometer. Even for rather longer wavelengths, there are potential advantages in getting away from the stringent requirements placed on the crystal that is used as the beamsplitter. An all-reflection interferometer with a division-of-wavefront beamsplitter, using grids or apertures, is relatively simple to design (in principle, anyway), but considerable effort has gone into a division-of-amplitude reflecting system in order to maintain the throughput advantage.

The concept of a reflection diffraction grating as a division-of-amplitude beamsplitter was suggested as early as 1959—see references given in the paper by Kruger *et al.* [33], who proposed a practical form of scanning interferometer and tested it in the UV, although not in the VUV. The wavelength dispersion is an inevitable, but undesirable, property of the grating when it is used as a beamsplitter; in order to maintain coherence of the recombining wavefronts over a wavelength range wider than the resolution of the grating, it is necessary to cancel the dispersion. Kruger *et al.* [33] did this with an additional pair of gratings, one in each arm of the interferometer. In one of their designs, these gratings were additional to the plane mirrors, one of which was scanned [34]; in another design the plane mirrors were eliminated to reduce the number of reflections, and one of the gratings was scanned [33]. This second design introduced a wavelength-dependent path difference, which distorted the wavelength scale in the transformed spectrum and had to be corrected for.

The availability of UV-sensitive array detectors and the development of the SHS technique combined to make feasible an all-reflection nonscanning interferometer, in which the dispersion of the grating is actually used to produce the required path difference. Recent work on all-reflection normal incidence FT spectrometers appears to have been confined to developing different configurations for SHS. The simplest version, as proposed by the Wisconsin group [30, 32], is shown in Fig. 10. The grating acts as a beamsplitter and at the same time introduces a wavelength-dependent wedge angle between the recombining wavefronts. Zero spatial frequency for the interference fringes corresponds to the wavenumber σ_0 for which the diffracted rays are normal to the mirrors, and the interferogram has exactly the same characteristics as that produced by the transmitting beamsplitter system. In plan, the output beam returns along the same path as the input beam, but it is shifted out of plane by a slight tilt, so that the two beams use, respectively, the upper and lower halves of the aperture.

Figure 11 shows another design developed for the EUV region [32, 35], using a common path configuration in which one-half of the grating acts as beamsplitter and the other half as beam recombiner. The divided beams travel around the system parallel to one another but in opposite directions. This design is particularly robust because any small disturbances to the components affect both beams equally, at any rate to first order. This design has been tested at Lyman α with a grating of 600 lines/mm in the second order and found to achieve 80% of

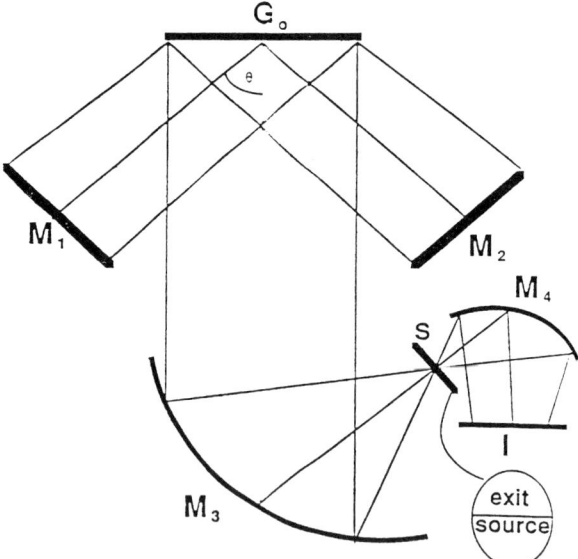

FIG. 10. Schematic diagram of the basic all-reflection SHS configuration. Light enters the system through the upper half of split aperture S and exits through the lower half. The mirror M_4 focuses the fringes onto the imaging detector I. The diffraction grating G_0 acts as both the beamsplitter and the dispersive element in the system (see Refs. 30, 32).

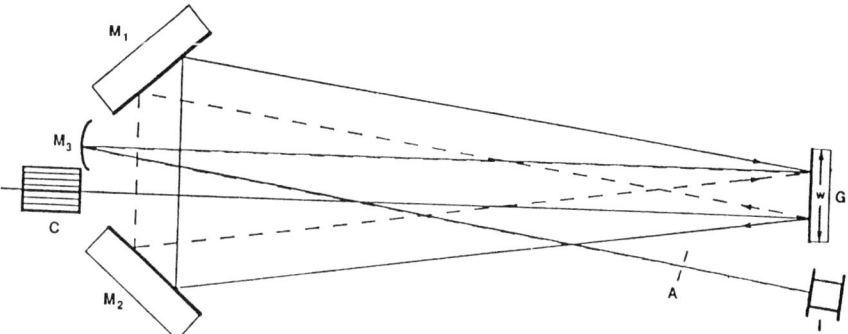

FIG. 11. All-reflection common path SHS configuration. Light enters through the collimator C, and the grating G splits and recombines the two beams (solid and dashed lines), which traverse the system in opposite directions. The converging mirror M_3 images the interference fringes onto the detector I (see Refs. 32, 35).

its theoretical resolution [35], which is set by the half-width of the grating that is illuminated by the incident beam. To emphasize the bandwidth restrictions, it should be noted that, at the resolving power of 120,000 offered by the grating, the free spectral range allowed by a detector of 512 pixels is only about 0.25 nm.

As with the transmission SHS described in Section 4.5, it is possible to double the free spectral range by introducing a small tilt of the mirrors to produce an unheterodyned interferogram in the y-direction. In the same way, it is also possible to exploit higher orders to give an echelle-like format [32]. The principal difference in the characteristics of the all-reflection and the transmission versions of SHS is the loss of a factor $\cos\theta$ in the throughput of the former, arising from the fact that the grating is not used in the Littrow configuration. The number of reflections in the two versions is similar, but the optical efficiency in the reflection interferometer is reduced by the need for the grating blaze to throw equal energy into the orders either side of the normal.

4.7 Soft X-Ray FTS by Grazing Reflection

In order to progress below about 40 nm, it is necessary to adopt a grazing-incidence reflection geometry and this is being attempted by a group at the Advanced Light Source synchrotron-radiation facility in Berkeley [36]. The construction of an interferometer in this spectral region poses significant new mechanical challenges, which are described later. The required technical solutions provide an indication of the type of effort involved in extending FTS to cover the entire VUV region. The present motivation is to achieve higher

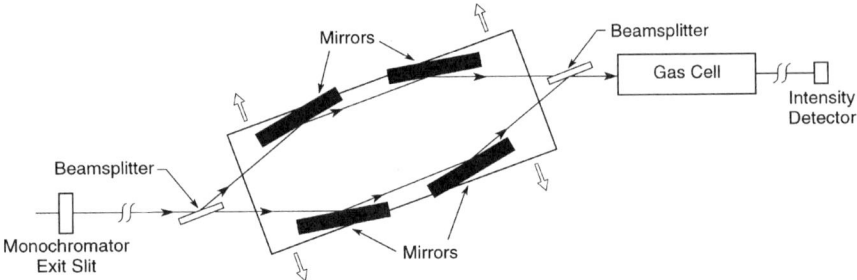

FIG. 12. Layout of the grazing-incidence Mach–Zehnder interferometer for use at 20 nm wavelength. The normal rectangular layout has been deformed into the rhombus ABCD, and four mirrors, all mounted to the same table, are used to enable the path difference to be scanned. This is accomplished by translating the table in the direction of the arrows. In this figure, the system is at zero path difference whereas in Fig. 14 it is shown at maximum path difference.

resolution than a grating for studies of two-electron excitations in helium near the double ionization region (60–80 eV). The original discovery, in 1963, of autoionizing series in the absorption spectrum of helium [37] was an early triumph of synchrotron-radiation spectroscopy, achieved at Gaithersberg. Subsequent work at Berlin [38, 39] (at 4 meV resolution) and later at Berkeley [40] (at 1 meV) has revealed further series, and these measurements (all by grating spectrometers) continue to excite strong interest in the theoretical-atomic-physics community.

At grazing incidence, it is essentially impossible to use the same beamsplitter for both division and recombination of the wavefronts, so the traditional Michelson geometry and its variants have to be abandoned in favor of a Mach–Zehnder type of layout [1, 41]. In the Berkeley scheme, the normal rectangular shape of the latter has been deformed into a rhombus, and extra mirrors have been added to allow the path difference to be scanned (see Fig. 12). The system presently under test has a path difference of ± 17 mm, which implies a potential resolving power of more than one million at 20 nm (see Section 4.3.2).

The spectrometer to be described here will be fed by a beam of synchrotron radiation from a low-resolution monochromator on an Advanced-Light-Source bending-magnet beam line. The light from such a monochromator will have a fractional bandwidth ($1/W_{in}$) of about 1/500, which is unusually narrow for an FTS source. This situation has some noteworthy features.

1. A grating spectrometer can often accept all the light from a high-brightness synchrotron-radiation beam. In such a case, neither the throughput (Jacquinot) nor multiplex (Fellgett) advantages exist, and it is again the resolution advantage that is being sought via FTS (see Section 4.1).

2. With a narrow-band input spectrum and a path-difference scanning rate v, the expected interferogram [see Eq. (15)] is close to a pure cosine wave of temporal frequency $v\sigma_m$. This cosine signal acts as a carrier wave on to which the spectral information is modulated as an envelope function that varies slowly in both amplitude and phase. The center (temporal) frequency of the incoming wave is down-shifted by a factor v/c (Section 4.3.1) so that for an experiment at 20 nm, an output at 1 kHz can be produced with $v = 20$ μm/sec. A frequency in this range is suitable for working with synchrotron sources that tend to suffer from small beam motions (noise) at frequencies below about 100 Hz.

3. Since only one W_{in} th of the spatial frequency range from zero to σ_m will contain any signal, the interferogram can be undersampled by a factor W_{in}, which is equivalent to sampling the envelope function at *its* Nyquist frequency. In the terms of Section 4.3.2, this means the system will be working in the W_{in} th alias, and the same conditions indicated there must be satisfied.

The first requirement for a grazing-incidence interferometer is a beamsplitter, and at 20° grazing angle and 20 nm wavelength a transmission device is impractical. The closest one can get to an amplitude-division beamsplitter is therefore a type of transmission grating used in zero order. Although such a device actually divides the *wavefront*, it may still produce a final result similar to amplitude division, provided it has a sufficiently small period and enough propagation distance is allowed for the dissected wavefronts to be smoothed by diffraction. Such a grating must be quite unusual, however, because the front faces of the opaque bars must form a planar reflecting surface of wavefront-preserving accuracy. Several transmission-grating beamsplitters of this type with 100 μm period have now been constructed by photolithography and directional etching of silicon [42] (Fig. 13). The reflecting surfaces obtained were flat within 0.6 μradian rms and smooth within 2 Å rms.

The flexural mechanism [43] used to translate the table carrying the four mirrors (Fig. 12) by ±7.5 mm has been constructed by electric-discharge machining (spark erosion) of a monolithic block of type C300 maraging steel. It includes a moving table that carries a special prism onto which all four mirrors are optically contacted. The design is a classical double-sided compound rectilinear spring [44, 45], and to obtain such a large motion with reasonable stresses, the hinges are implemented as monolithic crossed strips [46, 47]. The mechanism, which is about the size of a notebook computer, can be seen with the table at one end of its travel in Fig. 14. The rms tilt error of the table in the pitch direction over the 15 mm motion has been measured to be 0.38 μradian, leading to a wavefront tilt error of 0.57 μradian.

The possibility of realizing the theoretical resolution of soft x-ray FTS depends on being able to manufacture the spectrometer and optics within the

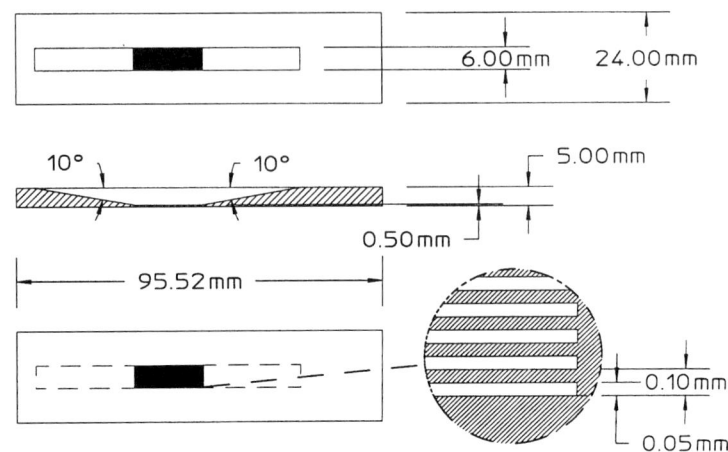

FIG. 13. Geometry of the transmission grating beamsplitter, made by optical polishing followed by photolithography and directional etching of silicon.

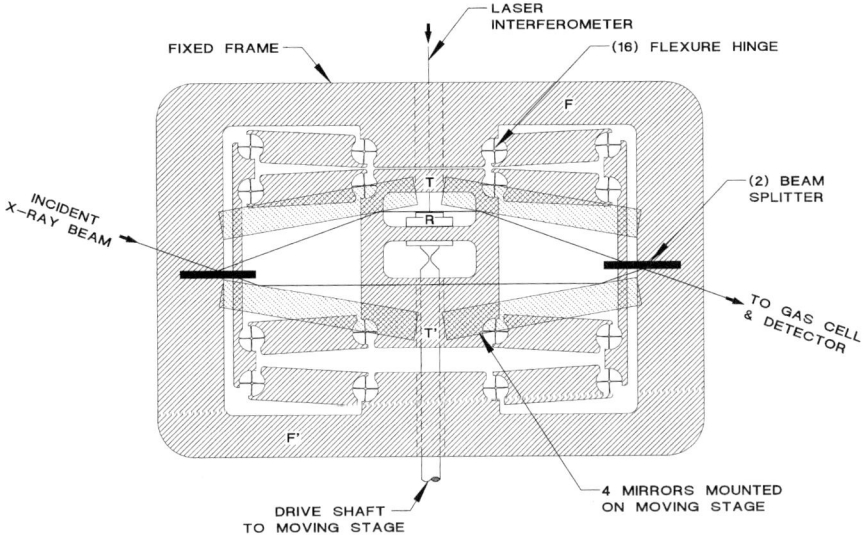

FIG. 14. The flexure mechanism for scanning the optical path difference. The outer frame of the mechanism FF' carries the two beamsplitters, which are fixed in position. The four mirrors are fixed to the moving table TT' here shown driven to one extreme of its travel. One side of the table is (kinematically) connected to a drive shaft activated by a double-sided hydraulic piston while the other side carries a retroreflector by which its position is measured with a laser interferometer [51].

required tolerances (see Section 4.4.1). Some of the tolerances and achieved values for the Berkeley system are summarized in Table I. The underlying goal is a relative tilt between the interfering wavefronts of less than 1.5 μradian, which should give greater than 90% of the maximum possible modulation.

The optics of Michelson and Mach–Zehnder interferometers are discussed in standard texts [1, 41, 48]. For a point source and a perfectly built system with no mirror tilts, the situation is quite straightforward. Viewed from the detector, there are, in general, two virtual point sources at different distances, and the observation space is filled with a single concentric ring pattern that is visible everywhere. Such a fringe system is said to be nonlocalized and an FTS experiment would proceed by placing a detector inside the center fringe of the system. On the other hand, if the source becomes extended, most of the pattern disappears and the fringes become localized, which means in this case that they can be observed only in the focal plane of a lens. Such "fringes of equal inclination" have already been discussed in Section 4.1.

The blurring of the nonlocalized fringe pattern by the increase of the source size can be understood as a degradation of the coherence of the optical field, and with this viewpoint one can calculate the degree of extension of the source that can be tolerated if nonlocalized fringes are to remain visible with a given contrast. Such calculations can be accomplished using the van Cittert–Zernike theorem [41], which provides a prescription for calculating the complex degree of coherence $\mu(P_1, P_2)$ between two points P_1 and P_2 (μ is defined, for example, by Born and Wolf [41]). According to the theorem, $\mu(P_1, P_2)$ is proportional to the amplitude that would be found at P_2 if a spherical wave converging toward P_1 were to be diffracted by an aperture function with the position and shape of the source intensity distribution. What is not always stated is that P_1 and P_2 need not be in the same plane, so that the theorem indicates the existence of a good-coherence volume determined by the wavelength and a "numerical aperture" equal to the half angle subtended by the source—that is, the angle θ of Section 4.1. For a uniform circular source, the distance from the center to the first zero of $\mu(P_1, P_2)$ is thus $0.61\lambda/\theta$ in the width direction and $2\lambda/\theta^2$ in the depth direction [49], similarly to a conventional resolution function.

TABLE I. Mechanical and Optical Tolerances

Quantity	Required	Achieved
Table rms tilt errors	*(quadratic sum	0.57 μrad
Mirror 1 rms slope errors	*(of all four	0.7 μrad
Mirror 2 rms slope errors	*(rms tilts	0.7 μrad
Beamsplitter rms slope errors	*(<1.5 μrad	0.6 μrad
Table motion rms indexing error	<8 Å	4 Å (noise limit)
Driver rms stick-slip (difference from linear motion)	<8 Å	4 Å (over 1 sec)

Both the width and the depth of $\mu(P_1, P_2)$ are important for interferometry because the detection point P, viewed now from the source via the two interfering beams, appears as two points (P_1 and P_2). These points in general have both a longitudinal separation because of the intentional path difference of the instrument and a transverse separation caused by mirror tilts. For the soft x-ray interferometer discussed here, the numerical aperture is defined by the size and distance of the monochromator exit slit to be about 30 µradian. At $\lambda = 20$ nm, this determines a depth resolution of many meters, which comfortably allows the required 17 mm path difference. It also defines a transverse coherence width at the interferometer of about 1 mm. The operational requirements are therefore that the period of the grating and any transverse separation of P_1 and P_2 (wavefront shears) that could be produced by mirror angle errors must be much less than 1 mm, both of which are easily achieved.

The conclusion from these arguments is that the soft x-ray interferometer will produce a single set of Haidinger fringes. This ring pattern will be nonlocalized and will fill the observation space, so that experiments should be possible without a "lens" in the detection region. Such a clean interference pattern results to a large extent from the fact that the incoming beam of synchrotron light has a high degree of spatial coherence. For other situations in which short-wavelength FTS might be interesting, such as soft x-ray emission spectroscopy, the coherence of the source is usually much worse and the experiment correspondingly more difficult.

The detection system may be one of two types: a photo diode downstream of the helium cell that would register the transmitted signal or an ion chamber of which the working gas would be the helium sample. The first would provide a transmission spectrum in which the SNR would be governed by Eq. (21), whereas the second would provide a signal analogous to that of an emission spectrum with the SNR governed by Eq. (22). In either case, the technical goal would be shot-noise-limited detection, which has been shown to be possible [50] at least in the case of a source with somewhat similar time structure to the subnanosecond pulses one gets from a synchrotron.

The ultimate uncertainty for applications of FTS to gas-phase soft x-ray synchrotron-radiation experiments is whether the attainable modulation and photon rate will allow an experiment in a reasonable time. This has been addressed by the Berkeley group by means of computer modeling of the Rydberg series that converges to the lowest excited state of He^+ as parameterized by Domke et al. [39]. The average modulation of the interferogram (the standard deviation of the recorded intensity divided by its mean), calculated by Fourier transforming the known spectrum, was found to be 0.005 [36]. This, combined with realistic estimates of the number of photons arriving in the assumed collection time of 10^3 sec, leads, via Eq. (21), to an SNR in the final spectrum of 330. Although this value was calculated without allowance for experimental error, it did use the conservative assumption of a transmission experiment.

This calculation provides a conclusion to this section on soft x-ray FTS experiments. We have shown that if a reasonable fraction of the maximum possible modulation can be achieved (and the mechanical requirements for that have mostly been met as described previously), then the system will yield good spectra, and we therefore look forward to a successful experimental program.

References

1. W. H. Steel, *Interferometry*, 2nd ed., Cambridge University Press, 1983.
2. P. Bousquet, *Spectroscopy and its Instrumentation*, Adam Hilger, London, 1971.
3. G. Hernandez, *Fabry–Perot Interferometers*, Cambridge University Press, 1986.
4. Y. Guern, A. Bideau-Méhu, R. Abjean, and A. Johannin-Gilles, *Opt. Commun.* **12**, 66–70, 1974.
5. A. Bideau-Méhu, Y. Guern, R. Abjean, and A. Johannin-Gilles, *J. Phys. E* **13**, 1159–1162, 1980.
6. A. Bideau-Méhu, Y. Guern, and R. Abjean, *J. Phys. E* **17**, 265–267, 1984.
7. T. Barbee and J. H. Underwood, *Optics Comm.* **48**, 161–166, 1983.
8. R. J. Bartlett, W. J. Trela, D. R. Kania, M. P. Hockaday, T. W. Barbee, and P. Lee, *Optics Comm.* **55**, 229–235, 1985.
9. M. P. Bruijn, J. Verhoeven, and M. J. Van der Wiel, *Optical Engineering* **26**, 679–684, 1987.
10. L. Mertz, *Transformations in Optics*, Wiley, New York, 1965.
11. R. N. Bracewell, *The Fourier Transform and its Applications*, McGraw-Hill, New York, 1965.
12. J. Chamberlain, *The Principles of Interferometric Spectroscopy*, John Wiley, 1979.
13. P. R. Griffiths and J. A. de Haseth, *Fourier Transform Infrared Spectrometry*, John Wiley, 1986.
14. G. A. Vanasse and H. Sakai, in *Progress in Optics*, E. Wolf (Ed.), Vol. VI, 259–330, North-Holland, Amsterdam, 1967.
15. H. Sakai, in *Spectrometric Techniques*, G. A.Vanasse (Ed.), Vol. I, 1–70, Academic Press, 1977.
16. G. Guelachvili, in *Spectrometric Techniques*, G. A. Vanasse (Ed.), Vol. II, 1–62, 1980.
17. J. W. Brault, *Fourier Transform Spectrometry*, in Proc. of 15th Advanced Course of the Swiss Society of Astronomy and Astrophysics, A. O. Benz, M. C. E. Huber, and M. Mayor (Eds.), Saas Fee, 1985.
18. R. C. M. Learner, A. P. Thorne, I. Wynne-Jones, J. W. Brault, and M. Abrams, *J. Opt. Soc. Am.* **A12**, 2165–2171, 1995.
19. J. W. Brault, *Mikrochim. Acta* **3**, 215–225, 1987.
20. Bomem DA8, Hartman & Brown, Quebec, Canada.
21. Bruker IFS 120 HR, Bruker-Franzen Analytik Gmbh, Bremen, Germany.
22. J. W. Brault, *Oss. Mem. Osservatorio Astrofisico di Arcetri* **106**, 33–50, 1978.
23. B. A. Palmer, *Opt. Soc. Am. Technical Digest Series*, Vol. 6, 52–55, 1989.
24. A. P. Thorne, C. J. Harris, I. Wynne-Jones, R. C. M. Learner, and G. Cox, *J. Phys. E* **20**, 54–60, 1987.
25. Chelsea Instruments, Ltd., East Molesey, Surrey KT8 0QX, UK.

26. A. Thorne, *Physica Scripta* **T65**, 31–35, 1996.
27. G. W. Stroke and A. T. Funkhouser, *Phys. Lett.* **16**, 272–274, 1965.
28. T. Dohi and T. Suzuki, *Appl. Opt.* **10**, 1137–1140, 1971.
29. S. Fransden, N. G. Douglas, and H. R. Butcher, *Astron. Astrophys.* **279**, 310–321, 1993.
30. J. Harlander, R. J. Reynolds, and F. L. Roesler, *Astrophys. J.* **396**, 730–740, 1992.
31. J. Harlander, F. L. Roesler, R. J. Reynolds, K. Jaehnig, and W. Sanders, *EUV, X-Ray and Gamma Ray Instrumentation for Astronomy V*, O. Siegmund (Ed.), SPIE 2280, 1994.
32. J. Harlander and F. L. Roesler, *Instrumentation for Astronomy VII*, D. L. Crawford (Ed.), SPIE 1235, 622–633, 1990.
33. R. A. Kruger, L. W. Anderson, and F. L. Roesler, *Appl. Opt.* **12**, 533–540, 1973.
34. R. A. Kruger, L. W. Anderson, and F. L. Roesler, *J. Opt. Soc. Am.* **62**, 938–945, 1972.
35. S. Chakrabarti, D. M. Cotton, J. S. Vickers, and B. C. Bush, *Appl. Opt.* **33**, 2596–2607, 1994.
36. M. R. Howells, K. Frank, Z. Hussain, E. J. Mohler, T. Reich, D. Möller, and D. A. Shirley, *Nucl. Instrum. Meth.* **A347**, 182–191, 1994.
37. R. P. Madden and K. Codling, *Phys. Rev. Lett.* **10**, 516–518, 1963.
38. M. Domke, C. Xue, A. Puschman, T. Mandel, E. Hudson, D. A. Shirley, G. Kaindl, C. H. Greene, and H. R. Sadeghpour, *Phys. Rev. Lett.* **66**, 1306–1309, 1991.
39. M. Domke, G. Remmers, and G. Kaindl, *Phys. Rev. Lett.* **69**, 1171–1174, 1992.
40. K. Schultz, G. Kaindl, M. Domke, J. D. Bozek, P. A. Heimann, and A. S. Schlachter, *Phys. Rev. Lett.* **77**, 3086–3089, 1996.
41. M. Born and E. Wolf, *Principles of Optics*, 6th Edition, Pergamon, Oxford, 1980.
42. Boeing North American, 2511C, Broadbent Avenue Parkway NE, Albuquerque, NM 87107.
43. S. T. Smith and D. G. Chetwynd, *Foundations of Ultraprecision Mechanism Design*, Gordon and Breach, Reading, 1993.
44. M. R. Howells, Advanced Light Source Technical Report, LSBL 212, Lawrence Berkeley National Laboratory, 1994.
45. Smith & Chetwynd, *op. cit.*, Fig. 4.8.
46. M. R. Howells, R. Duarte, and R. McGill, Advanced Light Source Technical Report, LSBL 213, Lawrence Berkeley National Laboratory, 1994.
47. Smith & Chetwynd, *op. cit.*, Fig. 4.16b.
48. J. Dyson, *Interferometry as a Measuring Tool*, Machinery Publishing Co., Brighton, 1970.
49. Born & Wolf, *op. cit.*, paragraph 8.8.2.
50. E. M. Gullikson, J. H. Underwood, P. C. Batson, and V. Nikitin, *J. X-Ray Sci. Technol.* **3**, 283–299, 1992.
51. Hewlett-Packard, Test and Measurement Division, Mail stop 51 LSJ, PO Box 58199, Santa Clara, CA 95052-9952.

5. GAS DETECTORS

J. B. West
Daresbury Laboratory
Warrington, United Kingdom

5.1 Ionization Chambers

Ionization chambers are based on so-called ion chambers, which have been in use for a considerable time for detection of ionizing radiation and particles. Ion chambers are in principle very simple devices (an early example is shown in Fig. 1) used for the detection and quantitative intensity measurement of x-rays [1]. The x-rays pass through the thin window W and ionize the gas contained in the chamber C. The current is collected on the wire A, carefully insulated from the chamber walls by the quartz plug I and the guard ring G, since the ion current is small. A voltage of around 100 volts is applied to the wire to ensure saturation of the ion current in the chamber; it is also positive so that electrons are collected rather than the ions. This makes use of the fact that the mean free path for electrons in the gas is greater than that for the ions. Originally, an electrometer measuring the voltage across a high value, high precision resistor would have been used to measure currents larger than 10^{-12} A. For currents less than this, the rate of rise of voltage on the electrometer would have been measured and the current deduced if the capacitance of the circuit were known. With modern electrometers, currents as low as 10^{-15} A can be measured.

In order to ensure complete absorption in the gas, high pressure and a gas of high density such as methyl bromide was used. Although the device shown in Fig. 1 does give a relative measure of x-ray intensities, it cannot be used for quantitative measurements in the VUV. For this, the ionization chamber is more appropriate, and has to be in a windowless form. The chamber shown in Fig. 2 was designed some time ago [2] but design principles have changed little since

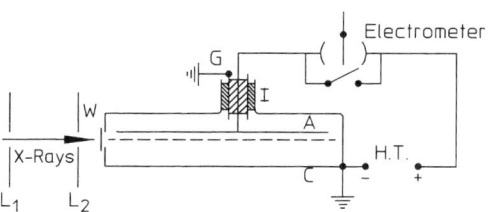

FIG. 1. An early design of ion chamber [1]. C: cylindrical wall; A: collecting wire; I: insulating quartz plug; G: guard ring; L_1 and L_2: collimators.

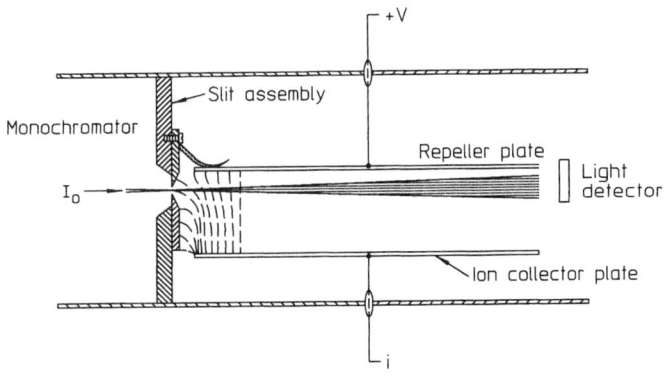

FIG. 2. Samson single plate ion chamber [2].

then. Plates rather than wires are used for the electrodes to increase the efficiency of collection for the lower field strengths, which are required to minimize further ionization. For linear measurements, it is necessary to ensure that all the ions generated along the path of the incident beam reach the collector plate without further collisions, and this limits the gas pressure used to 10^{-2} torr or less. The monochromator slit assembly must be held at the same potential as the repeller plate to ensure the efficient collection of ions generated near the entrance slit.

The following simple analysis shows how the ionization chamber can be used to measure absolute intensities: If the incident intensity is I_0 in photons/sec, the quantum yield is γ (number of singly charged ions generated per incident photon), and the intensity measured by the light detector is I, then the ion current i measured by an electrometer attached to the plate L is given by

$$i = e\gamma(I_0 - I) \qquad (1)$$

where e is the electronic charge.

Note that in this case, the ions rather than the electrons are collected. The electrons are drawn toward the repeller plate R, which is placed close to the light beam, to minimize the path of the electrons through the gas. This reduces the possibility of further ionization taking place because of electron collisions in the gas, so one must bear in mind that they will have kinetic energies of several tens of eV depending on the value of V, the voltage on the repeller plate. Even though their mean free path is greater than that for the ions, as stated previously, this places a further limitation on the gas pressure that can be used in the chamber.

In the region below their double ionization thresholds, the rare gases are known to have unit quantum yield and are the usual choice for these ion

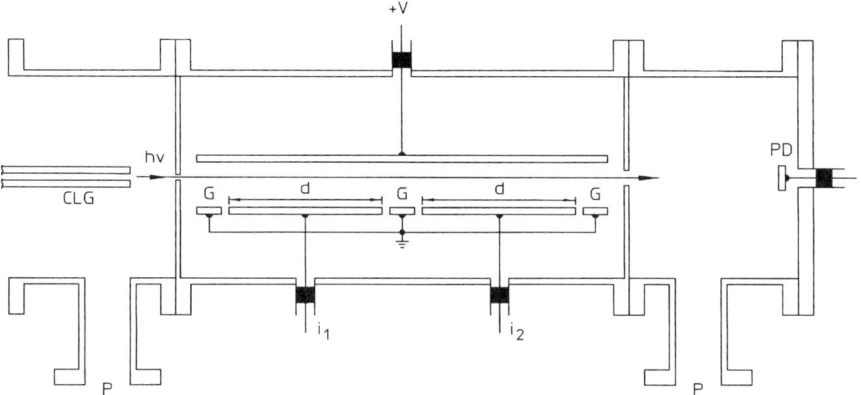

FIG. 3. Double ion chamber adapted for UHV use on a synchrotron radiation source. P: pumping ports; CLG: capillary light guide; G: guard plates; PD: photodiode detector.

chambers. Equation (1) can therefore be rewritten, with a little rearrangement:

$$I_0 = \frac{i}{e(1 - I/I_0)} \qquad (2)$$

The ratio I/I_0 (I_0 is the value of I with no gas in the chamber) can be measured using a detector, which must have a linear response with intensity, such as a sodium salicylate screen or aluminium photodiode. The ion chamber then becomes an absolute photon flux detector. In practice, it is more often used in its "double" form [2], with further modification to adapt it to the ultra high vacuum (UHV) environment of modern synchrotron radiation sources, where it may often be used. The layout is shown in Fig. 3. Since it is often inconvenient to have the exit slit of a monochromator at the repeller plate voltage, the chamber is separated from it by an intermediate differential pumping stage. The entrance to the chamber is a small hole rather than a slit, and the light from the source is brought to it by a capillary light guide as shown. This permits a vacuum of 10^{-4} torr in the differential pumping chamber, and 10^{-8} torr in the preceding monochromator because of the high impedance of the capillary light guide. Where it is being used to calibrate a secondary standard such as an aluminium photodiode, it is necessary for a further section of differential pumping to house this kind of diode detector. Without this, the photocurrent measured from the cathode of the diode will be enhanced by ionization in the gas near the cathode surface. If the length d of the collector plates is large in comparison with the distance of the light beam from the plate and the distance of the plate from the entrance or exit apertures, then end effects can be neglected. For accurate measurements, however, guard plates held at ground potential are usually fitted as shown in Fig. 3 to minimize ion losses in the collection plate regions.

The exit aperture must of course be large enough to allow all the light entering the chamber to leave it without scattering from its edges. It is also desirable for the hole at the entrance end to be large enough to avoid such scatttering, since the electrons generated will cause further ionization. Careful design is therefore needed to achieve this, bearing in mind the conflicting requirement that these apertures should be small for differential pumping.

The analysis of the double ion chamber follows from that for the single ion chamber, with the additional use of the Lambert–Beer law for photoionization:

$$I = I_0 \exp(-\mu d) \qquad (3)$$

where μ is the absorption coefficient of the gas at the pressure and temperature used in the chamber. If the currents in the two collector plates are i_1 and i_2, and the length of each plate is d, then this can be applied twice, giving

$$I_0 = \frac{i_1}{e[1 - \exp(-\mu d)]} \qquad (4)$$

and

$$I_0 = \frac{i_1 + i_2}{e[1 - \exp(-2\mu d)]} \qquad (5)$$

It is then straightforward to show that

$$I_0 = \frac{i_1^2}{e(i_1 - i_2)} \qquad (6)$$

and

$$\mu = \frac{1}{d} \ln\left(\frac{i_1}{i_2}\right) \qquad (7)$$

Thus, it is necessary to know only the values of the ion currents from the two plates in order to determine both the absolute intensity and the absorption coefficient. It must be remembered, however, that this can be used only where the Lambert–Beer law is valid, in particular where μ is independent of the pressure of the gas and of the absorption path length. Resonance regions, where the bandpass of the incoming light can be larger than the resonant line width, should therefore be avoided.

The main precautions to be taken in the design of the ionization chamber are concerned with elimination of pressure gradients in the chamber. This is likely to be a problem where differential pumping is used, and can be minimized by using both a low gas pressure and small apertures at the entrance and exit. A large diameter chamber, in comparison with the aperture diameters, will increase the volume of gas and thereby reduce the significance of the losses. A long chamber will compensate for the reduction in gas pressure, but one must bear in mind that a value of $\mu d \sim 1$ is the aim to give acceptable accuracy in the measurement of i_1 and i_2. With recent advances in picoammeter design, the highest accuracy is obtained when i_1/i_2 is between 2 and 3 [3]. The divergence of the light beam will limit what is acceptable for the chamber dimensions; an undulator beamline on a synchrotron radiation source, where the divergence is ~ 1 milliradian at most, is ideally matched in this respect. Realistic dimensions for the chamber are 20 cm diameter and 1 m long, with entrance and exit apertures of 1 mm and 5 mm diameter. This design is less suitable for laboratory sources, since they generally have higher angular divergence, but by collimating them and making a correction for the fraction of their output measured by the ion chamber, the design can still be useful.

The gas to be used in the chamber and the repeller voltage must also be chosen carefully. They will depend on the wavelength to be used and the chamber dimensions. It is important to ensure that the electrons ionized from the gas do not have enough energy to cause further ionization. The choice of the repeller voltage will also influence this, although, as pointed out previously, this effect is minimized by arranging for the photon beam to pass close to the repeller plate. Having chosen the gas, it will be necessary to establish a "plateau" region where the ion current does not depend on the voltage applied; it has already been pointed out [2] that for some combinations of incident wavelength and voltage, no plateau region will exist.

One advantage of the double ion chamber is that it is insensitive to variations of pressure, temperature, and source intensity because i_1 and i_2 can be measured simultaneously. Note also that for measurements of absolute intensity, there is no need for the gas pressure to be known. On the other hand, for measurements of ionization cross sections, the pressure and temperature do need to be measured, the former by an absolute gauge or a capacitance manometer calibrated against an absolute standard. It may also be desirable, where the highest accuracy is needed, to keep both temperature and pressure constant. In the case of temperature, a thermally controlled enclosure can be used; for pressure, regulated flow systems such as a needle valve controlled by a pressure gauge can be used. The cross section is related to the measured attenuation coefficient by replacing μ in the Lambert–Beer law in Eq. (3) by σn, where σ is the photoionization cross section, and n is the number of molecules/cc at the pressure P and absolute temperature T at which the measurement is made. The relationship between the measured attenuation coefficient and the photoionization cross section is therefore

$$\mu = \sigma N \frac{P}{760} \frac{273}{T} 10^{-18} \tag{8}$$

where N is Loschmidt's number, the number of atoms/cc at standard temperature and pressure, and σ is the cross section is expressed in Megabarns.

5.2 Proportional Counters

The proportional counter is a development of the ionization chamber, but it is used in a regime where gas amplification takes place. This is achieved by using a higher gas pressure and higher voltage in the chamber, so that ejected electrons have a much shorter mean free path in the gas and have sufficient energy to further ionize the gas, in distinct contrast to the ionization chamber. The construction is also different—usually symmetrical with a wire held at a high positive voltage being used as a central electrode, surrounded by the cylindrical wall of the counter. The mode of working depends on the electric field strength, itself being a function of the applied voltage and the relative radii of the central wire and the outer chamber wall. If the field is high enough, the ionizing particle generates electrons that are accelerated in the electric field and cause further ionization, leading to a Townsend avalanche. In the linear or proportional region, the magnitude of the current pulse depends on the number of initial ion pairs produced, multiplied by the gain, which results in the avalanche. This gain is usually $\sim 10^3 - 10^4$, depending on the voltage applied, and over a region of voltages (generally less than 1 kilovolt) varies linearly with it. At higher voltages, a discharge takes place that is independent of the initial number of ion pairs produced by the incident particle, and the device becomes a Geiger–Müller tube.

It will be appreciated that the proportional counter has a pulsed output, whose output pulse is proportional to the number of ion pairs originally generated, that is, it can be used to measure the energy of the incident particle. This pulse is generated by the electrons being collected at the anode wire. The much slower ions drift toward the counter walls, and would eject further electrons resulting in an after pulse if steps were not taken to prevent it. This is done by introducing a quenching molecular gas such as alcohol at a pressure of 1 torr; the ionizing gas is usually a rare gas at a pressure of ~ 10 torr. The ions exchange energy with the molecular gas, and this energy is dissipated through dissociation rather than by further ionization. Also, to maintain the efficiency of the detector, a flow system is generally used, whereby the gas in the detector is constantly being renewed. This removes photoproduced species that in general would quench the avalanche and lower the efficiency of the detector.

For purely statistical reasons, the energy resolution of the counter depends on the initial number of ion pairs produced; as a rough estimate, the energy required from the ionizing radiation to create one ion pair is ~ 35 eV. The statistical variation or the variance in this number has been calculated from first principles by Fano [4] and is known as the Fano factor F. Typically, for x-rays corresponding to the ^{55}Fe line with an energy of 5.9 keV, an energy resolution $\sim 14\%$ can be achieved. This can be improved by detecting the light emitted as the x-ray is absorbed; Fig. 4 shows the layout of such a detector in schematic form [5]. The electron cloud is localized around the initial x-ray direction and drifts toward the

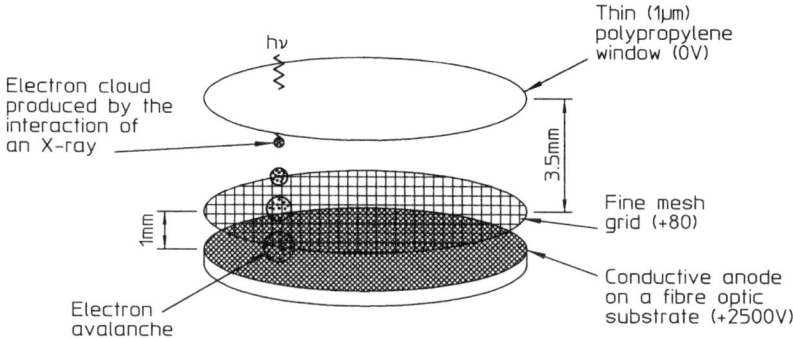

FIG. 4. The parallel plate imaging proportional counter used by Siegmund and colleagues [5].

grid. It is then accelerated toward the anode held at a high voltage by a transparent conductive layer, generating an electron avalanche in the gas mixture. Excited atomic and molecular states are generated in the gas decay, emitting light that can pass through the transparent anode. Either a photomultiplier is used to detect the emitted light, or an image intensifier to provide spatial information. Because of the improved efficiency, a resolution of ~9% can be obtained in this way for the ^{55}Fe line. In a further variant, the incoming radiation is absorbed in a rare gas such as xenon at relatively high pressure, and the VUV light emitted transmitted through an LiF window and used to ionize the lower pressure gas in the proportional counter. Such a system has also been used to provide spatial information in the multiwire proportional counter [6].

As far as detecting VUV radiation is concerned, however, the use of proportional counters is limited to the short wavelength end, generally described as the soft x-ray region. This is because a window is required both to contain the gas and thereby to limit contamination of the vacuum in the sample chamber, yet it must be thin enough to transmit the VUV radiation. For the region between 100 and 1000 eV, a parallel plate chamber has been shown to be the most effective design because of the large window-detector area possible, and has been described in detail by Smith and coworkers [7], from whom the following description is derived.

The design is shown in Fig. 5, and was operated at pressures of ~20 torr, using a 90% Ar/10% methane mixture. The window is typically 1-μ thick polypropylene, supported on a mesh and coated with an evaporated layer of Al on the inner surface to maintain ground potential across the window. The cathode mesh is held at ~200 V and the anode at ~900 V. The electrons ionized by the incident radiation drift toward the cathode and multiplication occurs between the cathode and anode. If the number of electrons passing through the cathode grid is n_c, then the number n reaching the anode and the gain M are

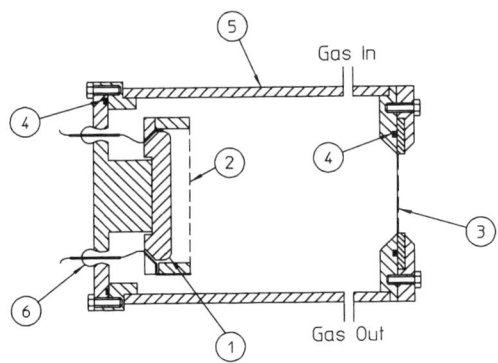

FIG. 5. Parallel plate avalanche chamber used by Smith and colleagues [7]. 1: anode; 2: cathode; 3: window with support mesh; 4: O-ring; 5: cylindrical body; 6: glass-metal feedthrough. Overall length: 60 mm.

given by

$$n = n_c \exp(\alpha d) \tag{9}$$

and

$$M = \exp(\alpha d) \tag{10}$$

where α is the first Townsend coefficient and d is the distance between the anode and cathode. To maintain constant gain, accurate machining is required to maintain d within an accuracy of $\pm 10\,\mu$, and the gas handling system maintains the pressure constant within ± 0.2 torr. The resolution of the detector was found to vary linearly between 40% at 1 keV and 100% at 100 eV, in reasonable agreement with the expression

$$\Delta E/E = 2.355[(F + f)/\langle n_c \rangle]^{1/2} \tag{11}$$

where F is the Fano factor, $\langle n_c \rangle$ the average number of initial electrons created by an absorbed photon, and $f = (\sigma_M/M)^2$ with σ_M the standard deviation in M. For argon, $F \sim 0.18$, $f \sim 0.6$ for M in the range 10^4 to 10^5, and the amount of energy required to produce one ion pair is ~ 25 eV [5].

A parallel plate detector of this kind has been used successfully at the Daresbury Synchrotron Radiation Source for extended x-ray absorption fine structure studies in the soft x-ray region. In this application, the fluorescence from the sample is measured in order to focus on the atom of interest [8]. The close spacing of the plates allows the high count rates (~ 1 MHz) necessary when used on a synchrotron radiation source, although care is needed in their construction to avoid electrical breakdown. The energy resolution at the higher end of the energy range is also useful in discriminating against unwanted fluorescent lines, some of which may be excited by higher order radiation from the primary monochromator. In cases where it is desirable to have a very low pressure in the experimental chamber, that is, $\sim 10^{-8}$ torr, a double window is necessary with differential pumping between the windows. The detector construction becomes complex, as does the associated pumping system, which must enable the user to control the differential pressures placed across the windows during pumpdown and venting.

References

1. J. Yarwood, *Atomic Physics Vol. II* (University Tutorial Press, Ltd, London, 1960) p. 153.
2. J. A. R. Samson, *Techniques of Vacuum Ultraviolet Specroscopy* (John Wiley, New York, 1967) p. 269.
3. J. A. R. Samson, *J. Opt. Soc. Am.* **6**, 2326 (1989).
4. U. Fano, *Phys. Rev.* **72**, 26 (1947).
5. O. H. W. Siegmund, S. Clothier, J. L. Culhane, and I. M. Mason, *IEEE Trans. Nucl. Sci.* **NS-30**, 350 (1983).
6. G. Charpak, *Nucl. Instrum. Meth.* **176**, 9 (1980).
7. G. C. Smith, A. Krol, and Y. H. Kao, *Nucl. Instrum. Meth. A* **291**, 135 (1990).
8. A. D. Smith, H. A. Padmore, and P. A. Buksh, *Rev. Sci. Instrum.* **63**, 837 (1992).

6. PHOTODIODE DETECTORS

L. R. Canfield

Electron and Optical Physics Division
Physics Laboratory
National Institute of Standards and Technology
Gaithersburg, Maryland

Abstract

Photodiodes of certain types are viable radiation detectors in the vacuum ultraviolet spectral region and are commonly used as the detectors of choice for radiometric applications. Some aspects of the historical evolution of the technology of these detectors are reviewed. The two general types of photodiodes most often used in this region, photoemissive and semiconducting, are discussed in detail. The most common materials and designs are given, with examples of their spectral response and applications shown. Special precautions necessary for proper use of photodiodes in the vacuum ultraviolet are identified and discussed.

6.1 Introduction

The photodiode is one of the simplest forms of radiation detectors suitable for use in the vacuum ultraviolet (VUV). This detector class is made up of two basic types, photoemissive (vacuum) photodiodes and semiconductor (solid state) photodiodes, which will be discussed individually in the sections that follow. The basic physical mechanism in the two types is quite different, but the end product of each is the production of measurable current that is proportional to the intensity of incident radiation, with an efficiency that is a function of wavelength. The process that occurs can, in either type, be characterized as the conversion of the radiant energy of incident photons to the measurable flow of electrons. Measured current represents the integration of this process and any event-related processes, with the photodiode efficiency (average electrons per photon) determined by the photon energy and the photodiode material(s).

We normally think of photodiodes as radiation detectors for relatively intense radiation levels, which they certainly are by comparison with detectors designed to give pulse amplification. Photodiodes are usually appropriate detectors when radiation intensity levels exceed those appropriate for photon counting techniques, although modern high-quality electronics have made it possible to operate some photodiodes (in a current-measuring mode) over a portion of the counting regime [1]. Photodiodes are widely used as radiometric standards for the VUV (and for other spectral regions) but are often useful for many much more mundane applications in which calibration is unnecessary. The discussion to follow will be biased toward the more demanding requirements of radiometry.

6.2 Radiometric Standards

Photodiodes have become the predominant detector standards for the VUV as well as for much longer wavelengths because of the simplicity of their basic operational mechanism, a primary dependence only on material properties, good temporal stability and spatial uniformity, and relative ease of use. Given a calibrated transfer standard photodiode and incident radiation with an appropriately narrow bandpass, one should be able to calibrate another detector or determine the flux incident on an experiment by direct intercomparison with the standard.

The ideal radiometric photodiode will have uniform response over its active area, will not exhibit changes in response over time, will give linear response over a wide range of intensities, and will be tolerant of imperfect conditions of use. These criteria are generally difficult to achieve in the VUV. One can approach the first three goals by design, materials selection, and preparation prior to radiometric calibration, but it is the responsibility of the end user to maintain conditions that will not degrade the calibration. Control of the vacuum environment is crucial to preserving the calibration of a VUV photodiode. Deposited contaminant films can absorb some of the incident radiation and/or change the photoemissive properties of the surface, and the material deposited may be decomposed by the radiation, leaving permanent surface films.

The intensity, $I(\lambda)$ (s^{-1}),* of an incident beam of monochromatic radiation can be easily obtained if one has a calibrated photoemissive or semiconducting photodiode, by an absolute measurement of the photocurrent, $i(\lambda)$ (A), produced by the radiation, and the simple relationship

$$I(\lambda) = \frac{i(\lambda)}{q\varepsilon(\lambda)} \qquad (1)$$

where $\varepsilon(\lambda)$ is the photodiode efficiency in average electrons per incident photon

---

* As an aid to the reader, the appropriate coherent SI unit on which a quantity should be expressed is included in parentheses when the quantity is first introduced.

at wavelength λ, and q is the electronic charge (C). Equation (1) can alternately be expressed as radiant power, $P(\lambda)$, and written as

$$P(\lambda) = \frac{i(\lambda)hc}{\lambda q \varepsilon(\lambda)} \qquad (2)$$

where h is Planck's constant and c is the velocity of electromagnetic waves in vacuum. One need not know the energy distribution of the photoelectrons in this application.

It is assumed throughout this discussion that radiation is incident normal to the detector surface (see Section 6.4.6). It is also assumed that the incident beam is of smaller cross-sectional area than the active area of the photodiode and falls totally within this area.

The spectral coverage of the two basic photodiode types will be discussed later, but in short, it is feasible to use either type, in one form or another, in any portion of the VUV. There generally will be a clear preference, determined by the operational characteristics of the detectors and the spectral region, among the types and configurations. The specific response of each type described later can be tailored somewhat in the fabrication process, and in some cases very restricted spectral response can be achieved in the design of the photodiode.

6.3 Photodiode Types

The two types of photodiode used in the VUV differ in a very significant way. Photoelectrons produced in photoemissive photodiodes must travel from the solid into vacuum to create a measurable current, whereas semiconductor photodiodes operate as a self-contained system, with (ideally) no charge flow into vacuum. The flow of current in a semiconductor photodiode is from one internal element to another, and will occur in response to appropriate radiation even without a vacuum environment. It would seem, from superficial considerations of the two types, that semiconductors should be superior detectors, but this is not always the case. The VUV segment of the semiconductor detector industry is relatively new and will surely mature, so one expects that eventually semiconductors will be the dominant detectors for this region, as they have become at longer wavelengths. There are many interesting and potentially useful directions in which designs could evolve, such as the application of high bandgap materials to exclude detection of visible radiation, the incorporation of on-chip electronic circuitry to produce amplified or digital output, or the use of internally multiplying semiconductor structures. The development process is relatively slow, however, because the major commercial markets for semiconductor detectors are concerned with other spectral regions.

6.3.1 Photoemissive Photodiodes

The external photoelectric effect was studied in the 1930s [2, 3], and a number of materials were characterized for use in detectors in and near the visible spectrum [4]. Many fundamental measurements in the VUV were made during the 1950s and 1960s, driven, in part, by the development of space technology. Several materials suitable for use at wavelengths longer than about 105 nm were studied during this period, and bandpass configurations useful in the emerging space program were developed by choosing window materials and/or emitters to adjust the spectral range [5, 6].

Photoemissive photodiodes emit electrons into vacuum in response to incident photons. The photocathode (or emitter) material must absorb photons sufficiently close to the surface, and the energy of the photons must be great enough to produce electrons that reach the vacuum interface with energies in excess of the work function of the photocathode material. An absorbed photon will produce an electron with enough energy to either escape directly or produce secondary photoelectrons, which may have enough energy to continue this process, or escape. An external electric field of sufficient strength to maximize the measured photocurrent generally needs to be provided. Electron transit times in a given material will set an upper limit to the rate at which incident photons can produce linear photocurrent. A more detailed description of the subject can be found in the literature (see, for example, references [4] and [7]).

Materials' properties produce the need for two subclasses within the photoemissive class in the VUV region for the following reasons. Materials with reasonable stability in air (such as certain metals) invariably have rather high work functions, and thus have little efficiency in the long wavelength portion of this spectral region (or beyond). Several materials, generally semiconducting intermetallic compounds, have lower work functions and much greater efficiency at, say, 200 nm wavelength, but these invariably are not stable when exposed to air. Such materials are practical only in sealed, evacuated photodiodes, in which the photocathode is formed *in situ* without subsequent air exposure, and an appropriate window is incorporated. (This is also the case for most photoemissive devices intended for use at longer wavelengths.) The shortest VUV wavelength at which operation of this photodiode type is possible will be established by the window's transmission limit, the shortest of which is nominally 105 nm (LiF). Photodiodes intended for use short of this wavelength must use a windowless configuration, and their short wavelength limits will be determined by the absorption characteristics of the cathode material.

6.3.1.1 Windowed Photoemissive Photodiodes. Windowed photoemissive photodiodes form a closed system in which the photocathode and the anode are located within an evacuated structure containing a window that transmits the radiation of interest. The photocathode can be either an opaque

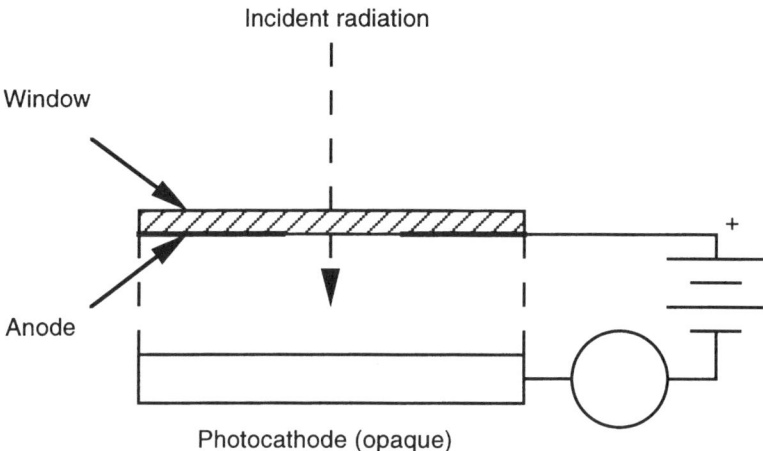

FIG. 1. Schematic of a windowed photoemissive photodiode with an opaque photocathode.

material opposite the window, or a semi-transparent thin film on the vacuum side of the window. Schematics of each type, with external biasing and current measuring circuitry shown, appear in Figs. 1 and 2. The distinction between many windowed photoemissive photodiodes for the visible/near ultraviolet, and those intended for the VUV lies primarily in the choice of window materials. The long wavelength response of these VUV photodiodes can be limited to some degree by selection of photocathode materials.

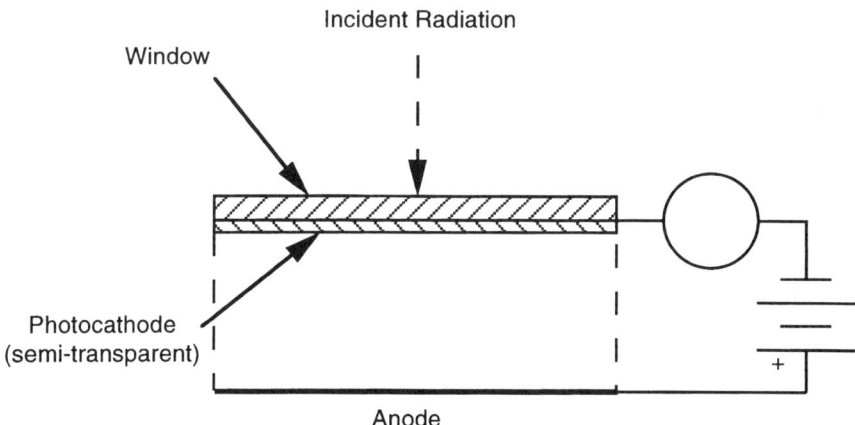

FIG. 2. Schematic of a windowed photoemissive photodiode with a semi-transparent photocathode.

Sealed vacuum photodiodes using reactive cathode materials will have good temporal stability only if the residual gas within the sealed volume is maintained at a very low level. Even better long-term stability can be achieved if the design also uses a very low operating potential. This will minimize the chance that emitted photoelectrons will ionize the residual gas, and that the ions will then impact the photocathode, possibly causing changes in the photoemissive properties of the photocathode. Another stability issue that may have to be addressed is the presence of residual unbound high vapor pressure materials, such as cesium, in the internal environment. Care in manufacture can adequately control this parameter.

Two basic methods, neither of which is well-suited to large-scale production, have been used to produce most sealed windowed photodiodes: appendage processing and remote processing. Both methods have been used successfully with reactive cathode materials, although remote processing appears to offer several advantages in the final product and is the more modern technology. The configuration can be as shown in either Fig. 1 or Fig. 2 with each of these schemes. Simplified descriptions of typical processing with each technique follow, with the assumption, for simplicity, that the photocathode is being formed on the window.

Appendage processing is basically an extension of glassblowing technology, and hence is relatively inexpensive to use. The processing of a photodiode typically starts with a fully assembled glass body with an anode, a sealed window, suitable electrode feedthroughs, and one or more glass tubing appendages attached to the body. The photocathode is formed by introducing a small evaporator through one of the appendages. The material in the evaporator is heated to a temperature sufficient to achieve the deposition of a film on the inner surface of the window. The evaporator is then disabled, and any necessary activating vapor (e.g., cesium) is allowed to diffuse into the tube body from an appendage. After appropriate activation of the original film, the vapor source is disabled, and the appendages are removed by fusing close to the tube body, leaving short residual sealed ports. Certain parameters, such as the spatial uniformity of the photocathode material and the re-evaporation of residual photocathode materials during the appendage tip-off procedure, may be difficult to control. It is possible to process photodiodes with acceptable cathode uniformity using this technique, but a great deal of artistry is generally required.

Remote processing of a photodiode with a semi-transparent photocathode typically starts with the placement of a fully assembled tube body and a separate window into a relatively large vacuum system. The body of the photodiode can be made of suitable materials other than glass, since no appendages will be required. High-density ceramic bodies with electrodes brazed to the ceramic are common. The photocathode and any necessary masking and conducting underlayer are formed on the window away from the body, after which the coated window can be joined to the body *in situ*. This technology permits the deposition of one or more carefully controlled films on the window with more or less conventional film techniques without residual contamination of the tube body. It also ensures that the gas remaining in the photodiode after sealing will be essentially that which existed in the processing system during the window bonding. Ultra-high vacuum conditions are realistically possible. The primary performance benefits realized from this technology are a high degree of both spatial uniformity and temporal stability of the photocathode. The major disadvantage is economic, in that specialized, rather expensive systems are required. A map of the spatial uniformity of a remotely processed photocathode at 121.6 nm wavelength is shown in Fig. 3, and includes, of course, any variations in transmission of this photodiode's MgF_2 window.

Other common window materials used with VUV photodiodes include LiF, CaF, Al_2O_3 (sapphire) and fused silica, each with a characteristic short wavelength limit in the VUV. The method of sealing the window depends on the window material and the tube body material, and may be fusing or chemical bonding. Silver chloride has often been used with fluoride windows, but it must be protected from exposure to visible radiation and moisture in order to reduce the rate of photochemical reaction and degradation of the seal. Indium alloys are convenient for the bonding of remotely processed windows to tube bodies while under vacuum, and have proven to produce durable seals [8]. One must, of course, avoid exposing photodiodes with window seals of this type to temperatures approaching the softening point of the alloy.

Proper operational conditions of a photoemissive photodiode can be confirmed by measuring the cathode photocurrent as a function of anode voltage. The voltage chosen should be well into the plateau of the measured function, but not unnecessarily high. A plot of the photocurrent of a particular windowed photoemissive photodiode as a function of anode voltage is shown in Fig. 4. The anode voltage chosen for operation in this case was 150 V. (The anode voltage used during the original calibration of a transfer standard should always be that used for radiometric measurements.) Photocurrent should be carried from the vacuum system by a low-noise coaxial or triaxial cable to the electrometer being used. (Radiometric applications will require an absolute measurement of the photocurrent, necessitating the use of a calibrated electrometer.) The vacuum feedthrough carrying the photocurrent should have high enough shunt resistance to prevent degradation of the signal, and the photocurrent should be kept well below the maximum specified by the device manufacturer to prevent premature degradation of the photocathode and loss of linearity. Stray charge appearing anywhere in the cathode circuitry will produce false readings of photocurrent, so it may be necessary to include a shielded enclosure for the photodiode, and to also shield the current-carrying wire attached to the cathode, and the vacuum side of the photocurrent feedthrough.

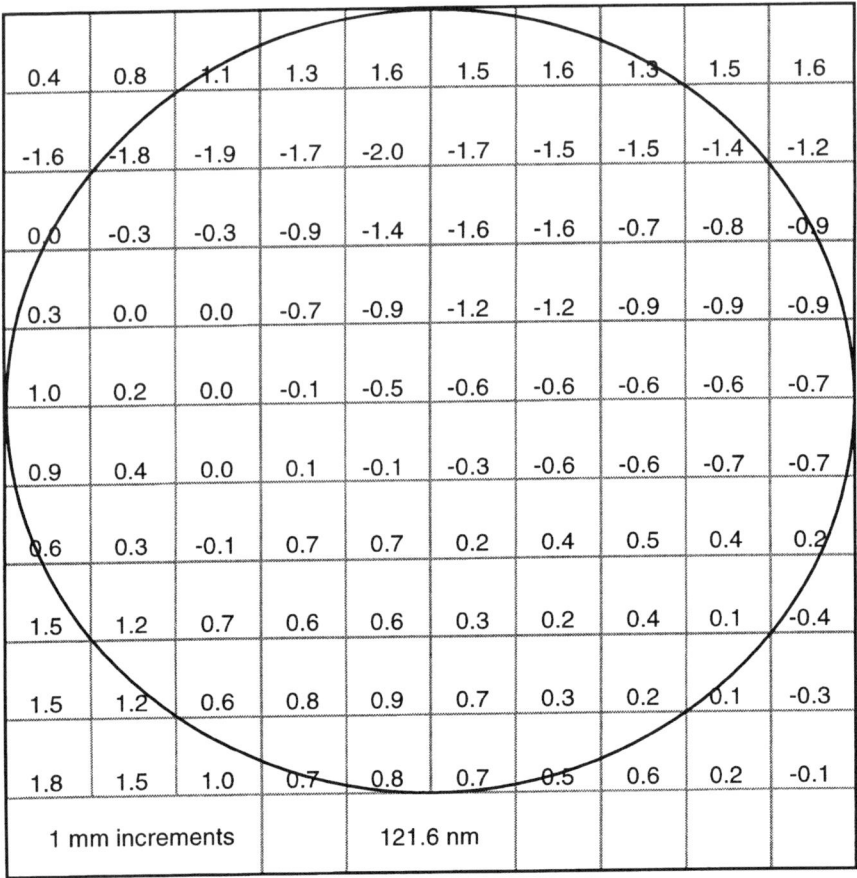

FIG. 3. Spatial uniformity map of a windowed photodiode with a remote-processed photocathode. Values are deviations from the mean, in percent.

6.3.1.2 Windowless Photoemissive Photodiodes.
Photodiode detection of radiation with wavelengths shorter than the cutoff of LiF requires an open, or windowless, configuration. Relatively few photodiodes of this type have been commercially produced. Advanced technology is not needed to construct an operational windowless photodiode, so for uncalibrated applications they can easily be constructed by the user. Generally a solid metal photocathode with an anode that is physically close to the cathode will operate well in the region from about 5 nm to about 150 nm. Radiometric transfer standards require more care in the choices of materials and construction, since the required open nature of the photodiode can easily compromise the temporal

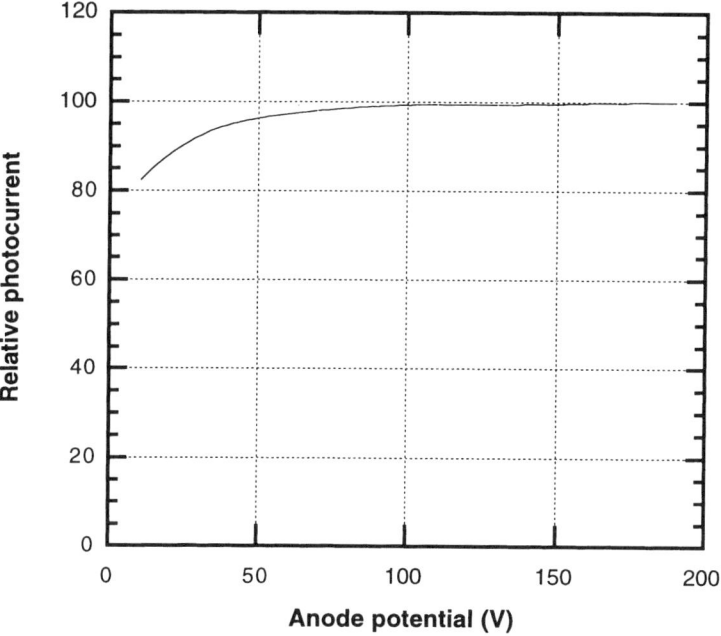

FIG. 4. Photocurrent of a photoemissive photodiode as a function of anode voltage.

stability of the photocathode. The previous comments on the selection of an anode voltage also apply to windowless photodiodes.

A study was conducted at the National Institute of Standards and Technology (NIST) in the late 1960s seeking a suitable photocathode material for use in NIST transfer standard windowless photodiodes for the VUV [9]. The primary criterion was stability of photoemission without the need for special storage. The photoemitter ultimately selected used Al_2O_3 artificially grown on a thin film of high-purity Al. This combination was the most stable of several candidate materials tested at NIST, and has been used in NIST windowless transfer standards since about 1970 [10]. A vacuum evaporated Al film about 150 nm thick is deposited on a quartz substrate, and then the thickness of the natural oxide is increased by anodizing [11] to a total oxide thickness of about 15 nm. The quantum efficiency of a typical photodiode of this type is shown in Fig. 5. The long wavelength calibration limit, imposed by the photocathode's relatively high work function, is normally 121.6 nm in NIST calibrations. Optical calculations predict that less than 1% of the incident radiation is transmitted into the aluminum at this wavelength, so the dominant photoemitter is the oxide. Note that the measured quantity must be photocathode emission current, rather than anode current. These two quantities are not, in general, equal in an open

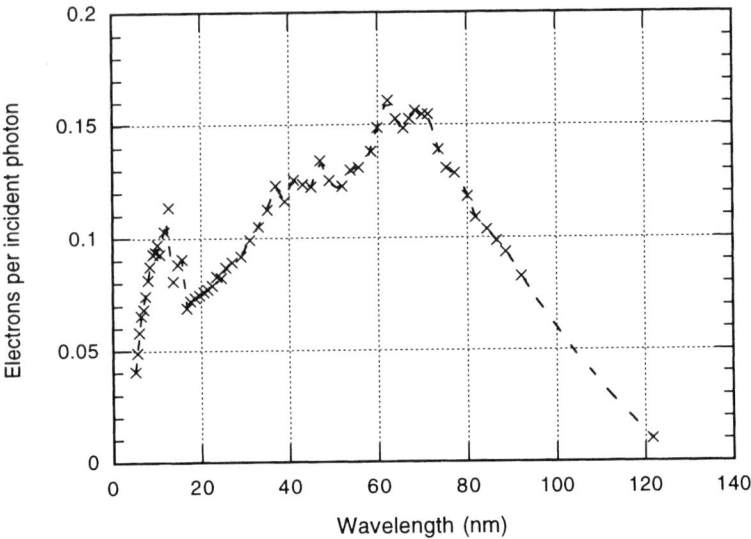

FIG. 5. Typical quantum efficiency of a NIST windowless transfer standard photodiode.

photodiode, since some portion of the emitted electrons will not reach the anode, depending on the electron energy, the electric field strength, and so forth.

6.3.2 Semiconductor Photodiodes

The first commercial semiconductor photodiodes appeared around 1960, and substantial technological development followed, particularly in processing for the visible spectral region. There was little, if any, movement toward devices specifically intended for the VUV, although it was shown in 1964 that operation of a silicon device was possible in the VUV [12]. Designs that were satisfactory in the visible spectral region proved, however, to have problems at somewhat shorter wavelengths [13].

Semiconductor materials are characterized by a relatively small energy gap between their conduction and valence bands. An absorbed photon with energy equal to or greater than this gap can excite a valence electron into the conduction band, creating an electron/hole pair. Suitable arrangements of n- and p-doped materials create internal electric fields, and, if the semiconductor's recombination length is large enough, photocurrent can be measured in an external circuit. There are additional considerations as the energy of incident photons increases well above the material's bandgap. For example, the penetration depth of the photons may exceed the thickness of the semiconductor chip, absorption in inactive surface layers may reduce the flux level reaching the active portion of the photodiode, and destructive interactions within either the semiconducting material or surface layers may result in unsatisfactory behavior.

The average pair creation energy has been measured for a variety of semiconducting materials, and is typically only a few eV [14]. The external efficiency, $\varepsilon(E)$, of an ideal semiconductor photodiode in the VUV can be expressed as

$$\varepsilon(E) = \frac{E}{w} F(E) \qquad (3)$$

where E is the energy of the incident radiation, w is the average pair creation energy for this semiconductor material, and $F(E)$ is a factor arising from absorption in any inactive ("dead") layer present and reflective losses. It is assumed here that there are no internal loss mechanisms, such as surface recombination, within the semiconductor material. It will be shown later that significant deviations from this simple model have been identified in silicon photodiodes in the VUV, particularly at relatively long wavelengths.

Equation (3) indicates that VUV photons may produce multiplication in semiconductor photodiodes if $F(E)$ is not too small. The average pair creation energy for silicon is given as about 3.6 eV [14, 15], suggesting that each incident photon (on average) with an energy of 100 eV (12.4 nm), for example, will produce almost 28 lower-energy electrons in the absence of optical losses. A typical value of $F(E)$ for this photodiode at this energy, where the loss is due almost entirely to absorption in the oxide, is 0.9, so the net efficiency would be about 25 electrons per photon.

A self-calibration model for semiconductor photodiodes has been published [16] wherein photodiode efficiencies can be determined at wavelengths shorter than 15 nm, with certain constraints, and without reference to a standard. Knowledge of the average pair creation energy for the semiconductor material is required, and measurements must be made at two or more angles of incidence at each wavelength.

Semiconductor photodiode technology offers major potential economic benefits over windowed photoemissive photodiode technology, since production can be at the wafer level. The same processes that would be necessary to produce one photodiode can be used to produce a large number of photodiodes on a single semiconductor wafer, and many techniques already developed for electronics fabrication can be advantageously adopted. A review of semiconductor detectors for the ultraviolet [17] gives some of the theoretical considerations that drive device technology, and discusses various configurations and semiconducting materials with potential applications in the VUV. We will discuss those that have received the greatest attention in this region thus far.

Two dominant issues must be considered when designing semiconductor photodiodes for the VUV: losses in "dead" layers and degrading interactions between the radiation and the detector materials. Significant progress has been made, but present devices could be substantially improved. Most of the commercially available devices that have proven useful in the VUV have been either Si (photovoltaic) or GaP/GaAsP (Schottky). Many designs using these and other semiconductor materials have not yet been fully explored.

The response time of semiconductor photodiodes optimized for the VUV is dependent on the area of the detector and the internal design, but tends to be relatively slow for pulse observations (typically microseconds). Techniques exist to improve on this (reverse bias, smaller active area, etc.) but generally they result in a compromise.

6.3.2.1 Silicon Photodiodes.

Silicon technology is the most advanced of the semiconductor disciplines, and silicon devices have dominated the electronics industry for decades. Silicon photodiodes, which can be designed to make use of common technology, have consequently been the most widely available of the semiconductor radiation detectors. The photovoltaic construction, dominant in the VUV, will be discussed here in some detail.

High-quality inexpensive silicon photodiodes for the near ultraviolet, visible, and near infrared have evolved from the first few decades of activity. The construction of a typical silicon photodiode designed for photovoltaic operation is shown in Fig. 6. (Either an n on p, or p on n configuration can be used; we show

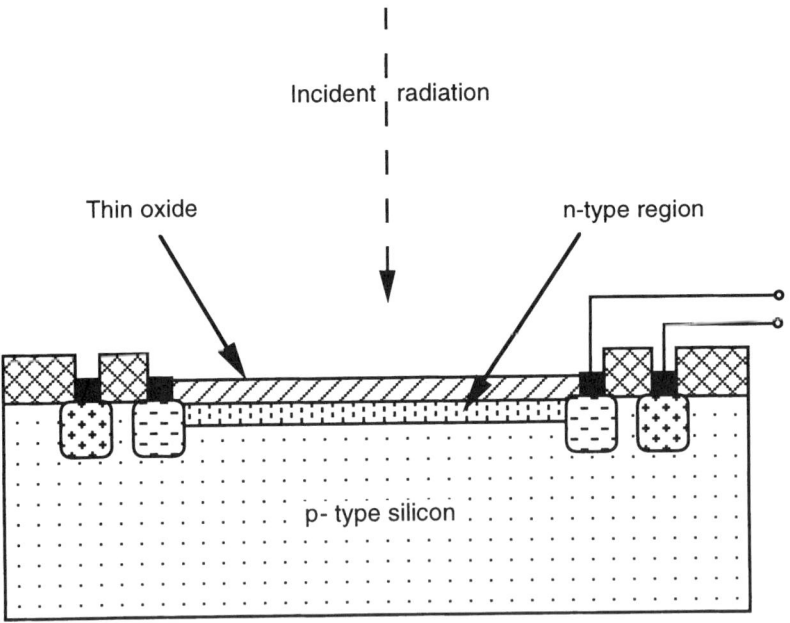

FIG. 6. Schematic of a VUV silicon photodiode.

only the former here.) An important development in detector radiometry was the discovery of a method to produce silicon devices in which nearly every absorbed photon produced an electron-hole pair [18] that could be measured by external circuitry. Surface recombination losses, which had been common in earlier devices, were essentially eliminated, and for applications in the visible (where absorption in the oxide is nil) there was effectively no longer a "dead" layer.

The oxide surface layer shown in Fig. 6 is usually grown under controlled thermal conditions, and serves to passivate the silicon. Detector configurations intended for longer wavelength applications usually use an oxide thickness that is chosen to minimize reflective losses in the spectral region of greatest interest by optical interference. Absorption losses in the oxide are negligible at wavelengths longer than about 160 nm [19], so control of reflective losses in this region by adjustment of oxide thicknesses is a viable technique. Absorption in the oxide is far from negligible in portions of the VUV [19], so much thinner layers, often much less than 10 nm, must be used when designing for this region.

A plot of the calculated radiation transmitted into the silicon of the photodiode of Fig. 6 with various oxide thicknesses is shown in Fig. 7. Clearly, if one wishes to design a reasonably efficient photodiode, particularly for use near 120 nm, a rather thin oxide must be used. There is a minimal practical oxide thickness of about 4 nm, below which operational instabilities have been observed [20]. Most successful VUV devices have had oxide thicknesses in the 5 to 15 nm range.

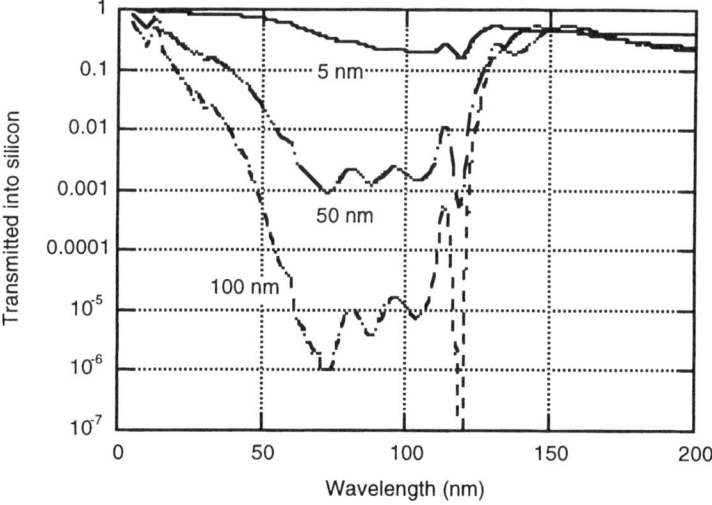

FIG. 7. Transmission into the silicon of a photodiode with the general configuration shown in Fig. 6, with three thickness of passivating oxide.

Photodiodes with very thin oxide layers have been proven to operate throughout the region of strong oxide absorption, but not strictly in accordance with a model based on calculated optical losses. A typical comparison between the calculated and measured efficiencies of a silicon photodiode is shown in Fig. 8. The published optical constants for Si [21] and SiO_2 [19] were used in a multilayer program to calculate the portion of the incident radiation entering the silicon in a photodiode with 9 nm of oxide on the surface. (This thickness was chosen to give a reasonable match of calculated and measured efficiencies at the shortest wavelengths shown.) It was assumed in the calculations that all photons entering the silicon were absorbed and created electron-hole pairs, and that there were no internal losses [18]. In fact, what one finds is that the measured efficiency is greater than that calculated in the spectral region in which there is strong absorption in the oxide, as seen in Fig. 8. It has been suggested [20, 22] that some charge produced by absorption in the oxide layer may appear in the silicon circuit and account for the observed increase in efficiency in the region of strong oxide absorption.

Early investigations of silicon photodiodes that had reasonable efficiencies in the VUV showed that there was a very significant degradation in the photodiode efficiency at moderate fluence levels in the presence of radiation strongly absorbed in the oxide [20]. It was found that the rate of degradation was very process-dependent, and that no processing then used yielded devices that were sufficiently stable to be used as detector standards in the oxide absorption region.

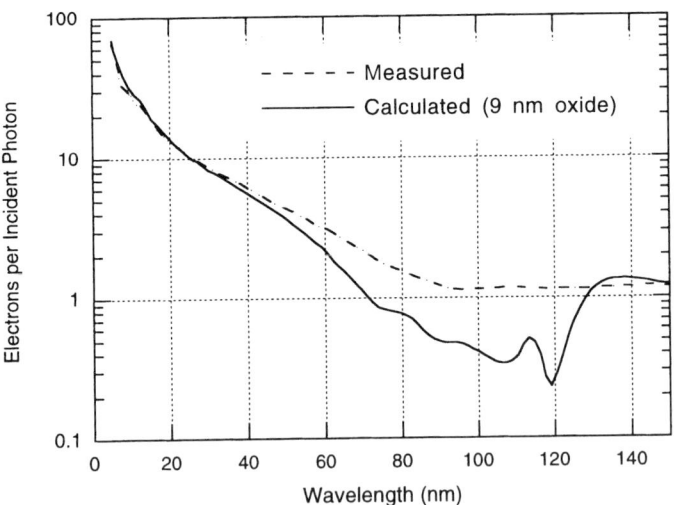

FIG. 8. Measured efficiency of a silicon photodiode compared with that calculated assuming 9 nm of passivating oxide.

6.3.2.2 Silicon Photodiodes with Nitrided Oxide.

A processing technique developed in the electronics industry was subsequently used to convert the oxide to oxynitride by rapid thermal nitridation [23]. The rate of degradation of devices thus processed was dramatically reduced, as is shown in Fig. 9, which compares the degradation of devices with and without nitrided oxides during exposure to 122 nm radiation. It was also found that the nitriding process greatly lessens the tendency of the unprotected oxide in a windowless configuration to absorb water vapor [23].

Silicon photodiodes with the oxide converted to oxynitride have proven stable and reliable in the VUV and are suitable for radiometric applications throughout the region under certain conditions. The high fluence degradation at some wavelengths that was mentioned previously, is balanced to some degree by a recovery in efficiency, as shown in Fig. 10, where a similar experiment was interrupted overnight. (Approximately 1×10^{11} photons/s/cm^2 was the average intensity level during this exposure.) The recovery process probably takes place during the exposure as well, leading to the conclusion that the reduction in efficiency at a given fluence is a function of the rate of exposure. This implies

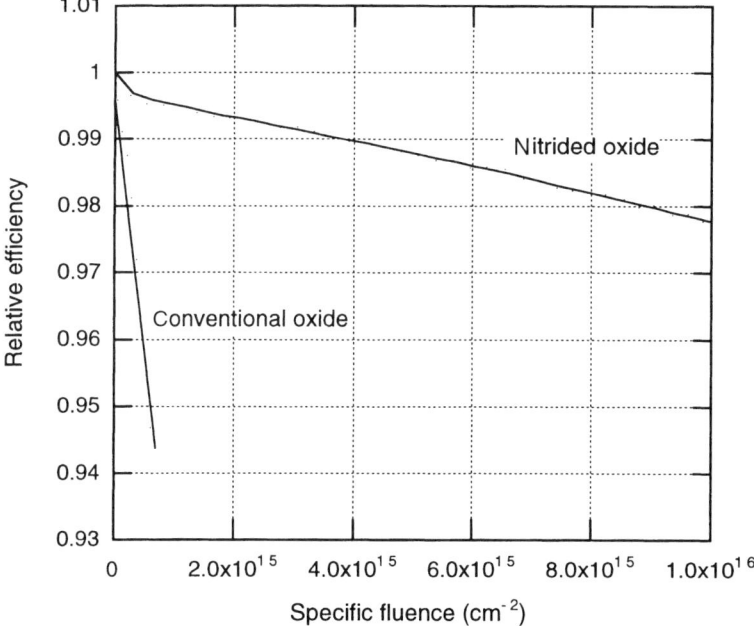

FIG. 9. Observed degradation of the efficiency of silicon photodiodes with conventional oxide and with nitrided oxide. Each sample was exposed to 122 nm radiation at about 2×10^{11} s^{-1} cm^{-2}.

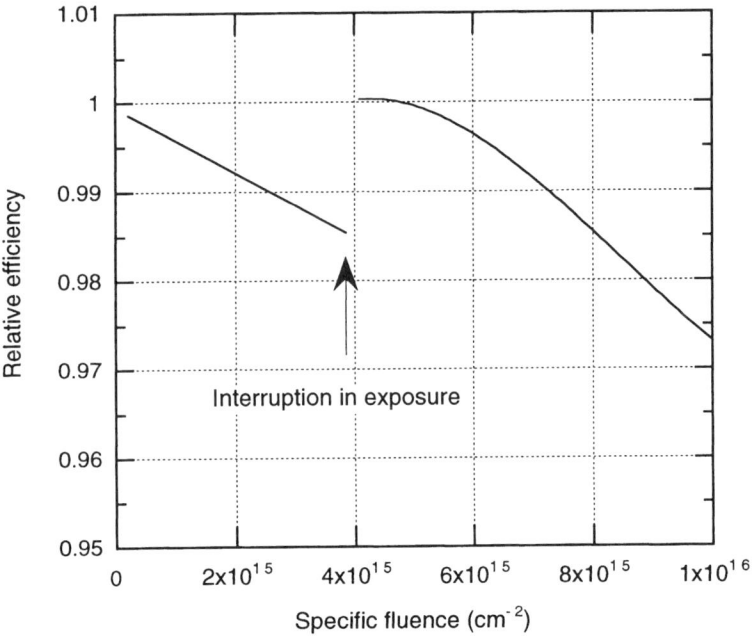

FIG. 10. Effect of interrupting a prolonged exposure of a silicon photodiode with nitrided oxide to intense radiation of 121.6 nm wavelength. The interruption lasted approximately 16 hours.

that the degradation will far outpace the rate of recovery at extremely high flux levels, and may also explain the lack of observed degradation shown in Fig. 11 at 10 nm wavelength with a flux level of about $10^9/(s \cdot cm^2)$, even though the calculated oxide absorption at this wavelength is about 0.1. Photoemission effects and oxide charging may be at least partially responsible for these phenomena.

6.3.2.3 Silicon Photodiodes with Integral Filters. A practical shortcoming of silicon photodiodes in the VUV is the broadband nature of their response, which extends from about 1.1 μm wavelength to the x-ray region. It is not unusual that out-of-band radiation may dominate experiments in the VUV. This situation can be alleviated to some degree with a relatively simple modification of the spectral response by depositing thin films of filtering materials (either metal or dielectric) directly on the surface of a photodiode [24]. (Freestanding thin film filters of various materials can be used with uncoated photodiodes, of course, but the films tend to be relatively fragile, and usually incorporate some form of adhesive, often undesirable in ultra-high vacuum applications.) The SiO_2 outer film of the silicon photovoltaic design seems to provide a sufficient barrier in typical thicknesses to prevent the short-term

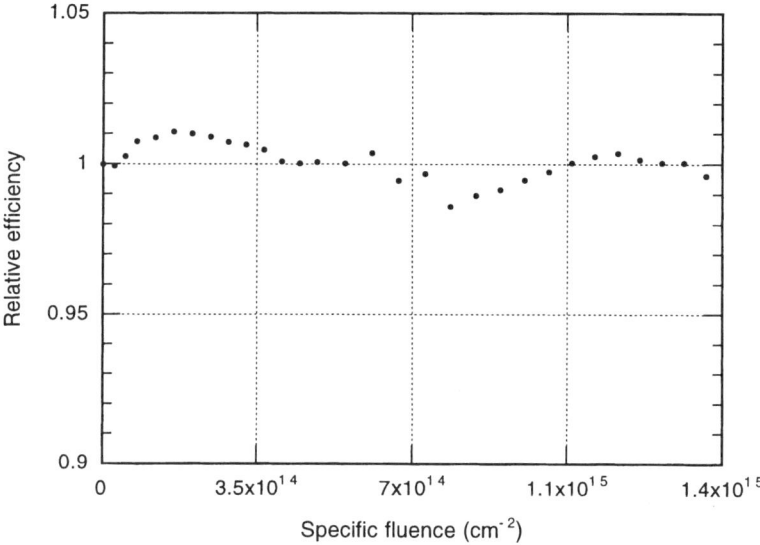

FIG. 11. Result of prolonged exposure of a nitrided silicon photodiode to 10 nm radiation. The average normalized intensity during exposure was about $8 \times 10^{10}\,\mathrm{s^{-1}\,cm^{-2}}$.

diffusion of the deposited filter materials into the silicon. A variety of materials can be applied to give filter radiometer-type response over selected spectral bands. An example of the calculated efficiency of a silicon photodiode coated with 200 nm of zirconium is shown in Fig. 12. Coated photodiodes have proven to be useful in such fields as solar physics [25] and plasma diagnostics [26, 27], where spurious radiation can impede the detection of the subject phenomena. Discrimination of nine orders of magnitude against radiation from the visible has been achieved for rocket-borne solar observations using metal filters on silicon photodiodes [28]. Radiation from the x-ray region can be rejected by insertion of a single reflector at an appropriate angle of incidence, or by reduction of the thickness of the silicon [24].

6.3.2.4 Avalanche Photodiodes. A modest amount of gain (with even low energy photons) can be achieved by designing silicon photodiodes for operation in the avalanche mode. It is also possible to achieve very short response times with reasonably large area devices with this approach. Application of a relatively high reverse bias voltage leads to the creation of secondary electron-hole pairs, with efficiency gains in the range of 10 to 100 being typical for devices now available for longer wavelengths. The use of an avalanche photodiode in the x-ray region has been reported [29], but avalanche technology has not yet merged with that used for the photodiodes described in the preceding sections to create devices suitable for the VUV. It will not be clear until this

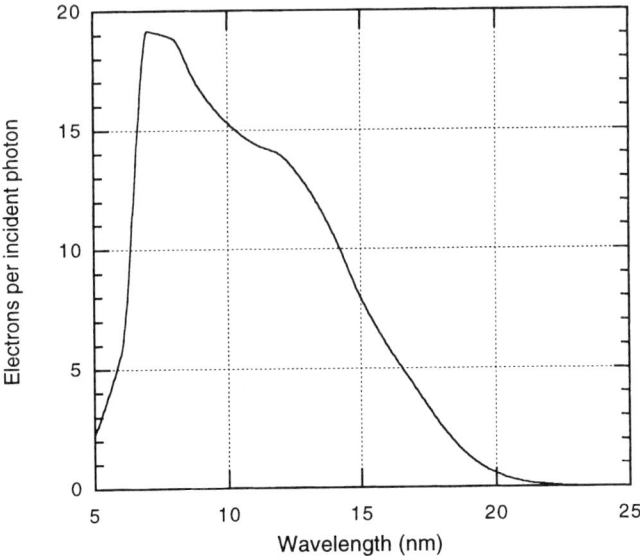

FIG. 12. Calculated efficiency of a silicon photodiode with a 10-nm thick oxide, coated with 200 nm of zirconium.

takes place whether a real improvement in signal-to-noise will have been achieved, since high-quality silicon photodiodes (non-avalanche) have been shown to be capable of photomultiplier-like detection of weak radiation sources in the visible [1].

6.3.2.5 Other Materials.
There are few materials other than silicon that have been developed commercially as VUV semiconductor photodiodes. Present offerings consist of gallium phosphide and gallium arsenide phosphide, both of which have larger bandgaps than silicon. The configuration with these materials that has been widely investigated is the Schottky diode, in which the outer (p) layer is a thin coating of gold, typically about 10 nm in thickness. There have been studies of these materials that showed them to be viable photodiodes for the VUV [30, 31]. They appear to be relatively immune to radiation damage, but have been found [32] to have significant temporal instability over part of the VUV. Several other materials with bandgaps much greater than silicon have been investigated [33, 34], but are not yet widely available as VUV detectors.

6.4 Proper Use of Photodiodes in the VUV

The use of photodiodes in the VUV is not as simple as it is at longer wavelengths, because of the energy of the radiation and the necessity of working in a vacuum environment. Conditions of use should be evaluated to avoid incorrect measurement results and/or degradation of the detector being used. Several of the more important of these are discussed briefly in the following sections.

6.4.1 Vacuum Conditions

It is obvious that condensables within the vacuum system in which radiometric photodiodes are used must be maintained at as low a level as possible. A surface film deposited on either a windowless semiconductor or a windowed photoemissive photodiode will, in general, absorb radiation, and may even fluoresce. A film on a windowless photoemissive photodiode will probably change the photoelectronic properties of the device. In either of these cases, a permanent film could result from interaction with relatively high energy radiation, leading to irreversible changes in device efficiencies. Should there be marginal vacuum conditions, either in terms of contaminants or operating pressure, a semiconductor photodiode would be preferred over a windowless photoemissive photodiode, because of the reduced sensitivity to surface conditions.

6.4.2 Intensity Levels

One should ensure that the anticipated radiant intensity levels are within the linear range of the detector, as is prudent in any radiation detection application. Extremely high intensity levels can cause degradation of the detector, and may be more appropriately detected with thermal devices. Extremely low intensity levels may be more appropriate for pulse counting devices than for photodiodes.

6.4.3 Spectral Purity

Large out-of-band contributions to the total radiation to be detected will make it reasonable to consider choosing a photodiode that discriminates against the unwanted radiation, if possible. For example, if one wishes to detect VUV emission in the presence of visible radiation, a silicon photodiode with its natural broadband response may not be the best choice. A better choice would be a photodiode that is relatively insensitive to the visible radiation, for example, a high work function photoemissive photodiode, a filter-coated silicon photodiode, such as described previously [24], or a high bandgap device, if available.

6.4.4 Magnetic Fields

Operation of a calibrated photoemissive photodiode in the presence of strong magnetic fields may call for caution. Electrons escaping the surface of the photodiode could, in the worst case, return to the surface, reducing the efficiency. One should reevaluate the photocurrent as a function of anode voltage to ensure that saturation is being achieved. Semiconductor photodiodes should not be as seriously affected.

6.4.5 Photoemission from Semiconductor Photodiodes

Semiconductor photodiodes will generally simultaneously behave as photo-emitters over much of the VUV. Photoelectrons leaving the surface, or collecting at the surface, can create anomalous effects if the surface is not a conductor, which is the case with photovoltaic silicon photodiodes. Localized collection of charge can have an unpredictable effect on the controlling electric fields within the photodiode. Schottky photodiodes will also emit photoelectrons, but there will be no local charge accumulation because of the conductive outer layer.

Photoemission as a loss mechanism might not be an issue if it were constant. A possible difficulty comes when external electric fields in the vicinity of the photodiode surface are not deliberate. The efficiency of photoemission could then be thought of as a day-to-day variable, depending on the random placement of nearby biased wires, and so forth. Possible solutions to this potential problem are either to eliminate or maximize photoemission from the photodiode. A simple method to accomplish this would be to attach an electrode near the outer surface of the photodiode, and to bias the electrode sufficiently with respect to the photodiode surface (normally at nearly ground potential) to control the escape of photoelectrons. The voltage necessary with a particular geometry can be determined empirically by measuring the semiconductor photocurrent as a function of voltage. A plot of the photoemissive efficiency of a GaAsP Schottky photodiode is shown in Fig. 13. The added electrode could, by the choice of polarity, inhibit or encourage photoemission and perhaps reduce surface charge buildup in the case of photodiodes with a dielectric surface.

6.4.6 Incidence Angle Effects

Both photoemissive and semiconductor photodiodes are sensitive to the angle of incidence of the radiation being detected. Reflective losses and photon penetration depths are both functions of the incidence angle, and will thus effect the net efficiency of a photodiode. The reflectance of a photodiode is also a function of the polarization of the radiation, except in the case of normally incident radiation. The sensitivity of windowed transmissive cesium telluride photoemissive devices to variations in the incidence angle has been reported [35].

The photoemission from windowless photoemitters will generally rise somewhat above the normal-incidence value as the angle is increased. The rise is caused by a decrease in the absorption depth relative to the surface, with a resulting increase in photoelectron escape efficiency, while the (unpolarized) reflectivity is fairly constant. Further increases of the incidence angle will

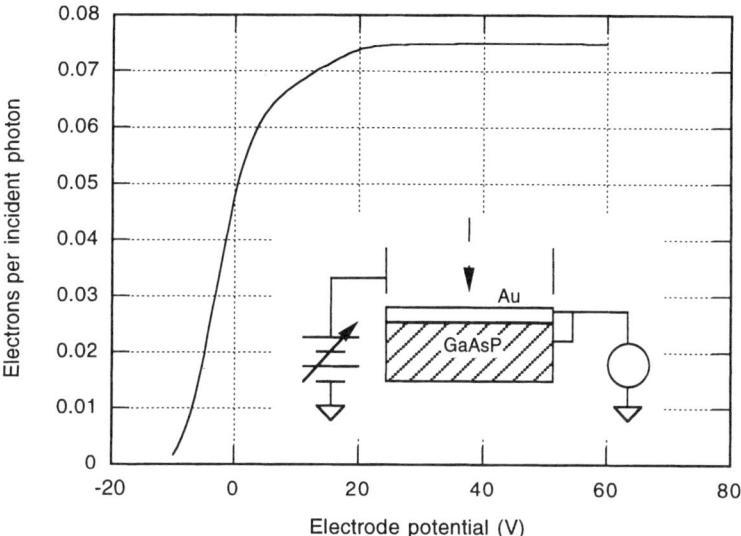

FIG. 13. Measured photoemission at 73.5 nm wavelength from a GaAsP Schottky photodiode with various external electrode voltages. The electrode was a cylinder coaxial with the incident radiation, extending about 2 cm from the front surface of the photodiode. Photocurrent was measured as shown in the schematic.

produce increasingly greater reflective losses and reduced photoemission efficiency.

Semiconducting photodiodes exhibit similar effects, although generally they are primarily caused by variations in the effective path length in surface dead layers and by reflective losses correlating to the angle of incidence. Angle-related changes in photoemission are relatively minor by comparison.

References

1. G. Eppeldauer and J. E. Hardis, *Appl. Opt.* **30**, 3091–3099 (1991).
2. C. Kenty, *Phys. Rev.* **95**, 891–897 (1933).
3. P. Görlich, *Z. Phys.* **101**, 335–342 (1936).
4. A. H. Sommer and W. E. Spicer. In *Methods in Experimental Physics* (K. Lark-Horovitz and V. A. Johnson, eds.), Vol. VI, part B, pp. 376–391, Academic Press, New York, 1959.
5. W. C. Walker, N. Wainfan, and G. L. Weissler, *J. Appl. Phys.* **26**, 1366–1371 (1955).
6. L. Dunkelman, W. B. Fowler, and J. Hennes, *Appl. Opt.* **1**, 695–700 (1962).
7. G. H. Rieke, *Detection of Light: From the Ultraviolet to the Submillimeter*, pp. 202–209, Cambridge University Press, New York, 1994.
8. U. Hochuli and P. Haldemann, *Rev. Sci. Instrum.* **43**, 1088–1089 (1972).

9. L. R. Canfield, R. G. Johnston, and R. P. Madden, *Appl. Opt.* **12**, 1611–1617 (1973).
10. L. R. Canfield and N. Swanson, *J. Res. Natl. Bur. Stand.* **92**, 97–112 (1987).
11. G. Hass, *J. Opt. Soc. Am.* **39**, 532–540 (1949).
12. A. J. Tuzzolino, *Rev. Sci. Instrum.* **35**, 1332–1335 (1964).
13. M. A. Lind and E. F. Zalewski, *Appl. Opt.* **15**, 1377–1378 (1976).
14. R. C. Alig, S. Bloom, and C. W. Struck, *Phys. Rev.* **B22**, 5565–5582 (1980).
15. R. D. Ryan, *IEEE Trans. Nucl. Sci.* **NS-20**, 473–480 (1973).
16. M. Krumrey and E. Tegeler, *Rev. Sci. Instrum.* **63**, 797–801 (1992).
17. M. Razeghi and A. Rogalski, *J. Appl. Phys.* **79**, 7433–7473 (1996).
18. R. Korde and J. Geist, *Solid State Electronics* **30**, 89–92 (1987).
19. H. R. Philipp. In *Handbook of Optical Constants of Solids* (E. D. Palik, ed.), pp. 749–763. Academic Press, New York, 1985.
20. L. R. Canfield, J. Kerner, and R. Korde, *Appl. Opt.* **28**, 3940–3943 (1989).
21. D. F. Edwards. In *Handbook of Optical Constants of Solids* (E. D. Palik, ed.), pp. 547–569. Academic Press, New York, 1985.
22. T. Saito and H. Onuki, *Metrologia* **32**, 525–529 (1995/1996).
23. R. Korde, J. S. Cable, and L. R. Canfield, *IEEE Trans. Nucl. Sci.* **40**, 1655–1659 (1993).
24. L. R. Canfield, R.Vest, T. N. Woods, and R. Korde, *SPIE* **2282**, 31–38 (1994).
25. H. S. Ogawa, L. R. Canfield, D. McMullin, and D. L. Judge, *J. Geophys. Res.* **95**, 4291–4295 (1990).
26. R. L. Kauffman, D. W. Phillion, and R. C. Spitzer, *Appl. Opt.* **32**, 6897–6900 (1993).
27. R. C. Spitzer, T. J. Orzechowski, D. W. Phillion, R. L. Kauffman, and C. Cerjan, *J. Appl. Phys.* **79**, 2251–2258 (1996).
28. J. R. Palmer and G. R. Morrison, *Rev. Sci. Instrum.* **63**, 828–831 (1992).
29. T. N. Woods, G. J. Rottman, S. M. Bailey, and S. C. Solomon, *Opt. Eng.* **33**, 438–444 (1994).
30. T. Saito, K. Katori, M. Nishi, and H. Onuki, *Rev. Sci. Instrum.* **60**, 2303–2306 (1989).
31. T. Saito, K. Katori, and H. Onuki, *Physica Scripta* **41**, 783–787 (1990).
32. R. E. Vest and L. R. Canfield, *Rev. Sci. Instrum.* **67**, 1–4 (1996).
33. M. Marchywka, J. F. Hochedez, M. W. Geis, D. G. Socker, D. Moses, and R. T. Goldberg, *Appl. Opt.* **30**, 5011–5013 (1991).
34. H. Morkoç, S. Strite, G. B. Gao, M. E. Lin, B. Sverdlov, and M. Burns, *J. Appl. Phys.* **76**, 1363–1398 (1994).
35. S. M. Johnson, Jr., *Appl. Opt.* **31**, 2332–2342 (1992).

7. AMPLIFYING AND POSITION SENSITIVE DETECTORS

Oswald H. W. Siegmund

Space Sciences Laboratory
University of California
Berkeley, California

7.1 Photon Detection

In the VUV region between 1000 and 3000 Å, the principal detection mechanism for amplifying detectors is the photoelectric effect. Normally a photocathode material is used to enhance the detection efficiency. Photocathodes may be opaque (reflection) photocathodes or transmission (semi-transparent) photocathodes. Photoemission from opaque photocathodes occurs from the front surface of the photocathode, while photoemission from transmission photocathodes occurs from the back surface. The general behavior of both types of photocathode are described in a number of articles [1–4].

Transmission photocathodes are deposited onto the entrance window (Fig. 1a) of a sensor. The window must be transparent to the wavelength of interest. Therefore, this technique cannot normally be used below $\lambda \approx 1000$ Å where thick transmissive windows are not effective (Volume 31, Chapter 16). Transmission photocathodes for imaging detectors are often mounted close, 0.2–1 mm, (proximity-focused) to the amplifying element. Photoelectrons emitted from the photocathode are accelerated to the signal multiplier by a high electric field (300–1000 V mm^{-1}) causing secondary emission. Opaque photocathodes are deposited directly onto the amplifying device, such as a microchannel plate, and photoemit into the multiplier, or away from it (Fig. 2). Forward emitted photoelectrons can be collected by applying a reverse bias (100–300 V mm^{-1}) between the sensor entrance window, or a mesh, and the multiplier.

In both configurations, a potential applied across the amplifying element then accelerates the secondaries forming a subsequent multiplicative cascade. In conventional photomultipliers, the dynode structure does not support position sensing. However, in other multipliers, such as microchannel plates, the photon position information is preserved allowing a position sensitive device to be used to determine the location of the detected photons. There are numerous techniques for photon image location that have been applied to this problem. The schemes used vary from discrete anode arrays, to charge division systems, signal timing techniques, and optical re-imaging.

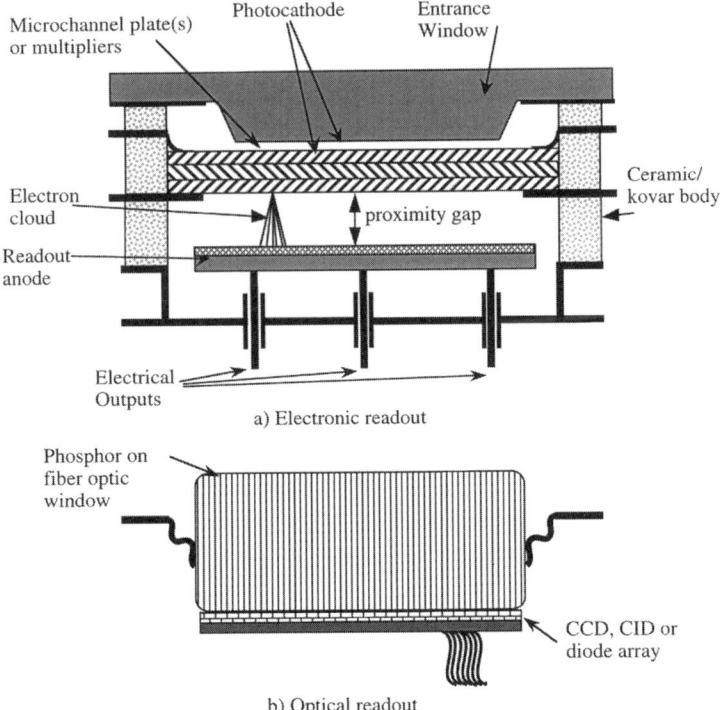

FIG. 1. Schematic of the components of a sealed tube imaging detector using a transmissive or opaque photocathode, microchannel plates for amplification, and electronic (a) or optical (b) readout techniques.

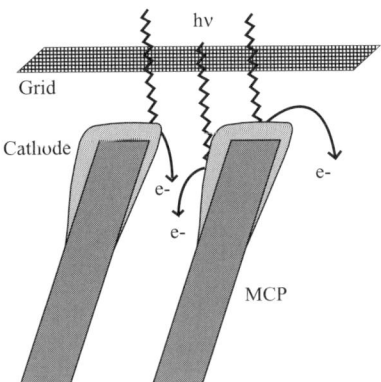

FIG. 2. Schematic of an opaque photocathode deposited onto an MCP with a retarding mesh.

7.1.1 Photocathode Types and Fabrication Methods

Commonly used photocathodes for 1000–3000 Å range are alkali halides and tellurides, such as CsI [5, 6], KBr [7, 8], MgF$_2$ [9], RbBr [10], KI [11], NaBr [11], CsBr [12], RbTe [4], and CsTe [4]. The photon energy must exceed the material valence band-conduction band gap energy, ϕ_g, plus the electron affinity, ϕ_a, which is analogous to the work function in metals. These photocathodes have larger work functions (4–12 eV) than the visible regime photocathodes ($\approx$2–2.5 eV). Photoelectrons emitted from a photocathode have a maximum kinetic energy that is given by the relation:

$$E = h\nu - \phi_g - \phi_a \tag{1}$$

RbTe and CsTe have sensitivity up to $\approx$3000 Å ($\varphi_g + \phi_a \approx 4$ eV), are fabricated by a process similar to that used for visible sensitive photocathodes (S20, bialkali) [3, 4, 13], and are unstable at poor vacuum pressures. These must be used in sealed tube devices, and are usually deposited as transmission photocathodes on the entrance window. CsI ($\phi_g + \phi_a \approx 6.4$ eV, 1940 Å), KBr ($\phi_g + \phi_a \approx 8.2$ eV, 1510 Å), MgF$_2$ ($\phi_g + \phi_a \approx 12.5$ eV, 1000 Å) and similar photocathodes are robust enough to withstand atmospheric conditions provided the humidity level is kept low. They may be used as transmission or opaque photocathodes and are deposited by evaporation. This is often done by resistive heating of powdered material in a vacuum bell jar, keeping the pressure below 10^{-6} torr and using evaporation rates <100 Å sec^{-1} [7]. The thickness of transmission photocathodes must be kept $\leq$1000 Å [4] to allow photoelectrons to escape, whereas opaque photocathodes are usually $\approx$10,000 Å thick [5, 7, 12]. Deposition rates should be monitored with a crystal oscillator, and an independent check can be made by observing interference colors of the photocathode layer (opaque cathodes are usually green or purple in color) (Fig. 3), or determination of the weight change of a foil witness sample. Electron microscope examination of such photocathode layers (Fig. 4) show that the layer is not uniformly smooth, but is in fact a granular structure with grain sizes of the order $\approx$0.5 μm.

7.1.2 Photocathode Detection Efficiency

Quantum detection efficiency (QDE) curves as a function of UV wavelength for two common photocathodes are compared with a bare microchannel plate in Fig. 5. Basic models [7, 14] interpret the characteristics of photocathode operation by using a number of factors including, the grazing angle of incidence of the radiation, the layer reflectivity, the linear absorption coefficient, the total number of photoelectrons produced and their surface escape probability, the photoelectron diffusion length in the material, and the probability of detection of the emitted photoelectrons (Section 7.2.4.1). Semi-transparent photocathodes are

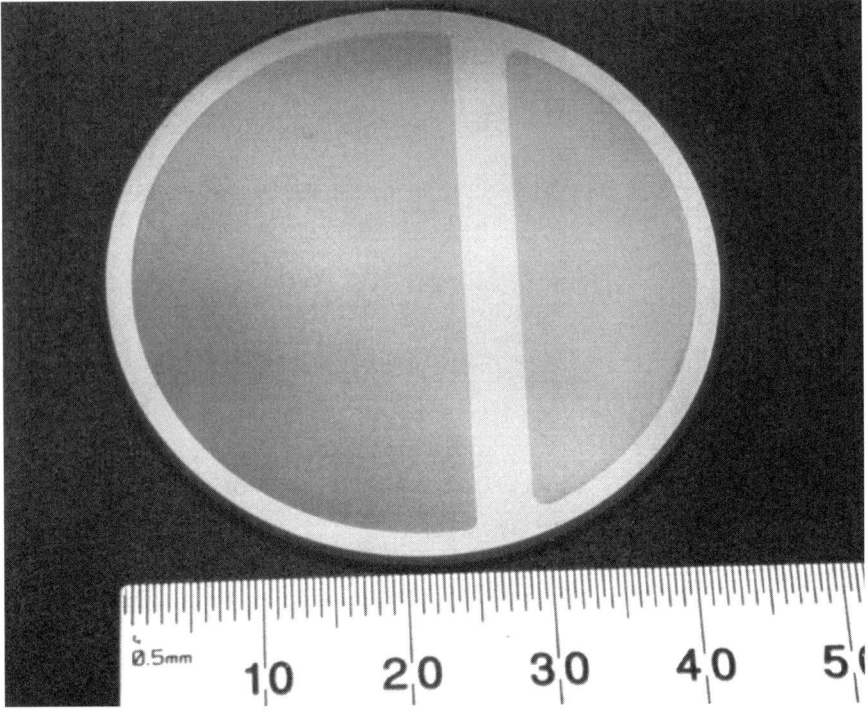

FIG. 3. Photograph of an opaque KBr photocathode deposited on an MCP, showing interference colors (bare MCP areas are white).

generally less efficient (10–15% maximum QDE) than opaque photocathodes because of a poorer geometry for photon absorption and photoelectron emission.

The alkali halide materials have similar QDE versus λ behavior (Figs. 5 and 6), often reaching 50% QDE and higher for opaque photocathodes. The QE for an opaque KBr photocathode on an MCP (Fig. 6) is a good illustrative case. The pore QDE is that caused by photoelectrons emitted from the inside surface of the MCP pores, and the web QDE is caused by photoelectrons emitted from the front surface of the MCPs and recovered by using a retarding field. The wavelength threshold for the onset of photoemission closely matches the material band gap energy (φ_g). The $\lambda \approx 1000$ Å QDE peak is associated with single photoelectron emission, and the broad minimum at $\lambda \approx 700$–800 Å marks the transition to production of two low energy photoelectrons [15]. The QDE peak at $\lambda \approx 400$–600 Å is caused by multiple photoelectron emission [15]. The QDE drops off below 400 Å (absorption coefficient declines) but KBr also has a large QDE peak at short wavelength ($\lambda \approx 70$–120 Å) as a result of atomic-like resonant absorption features [7].

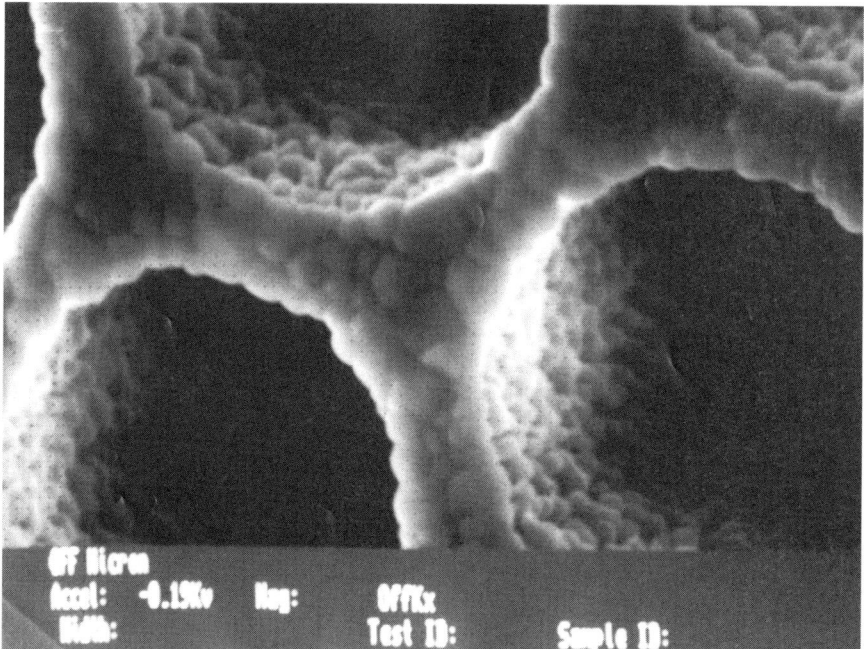

FIG. 4. Electron micrograph of an 8000 Å thick opaque KBr photocathode deposited on a 12.5 μm pore MCP, showing the granular KBr surface structure.

The angular dependence of the QDE is also important since UV optical systems can have widely differing angles of incidence. The angular response of the QDE is dictated by the ratio of the photoelectron escape depth to the photon absorption depth [7]. The angular response from normal incidence to ≈45° is usually fairly constant, but as the angles approach grazing, the QDE can change rapidly. For semi-transparent photocathodes the QDE will decrease as the the photons are absorbed closer to the input surface, thus reducing the backside photoemission probablity. Conversely, the opaque photocathode QDE increases (Fig. 7) at small grazing angles for the same reasons. For opaque photocathodes, there is also a sharp minimum in the QDE at very small graze angles caused by the high reflectivity of the photocathode layer [7]. Obviously, the QDE reaches its minimum when the incident angle is effectively straight down the pores for an opaque photocathode on an MCP (Fig. 7).

7.1.3 Photocathode Stability

The stability of a photocathode is critical to its application in long-term experiments. CsTe and RbTe photocathodes in ultra high vacuum sealed tubes

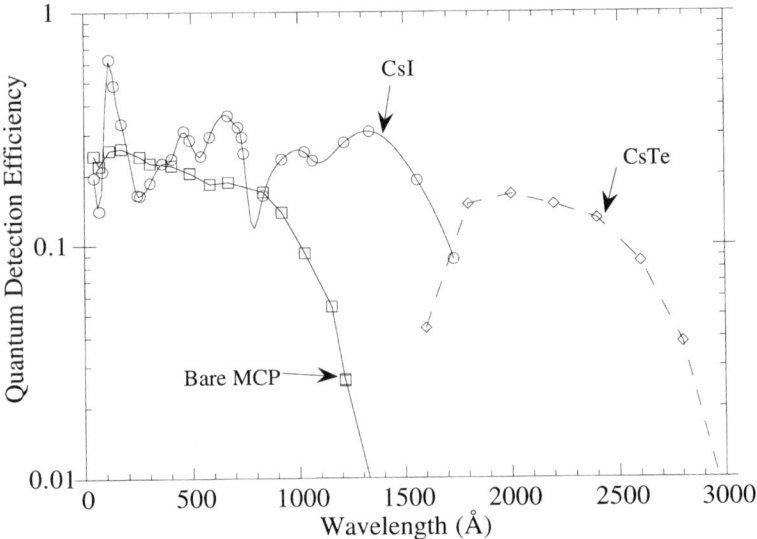

FIG. 5. Quantum detection efficiency versus wavelength for a bare MCP, a CsI coated MCP, and a transmissive CsTe photocathode.

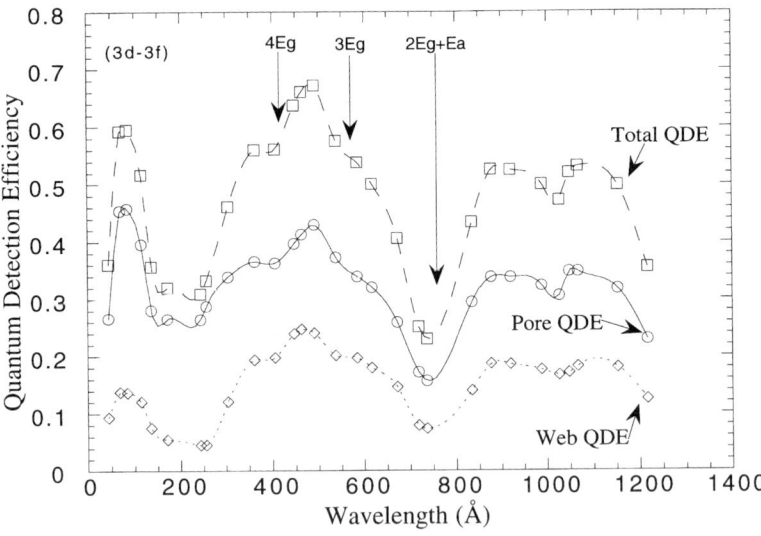

FIG. 6. Quantum detection efficiency of a KBr coated MCP as a function of wavelength. Web contribution measured with 100 V/mm retarding field applied.

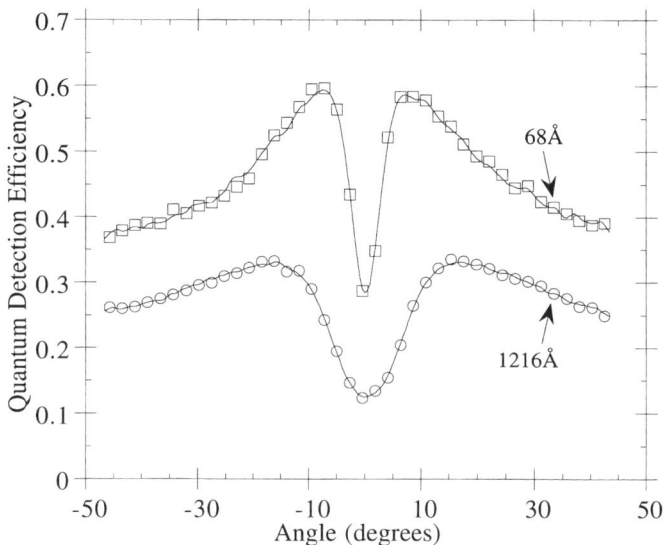

FIG. 7. Quantum detection efficiency versus incident angle (graze angle to MCP pore axis) at 68 Å and 1216 Å for an NaBr opaque photocathode on an MCP.

are normally stable over long periods if the integrity of the vacuum is held. This is usually ensured by inclusion of a getter inside the vacuum tube. For open-face detectors in the UV, stability often relates to the hygroscopic nature of the photocathode material. Hygroscopic materials such as CsI are known [16] to degrade in humid air, so caution is needed when handling these materials. Photocathodes of this kind should be stored and handled in dry nitrogen or kept in a pumped vacuum enclosure. Recent work on other alkali halides (KBr, RbBr) [10,17] show fairly good long-term stability (Fig. 8) while still providing high QDEs. If a photocathode has degraded because of exposure, it will usually have a "milky" looking appearance rather than clear interference colors.

7.1.4 Photocathode Dark Noise

There is no significant dark noise for far ultraviolet (FUV) photocathodes. The work functions are generally too high for any spontaneous electron emission as occurs for visible sensitive photocathodes because of thermionic emission. However, some alkali halide photocathodes may be "activated" [17] causing high background rates and enhanced long wavelength response after exposure to very high fluxes of photons or ions. Any wavelength below ≈ 3000 Å can activate the cathode in a fairly short time (≈ 10 minutes for an Hg vapor lamp). Although cathodes such as KBr show good solar blindness, ($<10^{-12}$ at 5000 Å) the response can be increased at longer wavelengths by

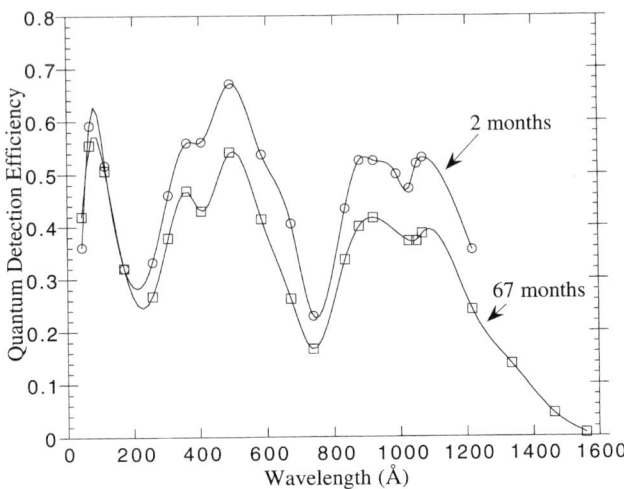

Fig. 8. Stability of quantum detection efficiency versus wavelength for a KBr coated MCP as a function of time under 5% relative humidity conditions.

more than five orders of magnitude [17] (Fig. 9), with QEs at wavelengths below 2000 Å hardly affected. Deactivation can be accomplished by re-exposure to air, or illumination with intense visible illumination. The activation effect is presumably caused by creation of low threshold energy electron trap sites in the photocathode.

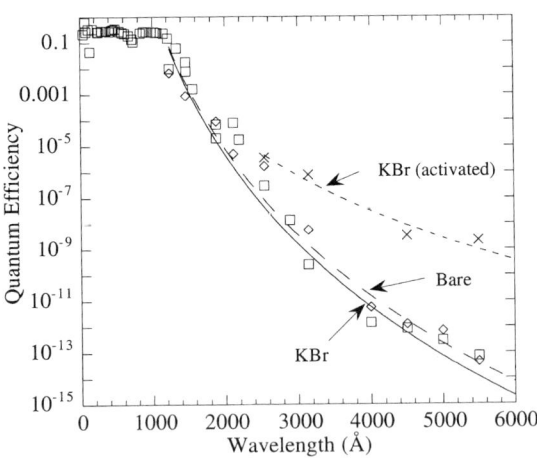

Fig. 9. Long wavelength quantum detection efficiency for a bare MCP, a KBr coated MCP, and an activated KBr coated MCP.

7.1.5 Photocathode Imaging Effects

The photocathode also has an effect on the position sensing capability for some detector configurations. In the case of semi-transparent photocathodes, the photoelectrons can spread laterally in the gap between the photocathode and multiplier element. In the case of opaque photocathodes with a mesh or equivalent photoelectron retarder, the ballistic trajectories of deflected photoelectrons will result in a halo around the origination point [18]. Both effects result in a spread of possible detection locations at the multiplier surface for a specific photon location at the photocathode. The full width at half maximum (FWHM) position error for semi-transparent photocathodes can be represented as a distribution with width, dx, given by [19]:

$$dx = 2.36d(2E/V)^{0.5} \qquad (2)$$

where d is the photocathode to multiplier distance, E (eV) is the mean photoelectron lateral energy, and V is the potential applied across the gap. The equivalent expression for opaque photocathodes is [20]

$$dx = 2.43dE/V \qquad (3)$$

The photoelectron spread for CsI semi-transparent and opaque photocathodes is compared in Fig. 10, showing that the opaque photocathode has significantly better resolution. The choice of gap distance and accelerating potential for each configuration requires evaluation of the specific application with respect to

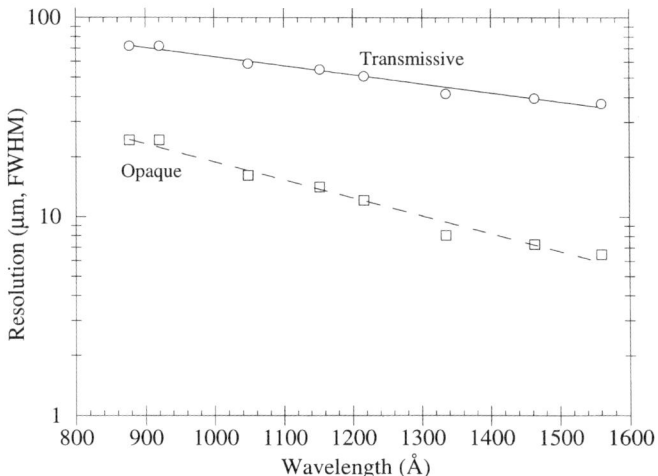

FIG. 10. Proximity focus contributions to image resolution for transmissive (250 μm gap, 400 V) and opaque (300 V/mm field) CsI photocathodes versus wavelength.

photocathode type, wavelength of operation, required position resolution, speed of optical beam, and other geometrical constraints [20].

7.2 Amplifying Detectors

7.2.1 Photomultiplier Tubes

Photomultiplier tubes (PMTs) have been used for many decades, and a wide variety are available commercially. Relatively few of these are optimized for UV wavelengths, but both transmissive [21, 22] and opaque photocathode PMTs are available [21] using fused silica or MgF_2 windows with CsTe or CsI photocathodes. Basically, two types of multiplication stages are used with UV PMTs. Circular-cage type dynodes are used for the opaque photocathode PMTs, and fast linear focused dynodes are used for the transmissive photocathode PMTs. Neither has imaging capabilities. Circular-cage type PMTs have opaque photocathodes deposited on the first dynode. Photoelectrons are then electrostatically focussed to a series (≈ 10) of uniquely shaped dynodes arranged in a compact circular fashion. This type of multiplier can have gains up to 10^7, with short pulse risetimes (1–2 ns). Fast linear focused PMTs with transmissive photocathodes electrostatically focus the photoelectrons onto the first dynode, which is followed by a linear array of curved dynodes (≈ 10). Gains are typically in the range 10^5 to 5×10^6, and pulse rise times can be very short (<1 ns). Both types of multiplier can be used in DC mode up to currents of ≈ 0.1 mA, and can also be used for photon counting, although the pulse amplitude distribution typically has a peaked but wide distribution. Many applications details are available in the standard catalogs [3, 21, 22] and PMT handbooks.

7.2.2 Focused-Mesh Multipliers

Focused-mesh electron multipliers are composed of stacks of perforated metal sheets [21, 22], with each sheet stacked so that the holes do not match with those of the succeeding sheet. The sheets are sequentially biased like PMT dynodes, so that a photoelectron incident on the first mesh will initiate an electron avalanche. Focused-mesh multipliers can have high gains, of the order of 10^5–10^9, with no ion feedback, even at high count rates. For single photon detection, the pulse amplitude distributions are poor ($\approx 200\%$ FWHM), but the rise times are quite fast (≈ 5 ns). Focused-mesh multipliers can be procured up to ≈ 100 mm formats [21], but the imaging characteristics are poor. Resolution of ≈ 3 mm FWHM has been achieved for a commercial focused-mesh multiplier with a wedge-and-strip anode [23], and ≈ 2 mm using a multiwire readout [24], with significant barrel distortion. The count rate linearity is, however, good and the background rate is very low ($<10^{-2}$ event cm^{-2} s^{-1}) compared with that of microchannel plates.

7.2.3 Channel Electron Multipliers

Channel electron multipliers (CEMs) have been widely used for amplification of the signals in analog and photon counting applications. CEMs [25] are tubes of reduced lead glass (Fig. 11) that operate when a high voltage is applied across the ends of the tube. Events striking the input end cause emission of photoelectrons, which are then accelerated down the tube. CEMs are typically curved or spiralled to eliminate ion feedback caused by ionization of liberated adsorbed gas. Electron [26] and ion [26] detection efficiencies for CEMs can reach $\approx 80\%$ (Fig. 12), although at different energies. CEMs can also be coated with an opaque photocathode as discussed in Section 7.1.1. Since the CEM tube is reduced in a hydrogen furnace, and the walls have a high secondary electron emission coefficient, electrons striking the walls are multiplied and accelerated. CEMs can be operated at high gains (10^8, Fig. 13) [27] and have narrow pulse amplitude distributions (<30% FWHM) because of space charge saturation, allowing excellent discrimination for single event counting. The CEM gain for ions typically drops by a factor of 5 as the ion mass is increased from 5 to

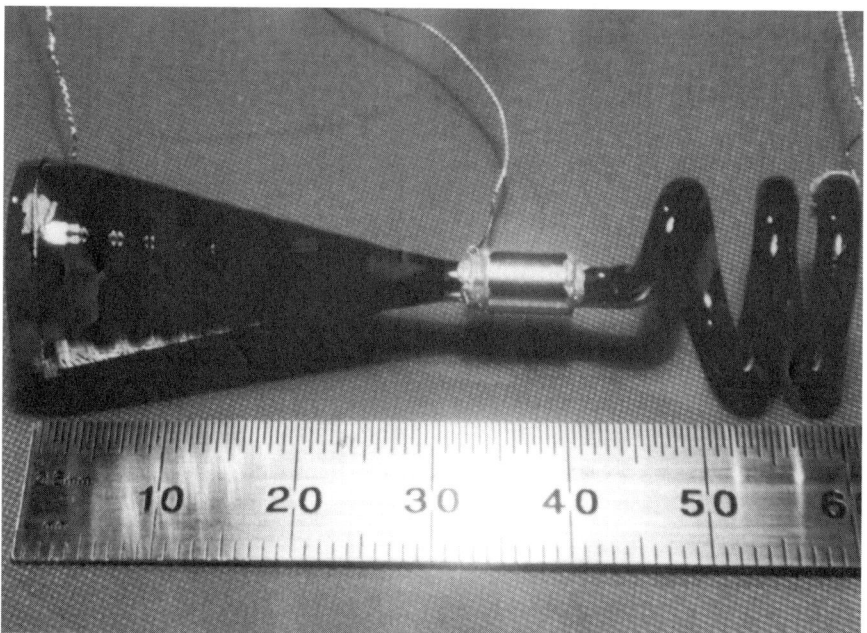

FIG. 11. Channel electron multiplier, Galileo model 4503.

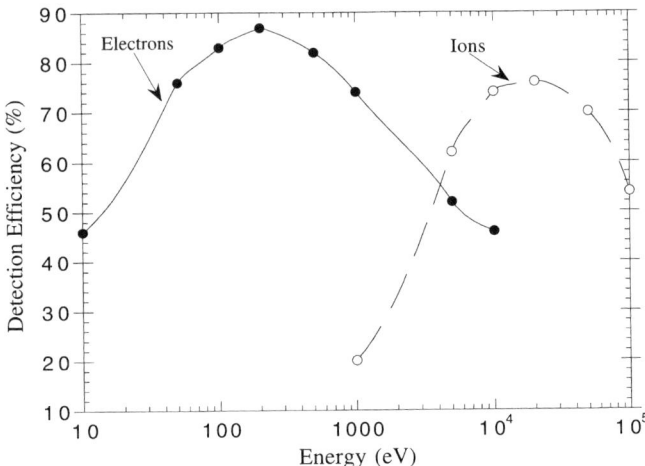

FIG. 12. Channel electron and ion detection efficiencies as a function of particle energy [26].

150 AMU [28]. In analog mode up to gains of $\approx 10^5$, the current output is linear up to $\approx 10\%$ of the strip current, with maximum output values of the order 10 µA. In high-gain photon counting mode, the gain begins to drop (Fig. 14) [26] if the event rate is too high. This occurs at about the same output current as the analog mode. Pulse rise times for CEMs are of the order 10 ns, and background rates are quite low (<0.05 events sec^{-1}). Lifetime data for CEMs

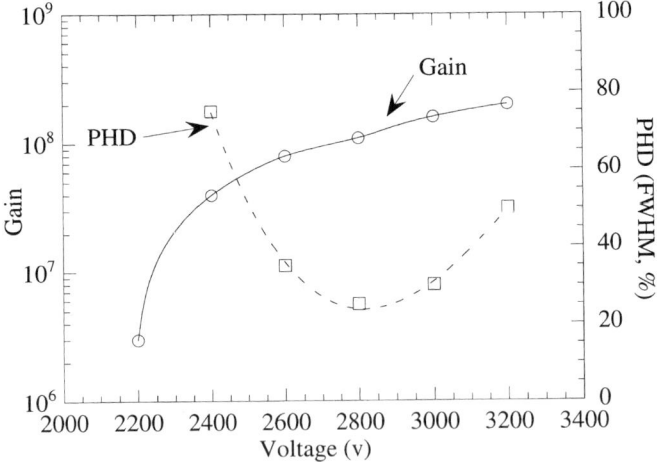

FIG. 13. Channel electron gain and pulse height distribution width as a function of applied voltage, Galileo model 4039 [27].

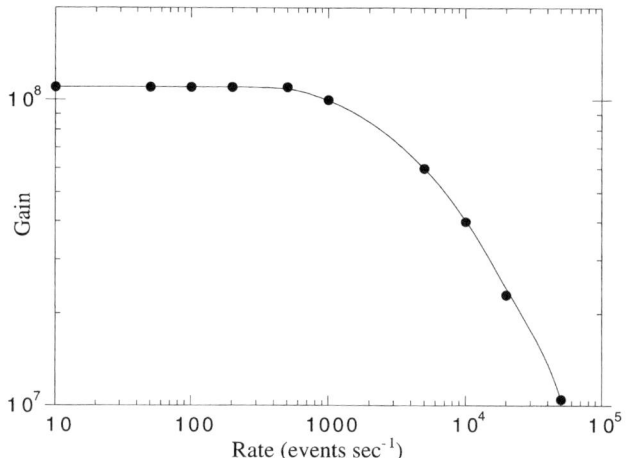

FIG. 14. Channel electron gain as a function of counting rate [26].

vary, but after an initial gain drop (Fig. 15) the gain gradually decreases as ≥10 coulombs are extracted. At some point, a significant drop in gain occurs because of chemical changes in the secondary emissive surface [29]. Lifetimes can be reduced in poor vacuum conditions or when organic contamination is present.

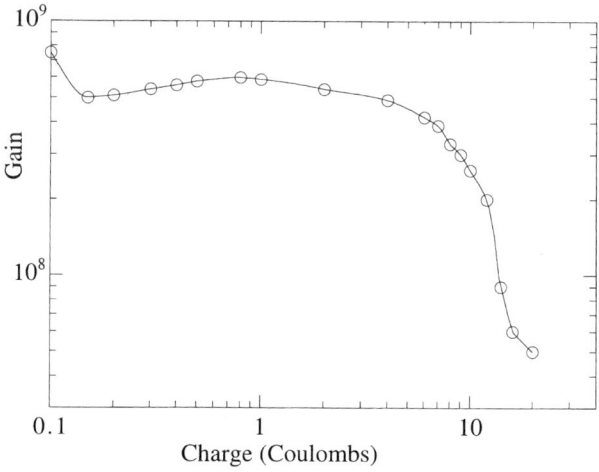

FIG. 15. Channel electron gain as a function of extracted charge [29].

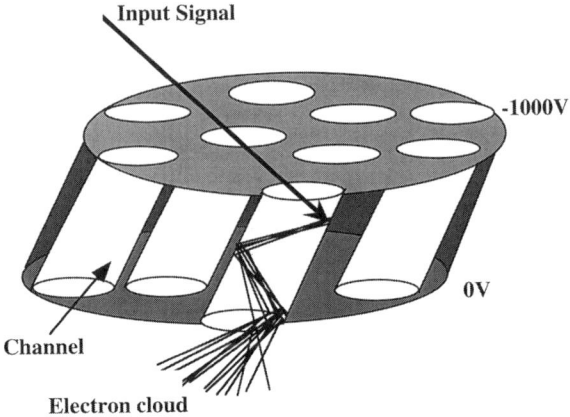

FIG. 16. Schematic of a microchannel plate electron multiplier array. When a potential is applied across the microchannel plate, an electron entering the input causes an electron cascade producing $>10^3$ at the output.

7.2.4 Microchannel Plates

MCPs are thin (≈ 1 mm), silicon-lead oxide glass wafers composed of a large number ($>10^5$ cm^{-1}) of small (≈ 10 μm) CEM tubes called channels, or pores [30] (Fig. 16). Radiation that strikes the MCP walls results in photoelectron emission. The photoelectrons are then accelerated and collide with the channel walls producing yet more electrons and an ensuing cascade. MCPs are fabricated in stages, first glass-filled lead glass tubes are reduced by drawing them from an oven, then stacked and fused to produce a solid fiber bundle (boule). The boule is then sliced and polished into wafers of the desired thickness and shape, and the glass cores are etched out. Reduction of the glass surface at high temperature ($\approx 450°C$) in a hydrogen atmosphere is subsequently performed to produce a high secondary emissive surface inside the pores. Application of thin NiCr electrodes to each face of the MCP then provides the electrical contacts. MCP resistances, in vacuum, are usually in the range 10 MΩ to 10 GΩ cm^{-2}.

MCP shapes are usually rectangular or circular flat wafers with a channel length-to-diameter (L/D) ratio of between 40:1 and 120:1. Pore sizes in the range 5–25 μm are available, but 10 μm and 12.5 μm are most common, with channel open areas of $\approx 60–70\%$. MCPs are normally cut so that the pores have an angle (0°, 6°, 8°, and 13° are common) relative to the MCP front surface. The most common MCPs are 25 mm diameter with 40:1 L/D ratio produced in large numbers for night-vision devices. Manufacturers of MCPs include Galileo [31], Philips [26], Litton [32], ITT [33], and Hammamatsu [34]. MCPs are also made for specialized applications with spherical (Fig. 17) or cylindrical surfaces [35], square pores [36], low resistivity [37], low background MCPs [38], and curved

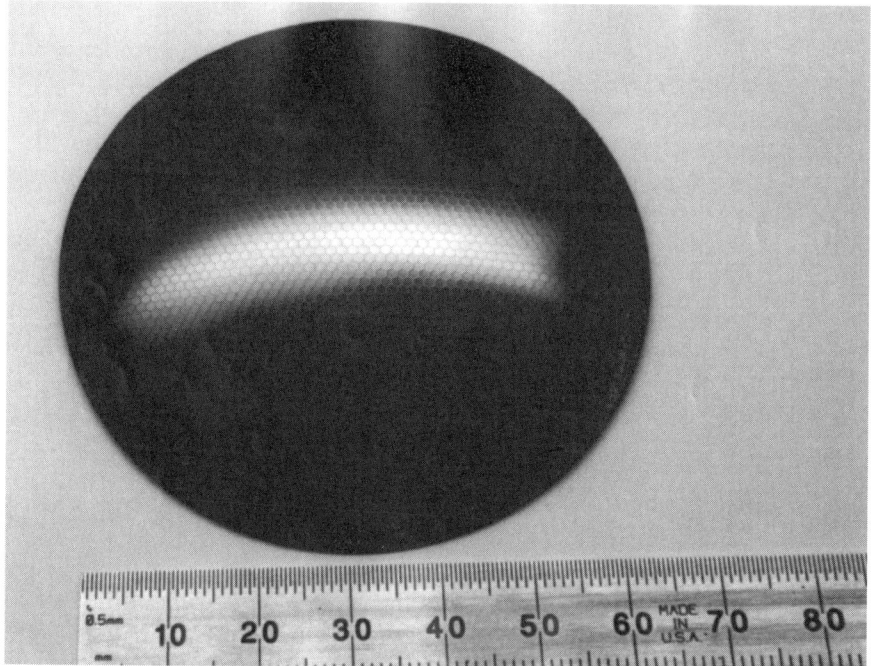

FIG. 17. Photograph of a spherically curved surface Philips 70 mm MCP.

channel MCPs [39]. Because of their small size, MCPs can withstand high magnetic fields (>0.5 T) without significant changes in performance [38]. The high spatial resolution (<10 µm), fast time response (<100 ps), and large areas (>100 mm diameter) achievable with microchannel plates make them a versatile amplification method.

7.2.4.1 Detection Efficiency. The photon detection efficiency of MCPs was described in Section 7.3.1.2 and is similar for MCPs from all manufacturers. The detection efficiency for photoelectrons and ions is a function of energy (Fig. 18) and incident angle. Electron detection efficiencies [40] are ≈80% for typical operating conditions (200–500v retarding voltage), whereas ion efficiencies [41] peak at ≈60% (no retarding voltage), and there is no obvious difference between H^+, He^+, and O^+ ions. The angular variation in electron and ion detection efficiency is similar to that for photons, peaking at ≈5–10° to the MCP pore axis, dropping sharply at smaller angles, and dropping more slowly at larger angles [41].

7.2.4.2 Gain and Pulse Height Distribution. Single straight channel MCPs (Fig. 19a) have gain curves that increase exponentially with applied voltage [42] up to a maximum of $\approx 10^5$, and the pulse amplitude distribution has

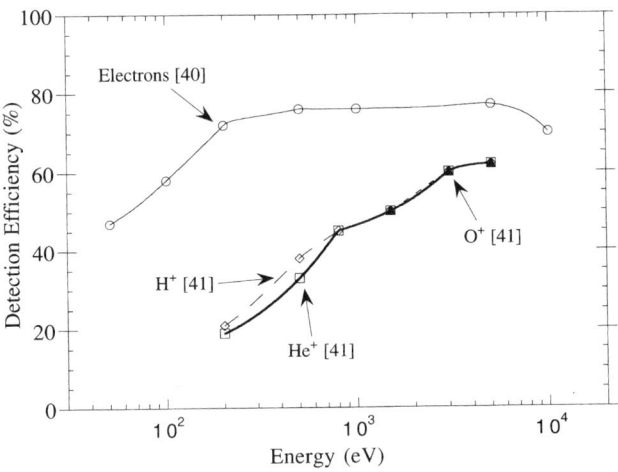

Fig. 18. MCP electron and ion detection efficiency as a function of energy [40].

a negative exponential shape [42] with no peak. In this analog mode, very high input fluxes can be accommodated, with maximum charge output of $\approx 10\%$ of the overall MCP strip current [43] up to $\approx 10\ \mu A$. However, at the highest gains, positive ions are produced by collisions of electrons with gas molecules released

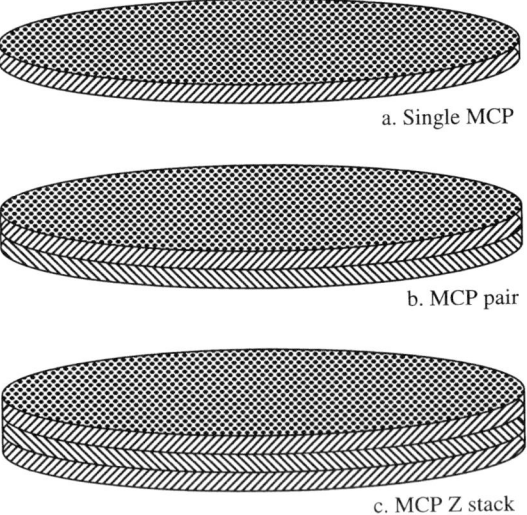

Fig. 19. Commonly used stacking configurations for MCPs. (a) A single straight channel MCP with $\approx 13°$ bias, (b) a pair of stacked straight channel MCPs, and (c) three straight channel MCPs in a Z stack configuration.

from the channel walls [43]. These ions are accelerated up the channel and can strike the MCP walls near the input face releasing electrons and causing a secondary electron avalanche (ion feedback).

Photon counting is achieved with MCP pairs, triplets (Zs) [44] (Figs. 19b, c), or quadruplets, or by using MCPs with curved channels [39]. MCPs can be stacked back to back, or with small gaps between them, with or without a gap potential. Stacks are oriented to maximize the angle between the channels at the stack interfaces (13–26°). This causes released ions to strike the channel walls close to their point of origin, minimizing ion feedback pulses. The gain of a microchannel plate stack is a function of the channel diameter, the L/D ratio, and the number and configuration of the MCPs [45]. Gains of 10^6–10^8 can be attained with no visible ion feedback effects. Figure 20 summarizes the gain versus voltage characteristics for several MCP back-to-back stacks (all Philips 12.5 μm pore). The larger the total L/D of the stack, the higher the achievable gain. Every gain curve asymptotically levels off at high gain because of saturation resulting from channel wall charge depletion at the output of the MCP stack. This causes a local reduction of the accelerating potential gradient, limiting the gain. Small channel diameters saturate at lower gains than those with larger channel diameters [45], in proportion to the channel wall area.

In back-to-back MCP stacks, small (≈ 10 μm) gaps between the MCPs, caused by MCP warpage, can increase the local gain [46] significantly. Therefore, care must be taken in configuring MCP stacks. Bonded MCP stacks

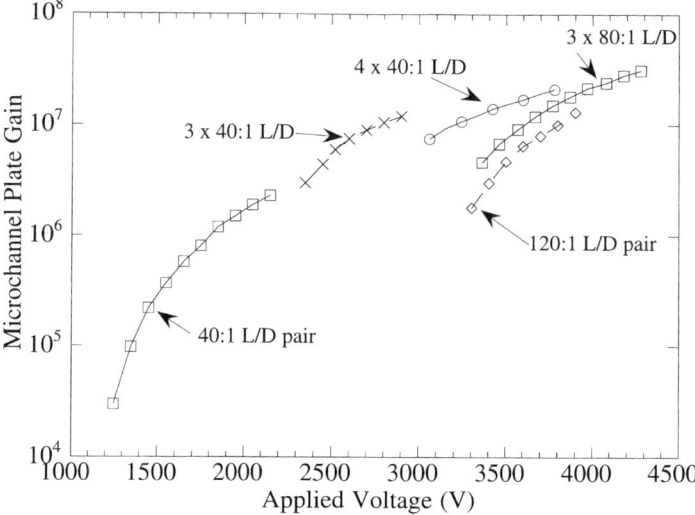

FIG. 20. Gain versus voltage curves for several common MCP back-to-back stacking configurations (Philips 12.5 μm pore MCPs).

attempted to eliminate this problem [47] but these have suffered from high noise caused by the bonding process. MCP stacks with an accelerating potential across gaps between each MCP [45] can successfully control the gain and voltage by changing the number of channels in the bottom MCP excited by the electron avalanche. However, this requires more high voltage taps on the MCP stack and can be inconvenient to implement.

The pulse amplitude distributions (PHDs) of high gain ($>10^6$) MCP stacks is Gaussian shaped (Figs. 21 and 22) because of the gain saturation effect. The full width at half maximum (FWHM) of the distributions decrease [45] as the MCP gain is increased. MCP stacks with shorter overall L/D ratios, and smaller channel diameters [48], achieve tighter PHDs at lower gains since they "saturate" at lower gains. Tight PHDs are useful both for discrimination of photon events from background and in reducing the required dynamic range of position encoding electronics. At high event rates, a progressive drop in the gain and a degradation of the MCP PHD is observed [49, 50]. Figure 23 shows the reductions in gain for high local count rate for several sizes of illuminated area. Studies [49, 50] have shown that during this type of illumination, the gain of the MCP is reduced for a significant distance around the affected spot. The limiting event rate for MCPs can be as low as <1 event sec^{-1} for high-resistance MCP stacks [46] to >100 events sec^{-1} for curved channel or low-resistance MCPs [51]. If high local event rates must be accommodated, low MCP gain and, or, low resistance MCPs [37] should be used.

Life tests [52] of MCPs show that the gain initially drops quite rapidly because of outgassing of the channel walls. Then there is a slow decrease toward

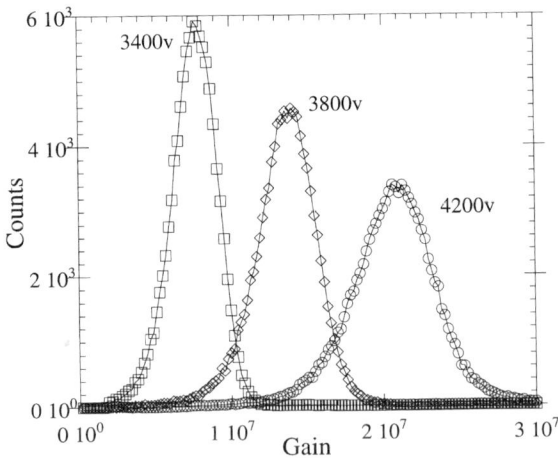

FIG. 21. Pulse height distribution as a function of gain for a back-to-back stack of four Philips MCPs, 12.5 μm pore, 40:1 L/D, 36 mm diameter.

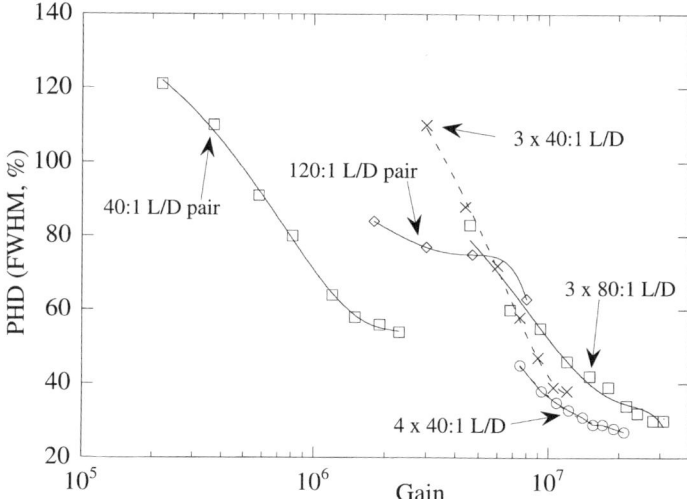

FIG. 22. Pulse height distribution versus gain curves for the MCP stacking configurations shown in Fig. 20.

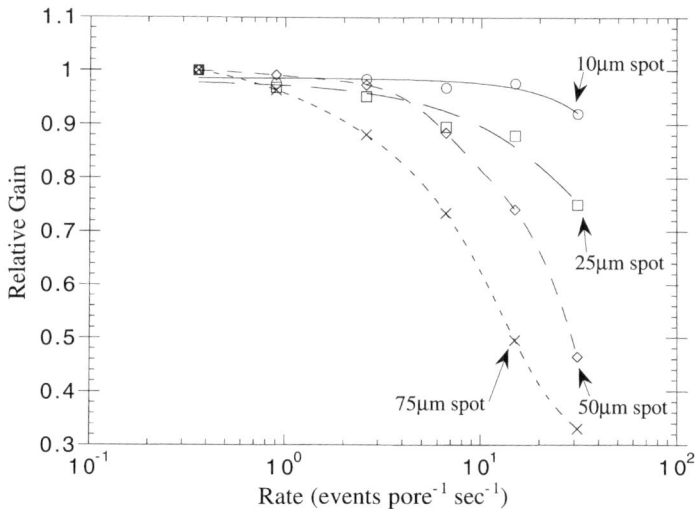

FIG. 23. Relative gain as a function of local counting rate for several illumination spot areas using a stack of four back-to-back Philips 12.5 μm pore, 40:1 L/D MCPs with 30 MΩ resistance each.

a plateau region of stable operation. Preconditioning [53] of MCPs helps to stabilize their performance. First the MCPs are cleaned ultrasonically (50:50 isopropyl/methyl alcohol) and baked (100–300°C) at ultra high vacuum to de-gas the MCP. Note that the MCP resistance decreases [53] as the temperature rises, with a temperature coefficient of ≈ 0.02. A "burn in" where the MCPs are run at low gain under high photon or electron flux (≈ 1 μA cm^{-2} output) stabilizes the gain after ≈ 0.1–0.5 C cm^{-2} is extracted [53]. This decreases the gain at a given voltage by a factor of 5–10×, but gain can be recovered by increasing the voltage. Tests have shown [54] that MCPs can endure the extraction of many C cm^{-2}, but the ultimate lifetime is not well known. However, the preconditioning process is partially reversed if the MCPs are re-exposed to air.

7.2.4.3 Background Characteristics.
The pulse amplitude distribution of intrinsic MCP background events has a negative exponential shape (Fig. 24) [55, 56] with overall rates <1 event cm^{-2} s^{-1}. This background has been shown [55, 56] to be caused by β decay from potassium (^{40}K, Philips MCPs) or rubidium (^{87}Rb, Galileo MCPs) in the standard MCP glasses. Although many of the events can be rejected by applying an amplitude discrimination threshold (Fig. 24), the need for low background has led to the development of MCP glass without radioactive isotopes [32]. Such MCPs have much lower background (<0.1 event cm^{-2} s^{-1}) (Fig. 25) [57]. Other background effects include "hot spots" associated with dirt or damage on, or in, the MCPs. Therefore, it is

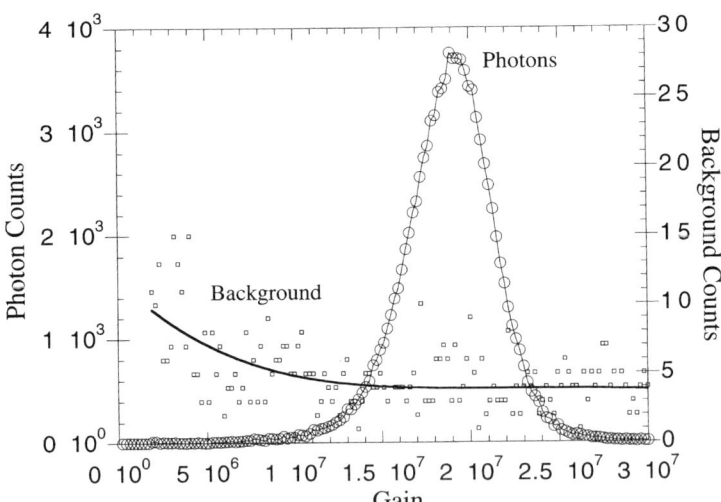

FIG. 24. Comparison of photon and background event pulse height distributions for a stack of four back-to-back Philips 12.5 μm pore, 40:1 L/D MCPs.

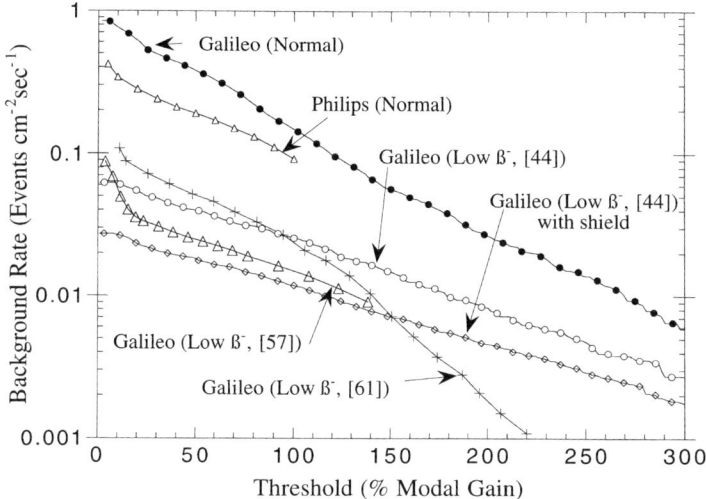

FIG. 25. Comparison of background event rates for several MCP types as a function of gain threshold.

important that care be taken in cleanliness and handling of MCPs to reduce the occurrence of hot spots.

7.2.4.4 Imaging Characteristics. The resolution of images amplified by MCPs is ultimately limited by the pore size and spacing. The effect of the pore size is illustrated in Fig. 26, showing that while 12.5 μm pore (15 μm spaced) MCPs can resolve to ≈40 line pairs/mm (group 5:3), 5 μm pores (6.5 μm spaced) can resolve to ≈80 line pairs/mm (group 6:3). This is a best-case

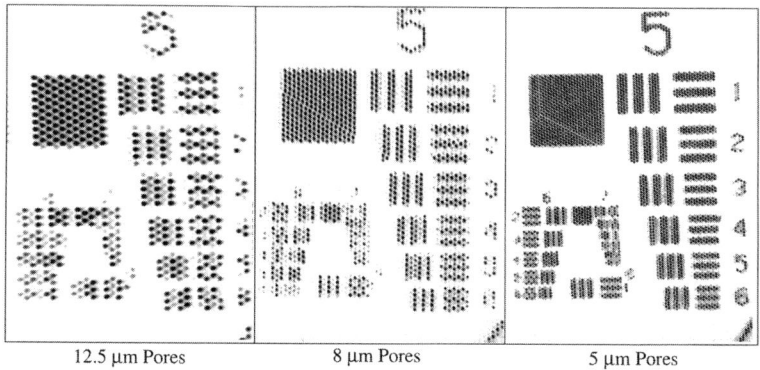

FIG. 26. Images of MCP pores taken with visible light projected through an Air Force 1951 test mask pattern in contact with the MCPs. Philips 12.5 μm pore MCP, Galileo 8 μm, and 5 μm pore MCPs.

scenario, because the MCP output charge can spread. Single and curved channel MCPs have the smallest charge footprints since photon events are always constrained to single channels. The electrode penetration depth in the output face of an MCP is an equipotential zone and the larger the electrode penetration depth and L/D ratio, the more focussed the electron cloud is as it leaves the MCP. However, if the charge density at the MCP output is very high, space charge effects will broaden the output distribution because of electron self-repulsion. MCP stacks [58] have larger charge footprints because of the increased number of channels that are excited in the output MCP. However, high saturated gains can also create an equipotential zone at the MCP output resulting in lower lateral electron energies. Studies of MCP charge clouds [58] indicate that the distributions have several components causing non-Gaussian charge distributions with wide wings. However, the important issue for most imaging systems is the consistency of the charge cloud centroid for MCP stacks, which permits the ultimate resolution to be determined by the pore dimensions for the top MCP (Section 7.3.2.1).

Because of the configuration and manufacture of MCPs, a number of effects are introduced in the image amplification process. Straight biased MCPs and MCP stacks, or curved channel MCPs, can displace the image by a significant amount ($\approx$400 µm). However, it is the inhomogeneities in the MCP fabrication that cause image distortions, particularly in the case of curved channel MCPs [59] where nonuniform curvature can result in small scale (<75 µm) local position errors. Several other effects can also be seen when a large statistical sample of events is integrated to give a flat field image. Dark spots the size of one MCP channel and up to one entire MCP multifiber [60] can occur because of blocked channels, or shear caused by fabrication problems. A consistent characteristic of flat field images is a hexagonal modulation (Fig. 27a) across the entire MCP [60]. This corresponds to the edges of the hexagonal multifiber bundles, and can modulate the image intensity and local gain by $\approx$10%. Channel deformations at the multifiber boundaries that occur during MCP fabrication affect the local gain and change the direction of the output charge clouds to give the effects seen. In the case of back-to-back MCP stacks, rotation of the MCPs with respect to each other can result in an interference effect that modulates the gain and intensity of the MCP output. This moire effect causes mottling of the image that can clearly be seen in Fig. 27b, so care must be taken with the rotational alignment of MCP stacks.

7.2.4.5 Time Response. The MCP is a very fast amplification system; output pulses can be extremely narrow (<1 ns) in duration [19]. The transit time for electrons through an MCP is about 250–800 ps (40:1–120:1 L/D MCP, 12.5 µm channels) with transit time spread of $\approx$50–140 ps [61]. Fast MCP photomultiplier tubes (PMTs) are commercially available [21, 48] for various applications including time of flight studies. Pulse widths of <200 ps can be

FIG. 27. Flat field images for Z stacks of 12.5 μm pore, 80:1 L/D back-to-back Philips MCP stacks. Multifiber modulation dominates when the multifibers are aligned (a), and moire mottling is dominant when the top and middle MCPs are rotated 12° with respect to each other (b).

achieved [21] with single MCPs, <300 ps for MCP pairs [62], and 300–500 ps with Z MCP configurations [43]. The transit time and transit time spread decrease with decreasing channel diameter [48, 61, 62] hence the best performance is obtained with the smallest channels (<6 μm). It is extremely important in these applications to establish an impedance matched system (typically 50 Ω) with short cable lengths and components mounted close to the device. With fast timing electronics, it is then possible to obtain event timing information that is of the order of <10 ps (see Section 7.3.2.3).

7.2.5 Microsphere Plates

Microsphere plates (MSPs) are a relatively recent innovation [63] for signal amplification. These devices are essentially a wafer composed of fused glass balls with high secondary emission coefficient coatings. MSPs have similar resistance ranges, sizes, and formats to MCPs. However, the multiplication is not restricted to a pore as in an MCP, so the charge avalanche can spread further, producing higher gains for the MSP (10^8). This also produces a larger charge cloud, introduces greater position uncertainty (>100 μm) in the centroid of the charge output, and results in poor PHD characteristics [63]. Background rates are similar to those of MCPs (0.5 events cm^{-2} sec^{-1}), but the QE is poor because of the variations in the effective input angles. Flat fields also have considerable modulations and gain variations caused by the irregularities of the MSP structure. Nevertheless, MSPs are a compact alternative to mesh grid multipliers.

7.3 Position Sensing Techniques

7.3.1 Discrete Element Position Sensors

7.3.1.1 Discrete Optical Position Sensors. Because of high QE, good spatial resolution (<10 μm pixels, >2000 × 2000 formats), low noise, and excellent spatial stability, the detector choice for most optical and near IR imaging is the charged-coupled device (CCD)[64]. However, direct illuminated CCDs are not commonly used in the VUV because of the poor QE obtainable with commercially available CCDs. The 1/e attenuation length of silicon is less than 100 Å between the wavelengths of 900–3600 Å and UV photons are absorbed in the CCD gate structure or in the surface dead layers. These problems can be minimized by coating the CCD with a phosphor (lumogen)[64] to convert the UV wavelength to a more penetrating optical photon, or to illuminate the backside of the CCD after the silicon substrate has been thinned to expose the active epitaxial layer [64]. However, the native silicon oxide layer that grows on exposure of the thinned backside to air is typically 50 Å thick and has an attenuation length of ~100 Å at wavelengths below 1200 Å. Left untreated, the native oxide-silicon interface also develops a positive charge depleting the nearby silicon and attracting the photoelectrons to the back surface. Various techniques can be used to reduce these problems. One is to use an intermittent UV flood that negatively charges the native oxide layer; another is ion implantation and laser annealing. QE response using these techniques plus lumogen can provide acceptable (30–40%) QE [64] in the range 1000–2000 Å, but because of optical signal losses in the lumogen the QE cannot be improved and is ineffective below 500 Å. Another problem with use of CCDs in the UV is their high QE at visible wavelengths. This requires the use of a Woods filter [65] to block out unwanted wavelengths.

Therefore, in the regime between 200 Å and 2000 Å, other techniques are usually used. The reconversion of UV photocathode photoemission back to visible photon output is the most well known method of observing an amplified image in an analog mode. A proximity-focused phosphor screen behind the signal multiplier is commonly used for second- (Fig. 1) and third-generation image intensifier tubes [32, 33, 66] and similar UV sensors [67, 68]. Multiplier-to-screen distances are of the order 0.2–1.0 mm, with accelerating fields of up to ≈10 kV/mm being fairly standard. The most common phosphor [21] is P20, but P43 is becoming widely used since it is less toxic. De-acceleration of the electron cloud at the phosphor screen produces additional light gain (2–50×). The overall resolution [67] is determined by the photocathode (7.1.5), the multiplier (7.2), the accelerating field and gap size, screen resolution, output window resolution, and image readout method. Current high-resolution MCP–based photon conversion devices can achieve 40–60 lp/mm [66] (7.2). Position sensing of the optical output images can then be accomplished with TV cameras [69], photodiode arrays [70], PMTs [71], CCDs [72, 73] or charge injection devices (CIDs) [67, 74] coupled to the phosphor screen by lenses, fiber optic windows, or fiber optic tapers. Optical output devices can also be used for individual photon position centroiding (7.3.2.1).

Observing the phosphor output with an array of lens-coupled PMTs, the PAPA [71] sensor images individual photon events. Each PMT has a unique mask of periodic black and clear stripes (gray code), vertical for X and horizontal for Y directions. Thus, the PAPA technique requires nine PMTs for each axis for 512×512 pixel imaging, plus a strobe PMT, and can accommodate $>10^5$ events sec^{-1} on 25 mm formats [75]. Because it uses PMTs, it is somewhat bulky. Care must also be taken in the optical alignment to avoid image modulations [75].

Photodiode arrays [70] provide another convenient method for imaging a phosphor screen, and can be fiber optic or lens-coupled. A recent development in photodiode arrays is the active pixel sensor (APS), which is a phototransistor device composed of sets of rows of photo-FETs. Large formats can be achieved ($\approx 2000 \times 1000$ [76]) with small pixels (7.5 μm) at high frame rates (30 frames/sec), and any row combination can be individually selected and read out at higher rates. As yet there has been little work in applying APS sensors for readout of UV intensifiers.

Intensified CCDs (ICCDs) where a CCD is coupled to an MCP–based intensifier, are commercially available for visible light applications, and have also been used with UV sealed tubes [73] and open tube devices [68]. Generally, the performance (resolution, QE) of the ICCD is determined by the intensifier, provided the lens, or fiber optic coupling is chosen appropriately. The CCDs can provide pixels <10 μm and array sizes $>2000 \times 2000$, but although smaller arrays can be read out at high frame rates (>100 frames/sec) most of the larger format CCDs have video frame rates. CIDs have also been proposed [74] in place of CCDs, and are useful where rapid addressing of specific areas of the image are required, or high radiation tolerance is needed.

In all of the techniques described previously, there are specific optimization steps required to provide best imaging performance. The amplification levels need to be matched to the sensitivity of the desired position sensor and observed light levels. The output window fiber optic fibers, optical coupler, and sensor pixel size need to be correctly matched to avoid aliasing effects such as moire modulations and other sources of fixed pattern noise. Also, in high counting rate applications, the phosphor decay times [77] must be chosen appropriately. These considerations are discussed in the references.

7.3.1.2 Discrete Electronic Position Sensors. A simple discrete electronic position readout scheme is a rectangular array of individual square metal anodes [21, 22], placed close behind the signal multiplier, each connected to an amplifier. The position resolution is limited to the individual anode size (typically >1 × 1 mm). Because of the potentially large number of connections and preamplifiers required, this kind of anode array has not been extended to give large numbers of pixels over large areas. Commercial optical photocathode, multianode devices are available with ≈10 × 10 pixels using MCP [22] or grid mesh multipliers [21, 22]. Multianode arrays are, however, useful as high-speed analog or photon counting sensors since the metal anodes permit signals at subnanosecond time scales [21] to be processed independently on each anode.

There are several multianode readout schemes that can encode larger formats at high speed and high resolution. One of these, the CODACON (Fig. 28) [78], uses an array of thin (15 μm wide) metal strips on a thin dielectric substrate placed close (50 μm) to the output of an MCP. Each photon event is collected on a strip, which then induces a signal on pairs of binary code tracks on the opposite side of the substrate underneath or to one side of the active area. The distribution of signals on the binary code tracks, detected by differential amplifiers, gives a unique gray code binary position readout. In one dimension, n

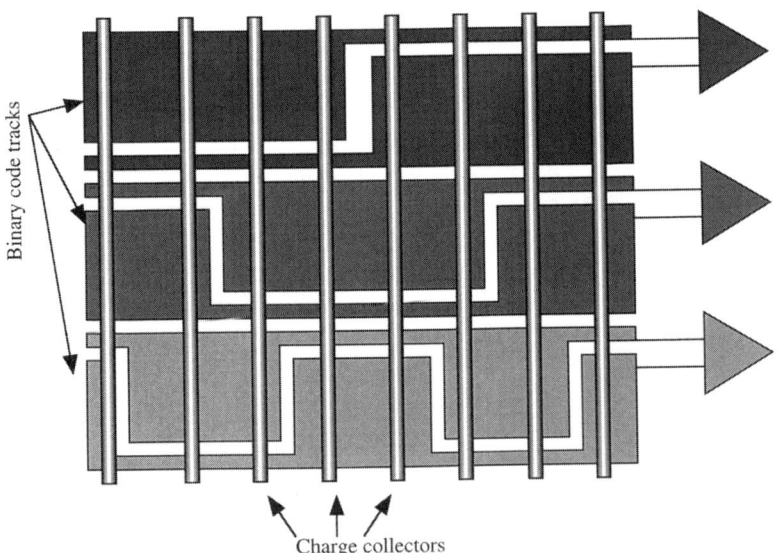

FIG. 28. Schematic for a three-bit, one-dimensional CODACON position readout system.

amplifiers are needed for 2^n position locations. Two-dimensional CODACON arrays can be constructed using two sets of orthogonally-aligned metal strips placed on top of one another with an insulator layer between the crossovers. Two independent sets of binary code tracks off to the sides of the array then encode the x and y event positions. Resolution of ≈ 25 μm FWHM has been achieved with 1024×64 arrays being used for space applications [78]. Image linearity with CODACON schemes is potentially very good if the MCP does not introduce displacements (Section 7.2.4.5). Because the anode pattern period and the MCP pores can be of similar size, care must also be taken to avoid significant moire modulation of the flat field response [79]. Counting rates of the order 10^6 sec^{-1} can be achieved with the CODACON, but because of increasing interanode capacitive cross coupling, the size of the arrays cannot be easily increased.

The multianode microchannel plate array (MAMA) [80] has two sets of orthogonal thin (≈ 10–20 μm wide) metal strips, similar to the CODACON, to collect the charge from MCPs. This anode is placed close (50–100 μm) to the MCP output, so that charge from the MCP is allowed to spread over at least two or three of the metal strips in each axis. The strips of each axis are connected to two sets of bus lines in a scheme similar to that illustrated in Fig. 29 and fed to fast amplifiers and discriminators. Each photon event position results in a unique signal combination on the top and bottom bus lines, which requires $2 M^{0.5}$ amplifiers for each axis to obtain M pixels. Therefore, 128 amplifier channels

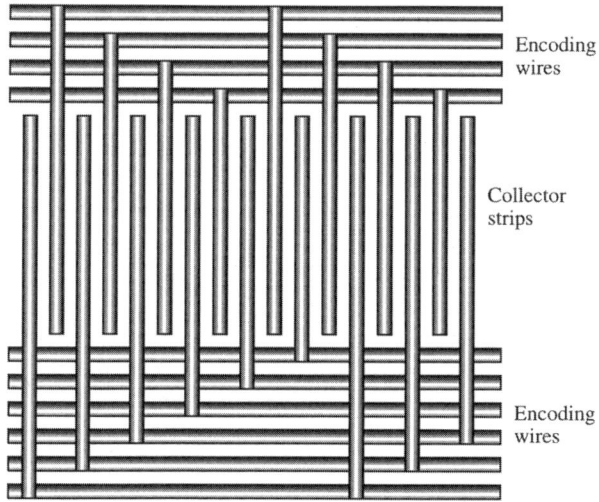

FIG. 29. Schematic for a segment of a fine-fine array, one-dimensional MAMA anode readout.

are required for a 1024 × 1024 format. Event positions are then decoded in both axes by high-speed address encoding electronics [80]. MAMA arrays from 1 × 512 to 1024 × 1024 [81] have been built with ≈25 μm pixels, and interpolation to half a pixel width has been demonstrated [80]. Counting rates up to 10^6 counts s^{-1} can be accommodated with good image linearity and position sensitivity.

Another image intensification method that uses CCDs for position sensing is the electron bombarded CCD (EBCCD) [82]. In this device, photoelectrons from a photocathode (either opaque or transmissive) are magnetically/electrostatically focussed onto a CCD. Electron energies of 10–20 keV are required to overcome the CCD dead layer and produce a sufficiently large signal in the CCD for photon counting applications. EBCCDs can also be used in analog mode if the CCD full well capacity is not exceeded. For applications using opaque cathodes, the EBCCD focus geometry can be somewhat bulky [82] and requires high voltages and magnets. Pixel formats are defined by the choice of CCD, as are the counting rates (frame rate, full well capacity).

7.3.2 Photon Centroiding Position Sensors

Position sensing event centroiding techniques derive individual photon event positions either with continuous readout anodes, or discrete element anodes with interpolation to provide essentially a continuous position readout.

7.3.2.1 Optical Centroiding Schemes.
ICCDs have been used as photon counting position centroiding sensors [72]. If the signal multiplier provides a high gain ($>10^5$), the statistical variation of the event centroid position is much smaller than the overall event spot size. If the spot of light from the intensifier is made large enough, event centroid positions on a CCD can be found by calculating the center of gravity of the charge signal levels on adjacent pixels. It is possible to achieve position determination to fractions of the CCD pixel size [72], and even resolve 6 μm MCP pores across an 18 mm intensifier [67]. One of the problems, however, is the residual nonlinearities of the image because of position interpolation errors. The overall counting rates of ≈10^5 s^{-1} can be obtained depending on the CCD frame rate and speed of the interpolation algorithm calculation. Counting rates within any zone the size of the interpolation area are ≈1 event per 10 frames for a 10% deadtime because of position confusion when two events fall in similar positions in the same frame. CIDs can also be used [74] instead of CCDs to enhance local counting rates, although their higher noise can be disadvantageous for low signal to noise situations.

7.3.2.2 Charge Division Centroiding Schemes.
The resistive anode technique [83], consisting of a uniform resistivity coating (100 kΩ–1 MΩ) on an insulating substrate, has been widely used for position encoding. A resistive

anode may be a wire (1 dimensional) or sheet [83] with contacts at the corners (Fig. 30) mounted close (<1 mm) to the signal multiplier output. Propagation of charge to the contacts from the event location gives pulse amplitudes and rise times that are proportional to the distance from the contact. Event positions can be determined from charge signal ratios, or difference in signal timing at opposing pairs [83] of the four contacts. For example, the centroid location of a photon event, dX, in one dimension is given by

$$dX = \frac{fQ_a}{(Q_a + Q_b)} \quad (4)$$

where Q_a and Q_b are the charge signals on the opposing anode corners, and f is a normalization factor. Amplifiers are connected to each anode corner, and each signal is usually digitized to several bits better than the expected resolution of the anode. This accommodates for the dynamic range of the multiplier signals and anode and allows characterization of the point spread function of the readout. To avoid considerable image distortion because of anode boundary signal reflection effects, resistive anodes with low resistivity borders [83] are commonly used (Fig. 30). High gain ($\approx 10^7$) MCP stacks are needed for resistive anodes to provide the high signal-to-noise ratio required to achieve good

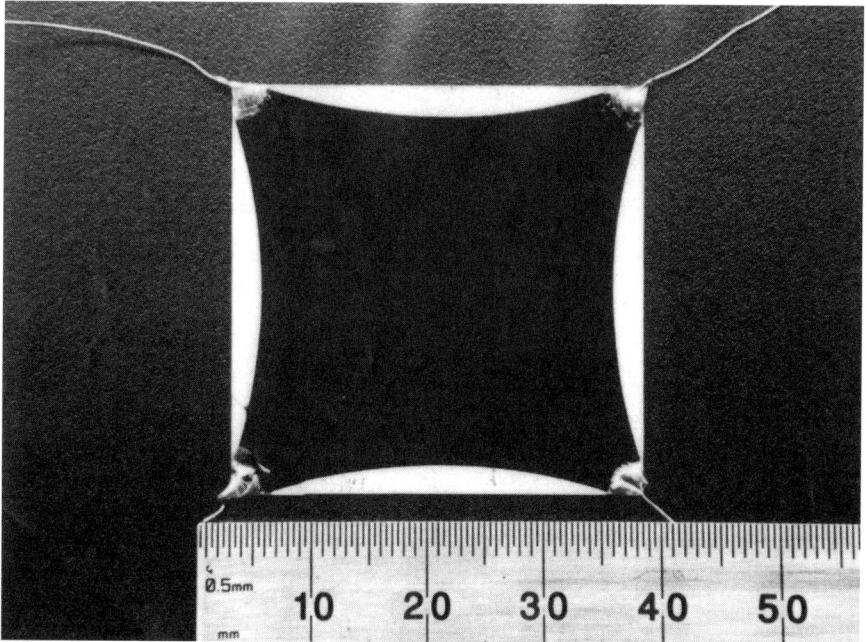

FIG. 30. Resistive anode position readout, 25 mm active area.

position resolution above the thermal resistive noise (KT_e noise) of the anode [83]. Resolution of ≈ 50 μm FWHM over a ≈ 25 mm anode is possible [84] with statistical limit flat field performance if the anode is of uniform resistivity. Using conventional encoding electronics [84], photon counting rates of up to $\approx 10^5$ events sec^{-1} can be achieved.

Crossed multiwire arrays (crossed grid, Fig. 31) that are resistively coupled can also be used as a position sensor system [85]. A two-dimensional readout [85] consists of two orthogonal planes of wires ≈ 0.1 mm diameter, closely spaced (≈ 0.2 mm intervals), with resistors interconnecting the wires in each plane. The multiplier (MCP or grid mesh [21, 22, 30]) charge output is spread over a number of wires. Preamplifiers, connected to every eighth wire, detect the signals, and the coarse event position is found by locating the preamplifier with the greatest signal. The centroid position is then determined from the charge ratios between adjacent preamplifiers. Crossed grid readouts up to 10 × 10 cm [85, 86] have been made with resolution of ≈ 25 μm FWHM. Depending on the interpolation algorithm used, the crossed grid readout may have small periodic image nonlinearities [85] or loss of some spatial resolution caused by the periodic amplifier taps. Because of the relatively large time constants, the counting rates are of the order $\approx 10^3$ counts s^{-1}.

A frequently used readout method is proportional charge division of the multiplier output signal between sets of conductive anode elements. A simple one-dimensional system, the "backgammon" encoder [87] has two sets of

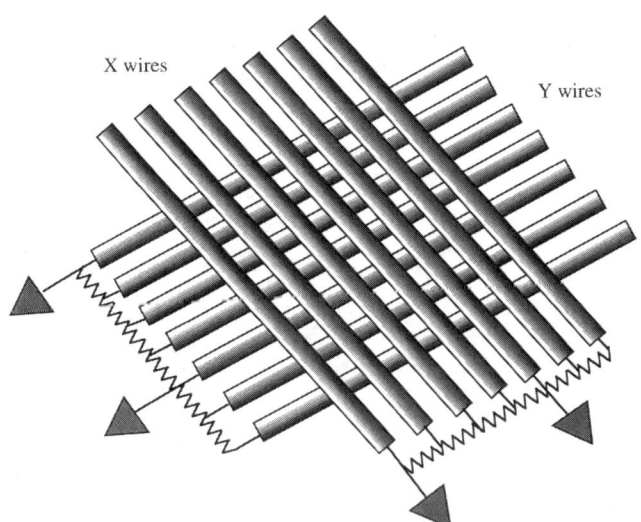

FIG. 31. Schematic of a section of a two-dimensional multiwire crossed grid sensor with resistive interconnections.

interleaved conductive wedge-shaped electrodes on a dielectric substrate, usually with a pattern repetition period of <1 mm. The signal is distributed over several periods of the pattern by choosing an appropriate multiplier-anode gap and accelerating voltage. The relative area of the electrodes varies linearly in one dimension so event centroid positions can be calculated from the charge ratio of the two electrode sets.

A two-dimensional position encoder of this type is the wedge-and-strip anode [88, 89]. The most common wedge-and-strip anode design has three electrodes (Fig. 32), a set of wedges, a set of varying width strips, and a zigzag pattern covering the area between wedges and strips. Wedge and strip anodes can be made a number of ways, but photolithographic etching of the pattern in a thin metallic layer deposited on a low dielectric constant insulator is often used. The wedge area varies linearly with position in one axis, and the strip area varies linearly with position in the other axis so the event centroid position dX in one axis is given by

$$dX = \frac{fQ_w}{(Q_w + Q_s + Q_z)} \qquad (5)$$

where Q_w, Q_s, and Q_z are the charge signals on the wedge, strip, and Z electrodes respectively, and f is a normalization factor. The expression for dY is

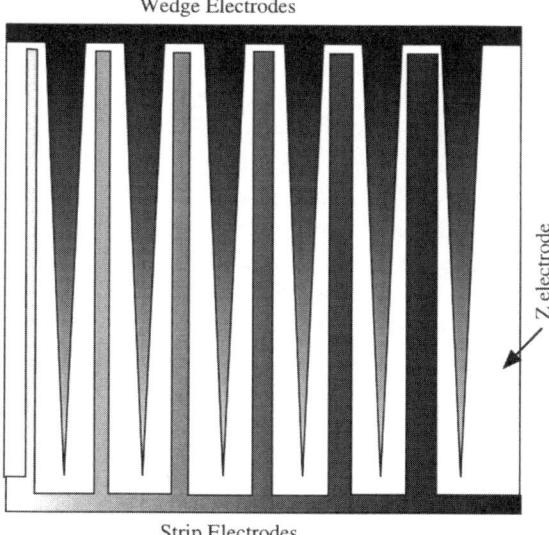

FIG. 32. Schematic of a three electrode, square, wedge, strip, and zigzag position sensitive anode.

equivalent. Wedge and strip patterns have repetition periods of ≈ 1 mm and are placed ≈ 1 cm behind the MCP output with an accelerating potential of ≈ 400 v. This ensures that the charge cloud from the multiplier (usually MCP) spreads over several pattern periods, avoiding potential flat field modulations [89]. Only three charge sensitive amplifers are required for two-dimensional imaging, and the signals are typically digitized and encoded by a hardware or software algorithm. Image distortion is low [89] if the anode is made larger than the required aperture to avoid edge effects. Spatial resolution is determined by the signal-to-noise levels, and by statistical variations in the charge division [88, 89]. Anodes with formats of ≈ 5 mm to ≈ 200 mm have been used with MCP stacks (gain $\approx 10^7$) giving $>512 \times 512$ resolution FWHM elements. Wedge and strip anodes with spherically curved surfaces have even been made [90]. The flat field response is limited by counting statistics and event rates of up to $\approx 10^5$ counts sec^{-1} can be achieved [20].

The spiral anode (SPAN) also uses charge division, but unlike wedge and strip, the electrode areas vary in a repetitive sinusoidal [91] fashion that decreases in amplitude from one side to the other. Three out of phase (60° or 90°) electrodes, connected to charge sensitive amplifiers, are required for each position axis. Only 8 bit digitization is required for high resolution ($>1024 \times 1024$ FWHM) with up to 50 mm anodes at rates of $\approx 10^5$ sec^{-1}. However, because of fixed pattern noise considerations [91], more bits are preferable. This system is also somewhat sensitive to the gain and charge cloud variations that can produce image nonlinearities [91].

7.3.2.3 Delay Line Centroiding Schemes.
Another method for position sensing uses delay line schemes to encode event position centroids by determination of the difference in arrival times of the event signal at the two ends of a high speed transmission line [92]. Delay line schemes can take a variety of forms including planar delay lines [92] and wire wound delay lines [93], and multilayer delay lines [94] that are placed behind (a few millimeters and ≈ 500 v) the signal multiplier. A planar double delay line (DDL) anode [95] (Fig. 33) has two sets of wedge-shaped electrodes that divide the charge deposited in a fashion linearly proportional to Y. Each wedge of a set is connected to an external serpentine delay line providing signal propagation delay linearly proportional to position. These anode patterns are photolithographically etched into a conductor deposited on a thin (0.25 mm) low loss microwave substrate (with high ε_r) that has a ground plane underneath. Typical delay times are ≈ 1 ns/mm, potentially enabling event rates $>10^6$ sec^{-1} to be accommodated. Determination of the Y event centroid coordinates is derived from charge division ratios, and the timing encoding electronics consists of an amplifier and discriminator for each end of the delay line, followed by a time to amplitude, and an analog to digital converter (ADC). Digital oversampling with the ADCs allows small electronic bin sizes (a few micrometers) to be used.

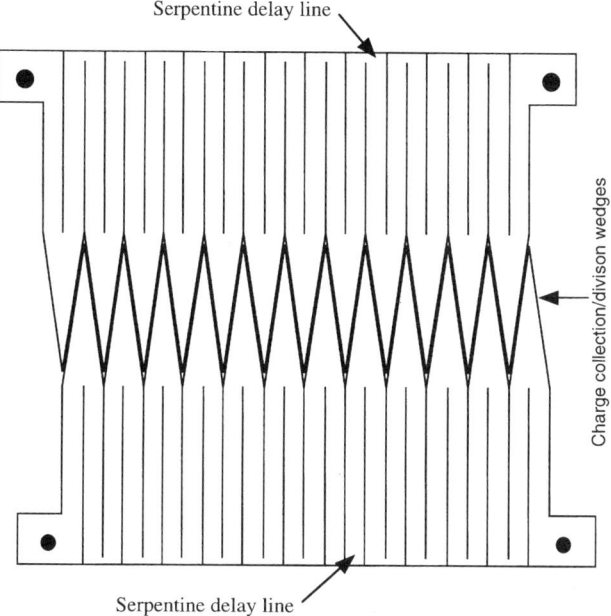

FIG. 33. Schematic of a planar double delay line anode with external serpentine delay lines and interleaved wedge charge collectors/dividers.

Crossed delay lines (XDL) have also been developed. One scheme places two serpentine delay lines above each other with insulating and ground layers in between. The upper (X) and lower (Y) delay lines are exposed so that each has a 50% share of the multiplier charge. Another scheme (Fig. 34) uses two sets of orthogonal charge collection fingers, one above the other with insulating and ground layers in between. Then external delay lines are attached as in the case of the DDL. The spatial resolution of the delay line technique is determined by the timing error (discriminator jitter and walk, ≈ 10 ps FWHM). DDL and XDL anodes [94, 95] have demonstrated resolution of <20 μm FWHM for formats of ≈ 10 cm, with counting rates up to $\approx 5 \times 10^5$ events sec^{-1}. Image distortions are small (<50 μm) and images are stable to the level of a few microns under varying operating conditions. Flat field performance is stable with characteristics that are in accord with poisson statistics. Wire wound delay lines use a pair of wires wound helically around a center plate to encode X positions, with a second pair wound around the other in an orthogonal direction for Y position encoding [93]. The wires must have a small dc bias to properly collect charge and are connected to differential preamplifiers. Performance is similar to planar delay lines, with resolution of <20 μm FWHM over ≈ 14 cm achieved.

FIG. 34. Photograph of a section of a cross delay line anode readout showing the multilayer charge collection fingers for X (lower) and Y (upper), and parts of the external serpentine delay lines.

References

1. Henke, B. L., Liesegang, J., and Smith, S. D., *Phys. Rev.* **19**, 3004 (1979).
2. Henke, B. L., *Proc. SPIE* **146**, 316 (1981).
3. Engstrom, R. W., *Photomultiplier Handbook*: RCA, 1980.
4. Sommer, A. H., *Photoemissive Materials*: J. Wiley and Sons, 1968.
5. C. Martin, S. Bowyer, *Appl. Opt.* **21**, 4206 (1982).
6. Siegmund, O. H. W., Vallerga, J., Everman, E., Labov, S., and Bixler, J., *Proc. SPIE* **687**, 117–124 (1986).
7. Siegmund, O. H. W., Everman, E., Vallerga, J.V., Sokolowski, J., and Lampton, M., *Appl. Opt.* **26**(17), 3607 (1987).
8. Fraser, G. W., *Nucl. Instrum. Meth.* **206**, 265 (1983).
9. Fraser, G. W., *Nucl. Instrum. Meth.* **206**, 251 (1983)
10. Siegmund, O. H. W., and Gaines, G., *Appl. Opt.* **29**(31), 4677–4685 (1990).
11. Marsh, D., Siegmund, O. H. W., and Stock, J., *Proc. SPIE* **2006**, 51–81 (1993).
12. Fraser, G. W., Pearson, J. F., and Lees, J. E., *Nucl. Instrum. Meth.* **256**, 401 (1987).
13. Kong, S. H., Nguyen, D. C., Sheffield, R. L., and Sherwood, B. A., *Nucl. Instrum. Meth.* **A358**, 276–279 (1995).

14. Pearson, J. F., Lees, J. E., and Fraser, G. W., *IEEE Trans. Nucl. Sci.* **NS-35**, 520 (1988).
15. Siegmund, O. H. W., and Gaines, G., *Proc. SPIE* **1344**, 217–227 (1990).
16. Whiteley, M. J., Pearson, J. F., Fraser, G. W., and Barstow, M. A., *Nucl. Instrum. Meth.* **224**, 287 (1984).
17. Jelinsky, S. R., Siegmund, O. H. W., and Mir, J. A., *Proc. SPIE* **2808**, 617–625 (1996).
18. Taylor, R. C., Hettrick, M. C., and Malina, R. F, *Rev. Sci. Instrum.* **54**, 171 (1983).
19. Eberhardt, E. H., *Appl. Opt.* **16**, 2127 (1977).
20. Siegmund, O. H. W., Lampton, M., Bixler, J., Vallerga, J., and Bowyer, S., *IEEE Trans. Nucl. Sci.* **NS-34**, 41–45 (1987).
21. Hamamatsu Photonics, photomultiplier catalog, 360 Foothill Road, Bridgewater, NJ 08807.
22. Photonis, Bp 520, F-19106, Brive, France.
23. Vallerga, J. V., Hull, J., and Lampton, M., *IEEE Trans. Nuc. Sci.* **NS-35**, 539 (1988).
24. Ziock, K. P., Hailey, and C. J., *Proc. SPIE.* **1159**, 280 (1989).
25. Eschard, G., and Manley, B. W., *Acta Electronica* **14**, 19 (1971).
26. Photonis, Bp 520, F-19106, Brive, France.
27. Galileo Corp, Model 4039 data sheet, Galileo Park, Sturbridge, MA 01566.
28. Galileo Corp, Channeltron electron multiplier handbook, Galileo Park, Sturbridge, MA 01566.
29. Galileo Corp, Technology seminar, 1989.
30. Adams, J., and Manley, B. W., *IEEE Trans. Nucl. Sci.* **NS-13**, 88 (1966).
31. Galileo Corp, Galileo Park, Sturbridge, MA 01566.
32. Litton Industries, Inc. 21240 Burbank Boulevard, Woodland Hills, CA 91367.
33. ITT, 7635 Plantation Road N.W., Roanoke, VA 24019.
34. Hamamatsu Photonics, image intensifier catalog, 360 Foothill Road, Bridgewater, NJ 08807.
35. Siegmund, O. H. W., Cully, S., Warren, J., Priedhorsky, W., and Bloch, J. J., *Proc. SPIE* **1344**, 346 (1990).
36. Siegmund, O. H. W., Marsh, D., Stock, J., and Gaines, G., *Proc. SPIE* **1743**, 274–282 (1992).
37. Feller, W. B., *Proc. SPIE* **1243**, 100 (1990).
38. Fraser, G. W., *Proc. SPIE* **1140**, 50 (1989).
39. Timothy, J. G., and Bybee, R. L., *Proc. SPIE* **687**, 109 (1986).
40. Galanti, M., Gott, R., and Renaud, J. F., *Rev. Sci. Instrum.* **42**, 1818 (1971).
41. Gao, R. S., Gibner, P. S., Newman, J. H., Smith, K. A., and Stebbings, R. F., *Rev. Sci. Instrum.* **55**(11), 1756 (1984).
42. Ruggeri, D. J., *IEEE Trans. Nucl. Sci.* **NS-19**, 74 (1972).
43. Wiza, J. L., *Nucl. Instrum. Meth.* **162**, 587 (1979).
44. Siegmund, O. H. W., Gummin, M. A., Stock, J., and Marsh, D., ESA symposium on detectors, **ESA SP-356**, 89 (1992).
45. Fraser, G. W., Pearson, J. F., Smith, G. C., Lewis M., and Barstow, M. A., *IEEE Trans. Nucl. Sci.* **NS-30**, 455 (1983).
46. Siegmund, O. H. W., Coburn, K., and Malina, R. F., *IEEE Trans. Nucl. Sci.* **NS-32**, 443 (1985).
47. Pearson, J. F., Lees, J. E., and Fraser, G. W., *IEEE Trans. Nucl. Sci.* **NS-35**, 520 (1988).
48. Laprade, B. N., and Reinhart, S. T., *Proc. SPIE* **1072**, 119 (1989).

49. Edgar, M. L., Smith, A., and Lapington, J. S., *Proc. SPIE* **1743**, 283 (1992).
50. Fraser, G. W., Pain, M. T., Lees, J. E., and Pearson, J. F., *Nucl. Instrum. Meth. A* **A306**, 247 (1991).
51. Siegmund, O. H. W., Gummin, M. A., Sasseen, T., Jelinsky, P., Gaines, G. A., Hull, J., Stock, M., Edgar, M., Welsh, B., Jelinsky, S., and Vallerga, J., *Proc. SPIE* **2518**, 344–355, (1995).
52. Fraser, G. W., Pearson, J. F., and Lees, J. E., *IEEE Trans. Nucl. Sci.* **NS-35**, 529 (1988).
53. Siegmund, O. H. W., *Proc. SPIE* **1072**, 111 (1989).
54. Laprade, B., *Proc. SPIE* **1072**, 102 (1989).
55. Siegmund, O. H. W., Vallerga, J., and Wargelin, B., *IEEE Trans. Nucl. Sci.* **NS-35**, 524 (1988).
56. Fraser, G. W., Pearson, J. F., and Lees, J. E., *Nucl. Instrum. Meth.*, **A254**, 447 (1987).
57. Siegmund, O. H. W., Gummin, M.A., Ravinett, T., Jelinsky, S. R., and Edgar, M. L., *Proc. SPIE* **2808**, 98–106 (1996).
58. Lapington, J. S., and Edgar, M. L., *Proc. SPIE* **1159**, 565 (1989).
59. Hassler, D. M., Rottman, G. J., and Lawrence, G. M., *SPIE* **1344**, 194 (1990).
60. Vallerga, J. V., Siegmund, O. H. W., Vedder, P. W., and Gibson, J., *Proc. SPIE* **1159**, 382 (1989).
61. Fraser, G. W., Pearson, J. F., Lees, J. E., and Feller, W. B., *Proc. SPIE* **982**, 98 (1988).
62. Oba, K., Kume, H., Wakamori, K., and Nakatsugawa, K., *Advances in Electronics and Electron Physics* **74**, 87 (1988).
63. Tremsin, A. S., Pearson, J. F., Lees, J. E., and Fraser, G. W., *Proc. SPIE* **2518**, 384 (1995).
64. Janesick, J., Elliott, T., Winzenread, R., Pinter, J., and Dyck, R., *SPIE* **2415**, 2 (1995).
65. Trauger, J. et al., *Ap. J.* **435**, L3 (1994).
66. Photonis, Bp 520, F-19106, Brive, France.
67. Vallerga, J., Jelinsky, P., and Siegmund, O. H. W., *Proc. SPIE* **2518**, 410–420 (1995).
68. Thompson, W. T., Poland, A. J., Siegmund, O. H. W., Swarz, M., Leviton, D. B., and Payne, L. J., *Proc. SPIE* **1743** (1992).
69. Boksenberg, A., *Image Processing Techniques in Astronomy*, Reidel: Dordrecht, Holland, 59 (1975).
70. Tennyson, P. D., Dymond, K., Moos, H. W., Feldman, P. D., and Mackey, E., *Proc. SPIE* **627**, 666 (1986).
71. Gonsiorowski, T., *Proc. SPIE* **627**, 626 (1986).
72. Read, P. D., van Breda, I. G., Norton, T. J., Airey, R. W., Morgan, B. L., and Powell, J. R., *Instrumentation for Ground-Based Optical Astronomy: Present and Future*, Robinson, L. B., ed., 528–538, Springer-Verlag (1988).
73. Torr, M. R., and Devlin, J., *Appl. Opt.* **21**, 3091 (1982).
74. Kimble, R. A., Chen, P. C. Haas, J. P., Norton, T. J., Payne, L. J., Carbone, J., and Corba, M., *Proc. SPIE* **2518**, 397–409 (1995).
75. Norton, T. J., Airey, R. W., Morgan, B. L., Fordam, J. L. A., and Bone, D. A., Conference on Photoelectronic Imaging Devices, IOP Publishing Ltd, 97 (1991).
76. Tanaka, N., Hashimoto, S., Shinoharo, M., Sugawa, S., Morishita, M., Matsumoto, S., Nakamura. Y., and Ohmi, T., *IEEE Trans. Elect. Dev.* **ED-37**, 964 (1990).
77. Torr, M. R., Torr, D. G., Baum, R., and Spielmaker, R., *Appl. Opt.*, **25**, 2768 (1986).

78. Lawrence, G. M., and McClintock, W. E. *Proc. SPIE* **2831**, 104–111 (1996).
79. Lawrence, G. M., *Appl. Opt.* **28**, 4337 (1989).
80. Kasle, D. B., and Morgan. J. S., *Proc. SPIE* **1549**, 52 (1991).
81. Morgan, J. S., and Timothy, J. G., *Instrumentation for Ground-Based Optical Astronomy: Present and Future*, Robinson, L. B., ed., 557. Springer-Verlag (1988).
82. Carruthers, G. R., Heckathorn, H.,M., Opal, C. B., Jenkins, E. B., and Lowrance, J. L., *Adv. Elect. Electron. Phys.* **74**, 181 (1988).
83. Lampton, M., and Carlson, C. W., *Rev. Sci. Instrum.* **50**, 1093 (1979).
84. Paresce, F., Clampin, M., Cox, C., Crocker, J., Rafal, M., Sen, A., and Hiltner, W. A., *Instrumentation for Ground-Based Optical Astronomy: Present and Future*, Robinson, L. B., ed., 542. Springer-Verlag (1988).
85. Murray S. S., and Chappell, J. H., *Proc. SPIE* **982**, 48 (1988).
86. Kenter, A. T., Chappell, J. H., Kraft, R. P., Meehan, G. R., Murray, S. S., Zombeck, M. V., and Fraser, G. W., *Proc. SPIE* **2808**, 626 (1996).
87. Allemand, R., and Thomas, G., *Nucl. Instrum. Meth.* **137**, 141 (1976).
88. Martin, C., Jelinsky, P., Lampton, M., Malina, R. F., and Anger, H. O., *Rev. Sci. Instrum.* **52**, 1067 (1981).
89. Siegmund, O. H. W., Lampton, M., Bixler, J., Bowyer, S., and Malina, R. F. *IEEE Trans. Nucl. Sci.* **NS-33**, 724 (1986).
90. Siegmund, O. H. W., Cully, S., Warren, J., and Priedhorsky, W., *Proc. SPIE* **1344**, 346 (1990).
91. Breeveld, A. A., Edgar, M. L., Lapington, J. S., and Smith, A., *Proc. SPIE* **1743**, 315 (1992).
92. Lampton, M., Siegmund, O. H. W., and Raffanti, R., *Rev. Sci. Instrum.* **58**, 2298 (1987).
93. Williams, M. B., and Sobottka, S., *IEEE Trans. Nuc. Sci.* **NS-36**, 227 (1989).
94. Siegmund, O. H. W., Gummin, M. A., Stock, J., Marsh, D., Raffanti, R., Sasseen, T., Tom, J., Welsh, B., Gaines, G., Jelinsky, P., and Hull, J., *Proc. SPIE* **2280**, 89–100 (1994).
95. Siegmund, O. H. W., Gummin, M. A., Stock, J., Marsh, D., Raffanti, R., and Hull, J., *Proc. SPIE* **2209**, 388–399 (1994).

8. ABSOLUTE FLUX MEASUREMENTS

S. V. Bobashev
A. F. Ioffe Physico-Technical Institute RAS
St. Petersburg, Russia

8.1 Introduction

In our institute's library, two books are very frayed. One of them is *Techniques of Vacuum Ultraviolet Spectroscopy*, by Samson [1] and the other is *Vacuum Spectroscopy and its Application*, by Zaidel and Schreider [2]. More than one generation of experimenters and engineers used these books very actively and even now the books are still in demand. It is convincing evidence of the fact that vacuum ultraviolet spectroscopy has been developing actively during the last several decades.

Intensive development of VUV and SXR spectroscopy, in the photon energy range from a few electron volts to a few thousand electron volts, during the past decade is closely linked with the progress in fundamental physics and astrophysics as well as technical applications such as hot plasma diagnostics, synchrotron radiation use, lithography and nanostructure materials, space research, biology, and practical medicine. Twenty years ago a 10–15% error in quantitative spectral radiation measurements was usual. Now technical advances require the accuracy of data to be of order 1% or even less. For instance, the absolute calibration of the sensitivity of a telescope equipped with a four channel spectrophotometer launched in the seventies by the European Space Agency for measurements of the ultraviolet star catalog (in the wavelength range 135–300 nm) was not better than 10–15% [3], whereas the ultraviolet equipment on the Hubble Space Telescope for the similar ultraviolet range required an absolute calibration accuracy of about 1% [4]. The need for high-quality radiometric standards, quantitative radiometric measurements, and dedicated facilities for the characterization of radiation sources, detectors and optical components in the VUV and the SXR is now apparent [5, 6]. These technological advances gave incentives to scientists to develop new measurement technology. Good examples of this progress are the outstanding results in radiator-based and detector-based radiometry at the PTB laboratory on the electron storage ring BESSY I, where there are six experimental stations on four beamlines optimized for quantitative radiation measurement in the spectral range from about 3 eV to 15 keV [7, 8]. Typical values of calibration uncertainty are less than 1%.

The object of this review is to consider detectors as devices for absolute flux measurements in the VUV and the SXR spectral range. Radiation generated by different sources in the VUV and the SXR has a very wide set of characteristics. The quantitative measurement of photon fluxes from an electron storage ring, a tokamak, a dense plasma discharge, a soft x-ray laser, or astrophysical objects demands essentially different approaches and instrumentation. Accordingly, many different kinds of radiation detectors have been developed and modified including thermocouples, bolometers, semiconductor and emissive photodiodes, photomultipliers, charge-coupled devices (CCDs), photographic film, scintillators, photoelectron spectrometers, and many others.

8.2 Primary Radiator Radiometry

In metrology, the radiation detector standards are divided into primary standards and secondary, or transfer, standards. Usually the basis of primary standard operation is some fundamental law or phenomenon. A major requirement for a transfer standard is a property to be transferred from one place to another without noticeable changes in stability, sensitivity, and reliability. Transfer standards of chosen types are calibrated by comparison with primary radiometric standards and are used as working transfer standards for the calibration of detectors of the same type, or other types, for a variety of customers. Transfer detector standards can be calibrated with the help of both primary detector standards and primary radiator standards. Therefore, the accuracy of absolute measurements is connected with the development of more accurate primary radiator standards, primary detector standards, transfer detector standards, and the calibration technique.

Although the advantages of synchrotron radiation (SR) from synchrotrons and storage rings as primary radiation standard sources were recognized many years ago [9, 10], the application of electron accelerating installations for quantitative radiometry became a widespread practice only during the last decade. This is because of the many complications for development of the radiometric technique and instrumentation compatible with the specific parameters of SR. At present, an electron storage ring is considered to be an almost ideal source for quantitative radiometry of electromagnetic radiation in a wide spectral range from the infrared up to far VUV and the hard x-ray spectral range [6, 11, 12].

Electron storage rings as dedicated sources of SR have been built at many places around the world. Some of them are used as radiometric facilities in the VUV and the SXR. Calibrations have been made in the VUV and the SXR at the National Institute of Standards and Technology (SURF II, Gaithersburg), the Electrotechnical Laboratory (TERAS, Tsukuba), the Budker Institute of Nuclear Physics (VEPP-2M and VEPP-3, Novosibirsk), and the Physikalisch-Technische Bundesanstalt (BESSY I, Berlin). A small synchrotron is used at the Russian

Research Institute for Optophysical Measurements (TROLL, Moscow) for calibration in the range 50–250 nm.

During the last few years, improvements in the measurement of storage ring parameters (current, energy of the orbiting electrons, as well as an accurate control of position of the orbital plane relative to the center of an aperture stop restricting the radiation flux used in calibration) provided high accuracy in determination of the absolute photon fluxes from storage rings. For all the previously mentioned electron storage rings, the spectral photon flux is greater than 10^{-6} W/cm^2·nm for electron currents of some hundred mA, corresponding to about 10^{11}–10^{12} stored electrons. This flux can be easily scaled down to that of a single orbiting electron by simply decreasing the stored electron current. Contemporary accurate control of the electron storage ring parameters and control of orbital plane have produced small relative uncertainties, usually less than 1%, with staff at BESSY reporting values of 0.18–0.35% for the spectral photon flux in the energy range 5–5000 eV. This spectral irradiation from the electron storage ring is used as a primary radiator standard to calibrate dispersive detectors. Using the SURF2, NIST has calibrated more than 150 spectral devices [13] up to 1992. In Russia, on the VEPP-2M storage ring beamline, a portable spectrometer designed to serve as a transfer detector standard has been calibrated with an uncertainty of about 1% at very small electron currents, when about 200 stored electrons were counted precisely as single units, observing steplike diminishing of photocurrent monitor signal about every 3 minutes [14]. Also, SURF2 in the range 116–254 nm and VEPP-2M in the range 50–550 nm [15] are used as a primary radiator standard to calibrate transfer detector standards, using a normal incidence monochromator with a single diffractive grating, whose reflectance is measured before the actual calibration procedure. Calibrations between 5 and 58.4 nm are carried out at NIST with SURF2. Using BESSY1 as a primary radiator standard, extensive work on the calibration of semiconductor detectors in the photon energy range 3–1500 eV and of energy dispersive semiconductor detectors, such as Si (Li) and high-purity Ge [16, 17], in the photon energy range from 100 eV up to 5000 eV has been performed by PTB. Monochromator-detector systems [18] and CCDs have also been calibrated at BESSY.

Thus, it is evident that accurately calculable spectral irradiation (or photon flux) from a storage ring is widely used not only for the calibration of transfer radiator sources but also for the calibration of transfer detector standard.

8.3 Primary Detector Radiometry

The straightforward technique is to calibrate a transfer detector standard by comparison with a primary detector standard using a tunable monochromatic radiation source [11, 19]. But even in this case, when only relative measurements are necessary, the absolute calibration procedure is not a trivial one because the operation mode and spectral range of the transfer detector standard differs from those of the primary detector standard.

For the photon energy range under consideration, three devices are now used as primary detector standards: the photoionization chamber (PC), photoionization quantometer (PQ) and cryogenic electrical substitution radiometer (ESR).

8.3.1 Photoionization Chamber

The photoionization chamber and its modifications are almost ideal primary photodetectors in the VUV. They have been used and are used in many measurements of the absolute photon flux intensity [1, 2, 20–23]. The operation of a photoionization device is based on the fundamental processes of photon interaction with isolated atoms or simple molecules. The double photoionization chamber [24], developed by Samson, made the problem of accurate absolute measurements in the VUV region a routine procedure in many cases. Photoionization chambers continue to be applied as a basis for metrological measurement [23, 25].

Basically, the idea of the photoionization chamber operation is in the fact that one absorbed photon produces one ion-electron pair [1]. Thus, measurements of the ion current and of the absorbed fraction of the incident radiation determine the absolute photon flux. The relation between the absolute photon flux and the ion current in a PC is given by the definition of the photoionization yield γ

$$\gamma = \frac{\text{total number of ions produced}}{\text{total number of photons absorbed}} = \frac{i/e}{N_\omega[1 - \exp(-\sigma n L)]} \quad (1)$$

where N_ω is the absolute photon flux,* σ is the total photoabsoption cross section, n is the number density of the gas, L is the path length in the ion chamber, e is the electron charge, and i is the total ion current.

In the wavelength range 102.2–25 nm, rare gases are the best gas targets if $\gamma = 1$. The long-wavelength limit is determined by the first ionization threshold of Xe. Thus, the knowledge of $\sigma n L$ and i defines the photon flux N_ω. The combination σn, which determines the accuracy of flux measurements, requires the detailed knowledge of photoionization cross section and gas density. In the double PC [1, 25] the σn value can be measured directly by means of recording electric currents i_1 and i_2 from first and second parts of the electrode system in the chamber and N_ω the number of photons can be calculated as a result of the double PC parameters.

The photoionization yield γ is an important characteristic of photon-atom interaction. Reliable information concerning the photoionization yield or mean

---

* In metrological nomenclature, we try to follow [11]. Photon flux N is a number of photons per time interval onto a given area, N_ω (photons/s-eV) = $dN/d\hbar\omega$.

ion charge makes it possible to extend the wavelength interval accessible for quantitative photon flux measurements by photoionization technique. Extensive study of γ values for rare gases has been made in [25] for Ne, Ar, Kr, and Xe over the photon energy range 44–1300 eV. The γ-value calculation is a result of experimental measurements of the yield and branching ratios for multiple charge ions produced by the photoabsorption of monochromatized synchrotron radiation. A time-of-flight mass spectrometer has been used for the ion yield and branching ratio measurements. The relative uncertainties in γ values claimed by authors in [25] are less than 1.5%.

A modified PC has been used recently for the development of a detector calibration facility at the Electrotechnical Laboratory (ETL) in Japan [23] using the synchrotron radiation from the storage ring TERAS. It represents a windowless system with a glass capillary array (of 93 capillaries, each with inner diameter 100 μm and length 2 mm) that provides the operation of the PC at a typical pressure around 2 kPa while the pressure in the electron storage ring is usually less than 10^{-6} Pa. In [23] the authors thoroughly discussed all the problems connected with accurate photon flux measurement in the wavelength range 10–92 nm. Some of them arise from the quality of the SR beam that passes through a monochromator and also depends on the beamline construction and operation characteristics of the accelerator. The most significant corrections resulted from the effects of secondary ionization of the gas by photoelectrons, multiple ionization above the multiple ionization threshold, and Penning ionization caused by impurities in the rare gases. Fig. 1 from [23] shows the computed pressure dependence of γ/γ_0 for neon, where γ is the apparent photoionization yield at a gas pressure p and γ_0 is the true photoionization yield at $p = 0$.

As a result, the detector standard based on a rare gas ionization chamber has been established in the wavelength range 10–92 nm with a maximum 1σ uncertainty of ±9%.

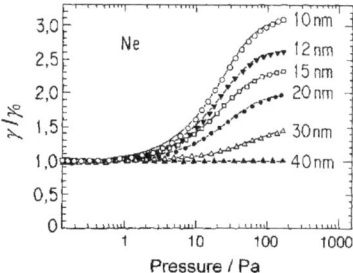

FIG. 1. Pressure dependence of multiplication factors caused by secondary ionization processes (including Penning ionization) for neon determined by ratios of the experimental-to-calculated ion currents for certain wavelengths [23].

From our point of view, 9% is one of the best values for the uncertainties of the absolute photon flux measurement at wavelengths ≥10 nm by a PC. The major drawback in using the PC for absolute flux measurements is a necessity to have enough photons absorbed by the gas to produce a sufficient number of ions for accurate electric current measurements. Also, too high a gas density leads to unknown secondary processes that cannot be taken into account properly.

8.3.2 Photoionization Quantometer

PQ as a new type of absolute detector for the spectral interval 100–0.1 nm has been proposed and demonstrated in [26–29]. The PQ is a photoionization chamber combined with a time-of-flight mass spectrometer. Its operation is based on the ionization of a gas successively by the photon flux under study and by an electron beam of known intensity and energy. Ions of different charge, produced by electrons and photons, are detected by the time-of-flight mass spectrometer (TOF). Thus, the PQ operation is based on the fundamental values of the cross sections for ionization of atoms by photons and by electrons. The development of this technique and the device in which this technique is implemented has been related closely to studies of the intensive impulse radiation from laser-produced plasmas [26, 29] and of absolute photoionization cross-section measurements in photon-ion interactions [30]. Subsequently, PQ was also used in experiments with other intense radiation sources such as x-ray tubes [31] and laser-produced plasmas for semiconductor detector calibration [32].

The essential parts of PQ are shown in Fig. 2. A diaphragm placed in the path of photon flux to be measured forms a beam of rectangular cross section passing between the electrodes of an ionization chamber. An electron gun produces an electron beam that passes through the ionization chamber in the direction perpendicular to the photon beam axis and is measured by a Faraday cup. The PQ operation with a single pulse of monochromatic radiation is as follows. The

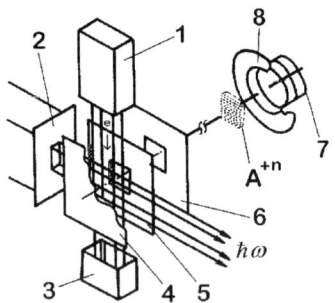

FIG. 2. Principle of operation and schematic of a photoionization quantometer (PQ): 1, electron gun; 2, aperture for photon beam; 3, Faraday cup; 4–5, electrodes of the ionization chamber; 5–6, ion accelarating space; 6–8, ion drift space; 7, ion detector [32].

photon flux produces ions in different charge states in the ionization chamber. The charge composition of the ions depends on the photon energy and the type of gas in the interaction region. It is supposed that the gas pressure is low enough to prevent ionization of atoms by secondary electrons. The separation and registration of ions are performed in the standard mode by the TOF mass spectrometer.

The number of n-charge ions A^{+n} produced in the gas chamber by a photon flux N_ω generates an electrical charge $q_{\omega n}$ in the ion detector of the TOF mass spectrometer that is proportional to the number of photons N_ω in the flux and the cross-section of n-fold photoionization of the gas $\sigma_{\omega n}(\lambda)$

$$q_{\omega n} = k_\omega N_\omega \sigma_{\omega n}(\lambda) \tag{2}$$

After a short time interval τ the gas in the chamber is ionized by a pulsed electron beam of energy E and intensity N_e. The duration of the pulse is set close to the duration of the photon beam. The ions of different charges, produced by the electrons, are separated in charge and detected in the same way as the ions produced by the photon flux. The TOF detector signal, in this case q_{en}, of the n-fold ionization of the gas by electrons depends on the electron impact cross-section $\sigma_{en}(E)$, that is

$$q_{en} = k_e N_e \sigma_{en}(E) \tag{3}$$

Practically, k_e and k_ω can easily be made equal and on this condition the number of N_ω is

$$N_\omega = \frac{\sigma_{en}(E)}{\sigma_{\omega n}(\lambda)} \cdot N_e \frac{q_{\omega n}}{q_{en}} \tag{4}$$

A comprehensive discussion has been carried out in [27] about the conditions when $k_e = k_\omega$. In short, both values are products of the filling gas density number and the detection efficiency of multicharged ions produced either by photons or electrons. As these detection efficiencies are maintained equal, so k_ω and k_e are equal accordingly.

Thus, the PQ operation is based on the ionization of a gas by photons, a separation of ions of different charges in the TOF mass spectrometer, and the calibration device using an electron beam. The complete set of data required for the absolute flux measurement are a recording of N_e and the signal ratio $q_{\omega n}/q_{en}$ combined with the partial cross-sections $\sigma_{\omega n}(\lambda)$ and $\sigma_{en}(E)$. The PQ spectral sensitivity depends on the photoionization cross section and the gas density in the photoionization volume of the device. The most complete information on cross sections for photoionization and electron impact ionization is presently available for the rare gases, which are also the most suitable for absolute measurements in the VUV and the SXR. The maximum gas pressure in the PQ depends on the filling gas and varies in the range $(0.1–1) \times 10^{-2}$ Pa. At these

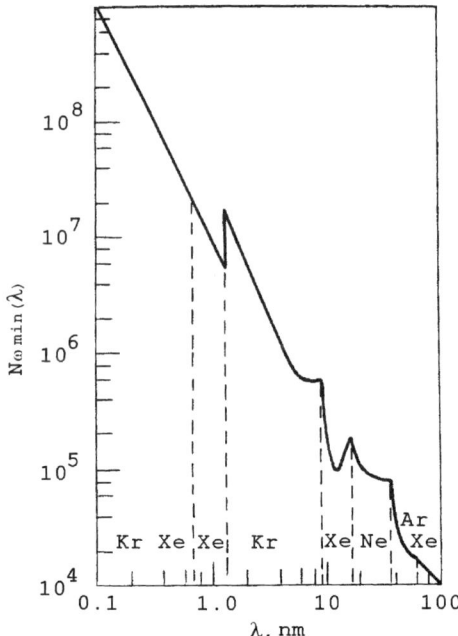

FIG. 3. Spectral sensitivity of a photoionization quantometer (PQ) as a function of wavelength λ for various filling rare gases. $N_{\omega\min}(\lambda)$ is the minimum number of photons that are necessary for flux measurement with uncertainty less than 10–15% [32].

pressures, the absorption of the photon flux in the active area of PQ is insignificant and does not exceed 0.1%. The active area of the device is practically limited to a few cm because of the size of the microchannel plates usually used as the ion detector.

It is well known that the photoionization cross sections fall off rapidly with increasing photon energy. Disregarding some subtle features of the photoionization processes connected with autoionization resonances [33] one can consider that the spectral response of the photoionization technique also drops rapidly with increasing photon frequency ω. Therefore, the application of this technique at $\hbar\omega \geq 10$ keV is not feasible.

Figure 3 illustrates the spectral sensitivity of the PQ, which has been used primarily as a detector standard for SXR intensity measurements from a laser-produced plasma in the spectral interval 1.5–0.35 nm [29, 31, 34]. The quantity $N_{\omega\min}$ corresponds to the minimum number of photons that can be detected by PQ with an accuracy of 10–15%.

A new version of PQ was developed [35] for photon flux measurements at synchrotron radiation facilities. It is based on the comparison of total ion yields of photoionization and electron-impact ionization. It consists of an ionization chamber (10 cm long), a microchannel plate (MCP) ion detector operated in the counting mode, an electron gun, and a Faraday cup for electron current measurements. The essential features of this new PQ are a hollow axis electron gun and a thin metal filter of known transmittance for soft x-ray radiation at the bottom of the Faraday cup. This design provides identical positions of the photon beam and the electron beam in the ionization chamber and, as a result, identical collection efficiency for ions produced by photons and by electrons. Moreover, the ions are accelerated up to kinetic energies high enough to obtain equal detection efficiency for differently charged ions.

With this PQ, the ratio of the number of impinging photons N_ω to the number of electrons N_e is given by:

$$\frac{N_\omega}{N_e} = \frac{\sigma_e}{\sigma_\omega} \cdot \frac{f_\omega}{f_e} \qquad (5)$$

where σ_e and σ_ω are the total ionization cross sections, f_e and f_w are the ion count rates for electron impact and photoionization, respectively.

The new PQ has not been used yet for photon flux measurements but it was used to determine the ratios σ_e/σ_ω of the ionization cross sections [35]. In these measurements the photon flux N_ω was obtained with a photodiode calibrated against an electrical substitution radiometer (ESR) with an uncertainty of 1% [12, 36]. The obtained uncertainties for the ratios of the ionization cross sections were as low as 1.5%. The measurements demonstrate that the new version of the PQ is very suitable for accurate photon flux measurements.

8.3.3 Detector Calibration with Electrical Substitution Radiometer

Since 1980, high vacuum grating spectrometers for normal incident radiation and grazing incident radiation have been developed that have low stray radiation and high suppression of radiation from higher diffraction orders. In addition, they can be rotated under high vacuum conditions to take into account the polarization of synchrotron radiation. A cryogenic ESR detector, about one thousand times more sensitive than existed earlier, has been developed for use as a primary detector standard. Highly sensitive semiconductor detectors with zones of uniform sensitivity, small dark current and high stability under irradiation, have been manufactured in Japan, the United States, and Switzerland. Armed with this new technology, the PTB radiometry laboratory developed a transfer detector calibration technique based on the ESR as the primary detector standard and the synchrotron radiation from the electron storage ring BESSY I [37] to attain the utmost possible accuracy of detector calibration in the VUV and the SXR spectral range. The monochromatized synchrotron radiation obtained from a monochromator and higher diffraction order cut-off filters, is refocused by a grazing-incidence ellipsoidal mirror at a fixed focal point where detectors can be placed with an error, less than 0.1 mm. For a spectral resolution of 3.3 nm and typical operation conditions of BESSY, the radiant power at the focal point is 1–15 µW [38]. More than 99% of this radiant power is found in an area of 2.8 mm horizontally by 1.8 mm vertically, which is less than the ESR entrance aperture (6 mm diameter) or the sensitive area of the photodiode transfer detectors (>4.8 mm diameter) to be calibrated. The ESR was manufactured by Oxford Instruments to meet specific requirements for operation at an electron storage ring, which include an ultra high vacuum environment, low photoelectron emission, and fluorescence of the absorber. The absorber cavity is made of copper foil with a wall thickness of 25 µm. This cavity is 60 mm long and has a 10 mm × 10 mm cross section. The thermal time constant is about 2 min. The cavity is thermally linked to the heat sink, the temperature of which is stabilized at 4.32 K. The temperature of the cavity is also actively controlled to better than 40 µK at a value of 6.8 K. With this cavity temperature, the ESR has a responsivity of 40 µK/µW, and is capable of measuring radiant powers up to 16 µW with a measurement time of 50 s. For a typical measurement time of 7 min, the total uncertainty of the radiant power measurement (for a power of 1 µW measured with a bias heater power of 16 µW) is 0.22%. Even for a radiant power of 100 nW determined with a bias power of 1.6 µW, the total uncertainty was below 0.5%. Two different types of photodiode (Hamamatsu G1127, GaAsP Schottky diode, and Hamamatsu S1337 silicon photodiode) have been calibrated with the relative uncertainty of the spectral responsivity less than 1% and a GaAsP diode did not show aging after exposure to radiant powers in the microwatt range over several hours. In contrast to this performance, silicon photodiodes have shown rapid degradation of the responsivity for wavelengths below 250 nm. Recently, the spectral range for the calibration of radiation detectors has been extended down to 0.8 nm using a grazing-incidence monochromator [12, 39] at BESSY.

Using intense radiation from the undulator installed in the NIJI-II storage ring at the Electrotechnical Laboratory, detector calibrations have been performed with a room-temperature operated ESR in the wavelength range 200–400 nm [40]. It has been shown that results of responsivity calibrations for Si photodiodes and GaP photodiodes agree within ±4% and ±5%, respectively, with the existing responsivity scale based on a thermopile as a wavelength-independent detector and silicon photodiode self-calibration for the absolute scale.

8.4 Transfer Detector Standards for Absolute Flux Measurement

The development of stable detectors for the VUV and the SXR spectral range is an important part of experimental instrumentation production. Photoemissive diodes with Al_2O_3, CsI, gold, and tungsten cathodes were developed and thoroughly investigated at NIST [41, 42] more than ten years ago. These detectors are very useful and reliable devices for absolute flux measurements. The calibrated transfer detectors, available from the NIST, are magnesium fluoride windowed CsTe-diodes for the wavelength range 116–350 nm, and windowless Al_2O_3-diode for the wavelength range 125–5 nm. They are highly linear, stable in intense fluxes, easy to operate, and solar blind. Their disadvantages are the low responsivity in SXR, spectral structure in the vicinity of absorption edges, and dependence on cathode surface conditions.

During the last ten years, photoemissive detector standards have been gradually replaced by semiconductor photodiodes. Silicon photodiodes with n-conducting front regions on top of a p-type epitaxial layer on a p^+-substrate [43, 44], as well as GaP, GaAsP, Schottky diode, and diffused junction GaAsP diodes have been shown by extensive study at PTB [45] to comply with requirements of stability, sensitivity, and time resolution. Semiconductor detectors are widely used in laboratories because of their high spectral responsivity, simplicity of operation, and reduced influence of surface condition. A self-calibration procedure enables absolute calibration as proposed in [45] and improved in [46]. This procedure provides absolute response measurements of high accuracy in the spectral range 150–2500 eV, with an error of about 2% [46].

For the SXR spectral range, the semiconductor detector response can be described by a simple model, because for this spectral range a semiconductor detector is a device with nearly 100% internal quantum efficiency. The spectral responsivity S (ampere/watt) is $S(hv) = (e/w) \cdot S_r(v)$ where e is the electron charge, w is the average energy in eV necessary for the production of one electron-hole pair, and S_r is a relative response that covers all loss processes. For the ideal device, without any dead layer and an infinitely large sensitive volume, S_r should remain constant over the entire spectral range. The quantity w has been found constant [46] at the room temperature with a value of 3.64 ± 0.03 eV. Fig. 4 taken from [46] presents the spectral response of n-on-p crystalline silicon photodiode measured with the ESR and the calculated value of S_r.

Methods and instrumentation for absolute radiometry of pulsed intense radiometric fluxes in the SXR range have been developed at the Ioffe Institute [32, 47] as part of a program to test multilayer polychromators for absolute spectral flux measurements for plasma diagnostics [48]. Diamond and p-on-n semiconductor detectors have been calibrated in the laboratory. The calibration setup for the photon energy range 0.1–1.0 keV is shown in Fig. 5. Two independent spectral channels for 0.1–0.4 keV and 0.4–1.0 keV have been used. Each channel is complete with a common laser-produced plasma SXR source with a quasi-continuous spectra, multilayer x-ray mirrors (MLM) for monochromatization, filters, and a PQ serving as a primary detector standard. A

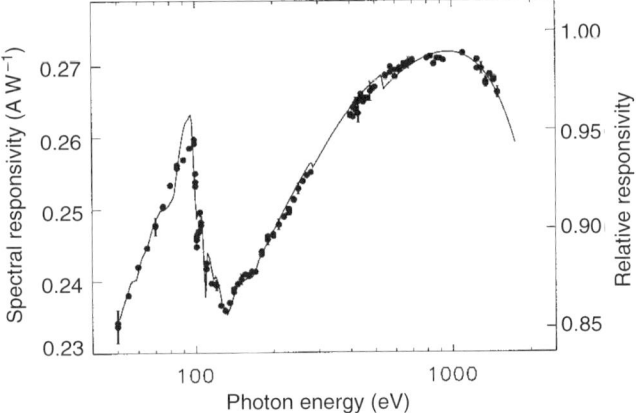

FIG. 4. Spectral responsivity $S(h\nu)$ (symbols, left axis) of *n*-on-*p* crystalline silicon photodiode (AXUV-100G) measured with the ESR and the relative responsivity $S_r(h\nu)$ (line, right axis) calculated by self-calibration procedure. Reprinted with permission from F. Scholze, H. Rabus, and G. Ulm, *Appl. Phys. Lett.* **69**, 2974–2976 (1996).

calibrated detector is placed after the PQ in the radiation flux so that the sensitive area of the detector is uniformly covered by the radiation flux.

Impulse radiation with typical intensities of 10^7–10^8 photons/pulse and duration ≤ 40 ns is directed into one of the spectral channels after the MLM

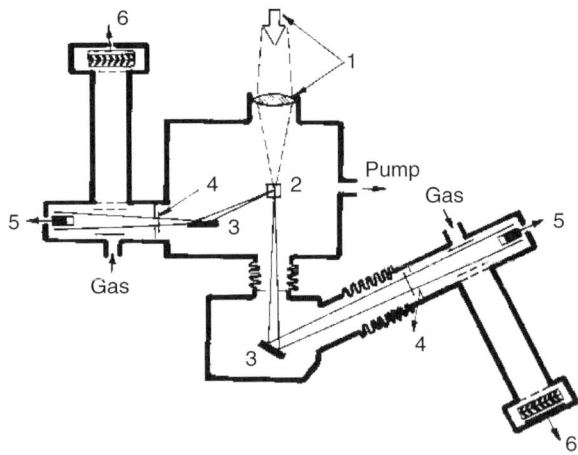

FIG. 5. Detector calibration setup using pulse SXR radiation (0.1–1.0 keV) from laser-produced plasma source: 1, focusing system for laser beam; 2, laser-plasma source target; 3, multilayer x-ray mirrors; 4, apertures and SXR radiation filters; 5, detector under calibration; 6, ion signal from PQ detectors [32].

monochromator with spectral resolution $E/\Delta E \approx 20$–60. The absolute responsivity S_d to photon fluxes has been measured with uncertainties of about 10%. A laser-produced plasma in combination with the MLM makes it possible to irradiate the detector with spatially uniform radiation. Thus, S_d (ampere · cm^2/watt) = $S(hv)$ (A/W) × (detector sensitive area, in cm^2) is an absolute detector sensitivity for photon fluxes. For certain applications S_d is more convenient to use than $S(hv)$ because it is independent of the detector's responsivity changes over its active surface.

The accuracy of the calibration procedure for PQ depends essentially on the accuracy of our knowledge of the cross sections for rare gas ionization by electrons and photons.

More accurate cross sections will make it possible in the near future to perform absolute calibrations of transfer detector standards using photoionization devices with uncertainties of about 2–3%. Relatively simple laboratory installations equipped with the photoionization technique will provide quick and reliable detector calibrations directly in laboratories.

8.5 Conclusion

Accurate calibrations of transfer detector standards can be performed in the VUV and the SXR range with synchrotron radiation from an electron storage ring by using two different calibration techniques. One technique is based on the cryogenic radiometer (ESR) as a primary detector standard. This method is insensitive to spatial nonuniformity and polarization of monochromatized radiation and enables one to reach the highest reasonable accuracy with an uncertainty less than 1%. The other is based on the calculable spectral flux of highly polarized synchrotron radiation emitted in the orbital plane of the electron storage ring and used in conjunction with a monochromator system of known resolution and transmittance. This allows one to calibrate the detector transfer standards with monochromatic radiation. Further, the radiation flux can be varied over ten orders of magnitude by simply reducing the electron beam current in the storage ring. This allows accurate calibrations of high sensitive detectors operating in the photon counting mode, such as photomultipliers, channeltrons, MCP and their assemblies with CCD and multianode systems (MAMA). The second technique is well suited for the calibration of detectors operating in very wide range of fluxes such as photoemissive diodes and semiconductor detectors, which can be used after calibrations for the accurate measurements of pulsed radiation. One may also calculate the responsivity of portable spectrometers serving as transfer standards of spectral irradiation detectors. Uncertainties of transfer detector calibration of less than 1% can be claimed and can be considered attainable in the VUV and the SXR range. It seems possible that further development of these methods can produce, in the next decade, routine procedures to make absolute measurements, based on calibrated transfer detector standards and ionization chambers, in the VUV and the SXR range with uncertainties of 2–3%.

Acknowledgment

The author would like to thank V. I. Ogurtsov for many helpful discussions.

References

1. J. A. R. Samson, *Techniques of Vacuum Ultraviolet Spectroscopy*. John Wiley and Sons, New York, 1967.
2. A. N. Zaidel and E. Ya. Schreider, *Vacuum Spectroscopy and its Application*. Science Publication, Moscow, 1978. [In Russian]
3. C. M. Humphries, C. Jamar, D. Malaise, and H. Wroe, *Astron. and Astrophys.*, **49**, 389 (1976).
4. F. Macchetto, *Proc. of Second ESO/ESA Workshop*, Munich, 15–21 (1981).
5. M. Kühne and B. Wende, *J. Phys.* **E18**, 637 (1985).
6. B. Wende, *Metrologia*, **32**, 419–424 (1995/96).
7. G. Ulm and B. Wende, *Rev. Sci. Instrum.*, **66**, 2244 (1995).
8. G. Ulm and B. Wende., *Optical and Quantum Electronics*, **28**, 299–307 (1996).
9. D. H. Tombulian and P. L. Hartman. *Phys. Rev.*, **102**, 1423 (1956).
10. J. Schwinger, *Phys. Rev.*, **75**, 1912 (1949).
11. E. Tegeler, *Physica Scripta*, **T31**, 215–222 (1990).
12. H. Rabus, V. Persch, and G. Ulm, *Appl. Opt.*, **36**(22), 5421–5440 (1997).
13. M. L. Furst and L. R. Canfield, *SPIE Proceedings*, **1764**, 278–284 (1992).
14. E. S. Gluskin and V. I. Ogurtsov, The Report on SR Activities of NPI, Novosibirsk, USSR, 135 (1981). [In Russian]
15. E. S. Gluskin, V. S. Kuzíminikh, O. A. Makarov, and V. I. Ogurtsov, *Nucl. Instrum. Meth.*, **A246**, 397–399 (1986).
16. M. Krumrey, E. Tegeler, and G. Ulm, *Rev. Sci. Instrum.*, **60**, 2287–2290 (1989).
17. F. Scholze and G. Ulm, *Nucl. Instrum. Meth.*, **A339**, 49–54 (1994)
18. K. Molter and G. Ulm, *Rev. Sci. Instrum.*, **63**, 1296–1299 (1992)
19. F. Tegeler, *Nucl. Instrum. Meth.*, **A282**, 706 (1989).
20. K. Watanabe, F. F. Marmo, and E. C. Y. Inn, *Phys. Rev.*, **91**, 1155 (1953).
21. E. B. Saloman and D. L. Ederer, *Appl. Opt.*, **14**, 1029–1034 (1975).
22. J. A. R. Samson and G. N. Haddad, *J. Opt. Soc. Am.*, **64**, 47–54 (1974).
23. T. Saito and H. Onuki, *Metrologia*, **32**(6), 525–529 (1996).
24. J. A. R. Samson, *J. Opt. Soc. Am.*, **54**, 6 (1964).
25. I. H. Suzuki and N. Saito., *Bull. Electrotechnical Lab.*, **56**, 688–711 (1992). [In Japanese]
26. V. V. Afrosimov, V. P. Belik, S. V. Bobashev, and L. A. Shmaenok, *Sov. Tech. Phys. Lett.*, **1**, 370 (1975).
27. S. V. Bobashev and L. A. Shmaenok. Preprint 634. A. F. Ioffe Physico-Technical Institute, Leningrad, 1–36 (1979). [In Russian]

28. S. V. Bobashev and L. A. Shmaenok, *Rev. Sci. Instrum.*, **52**, 16–20 (1980).
29. V. P. Belik, S. V. Bobashev, M. P. Kalashnikov, I. F. Kalinkevich, Yu. A. Mikhailov, A. V. Rode, G. V. Sklizkov, S. I. Fedotov, and L. A. Shmaenok, *Pis'ma Zh. Tekh. Fiz.*, **6**, 1273–1278 (1980).
30. M. Ya. Amusia, V. V. Afrosimov, V. P. Belik, S. V. Bobashev, S. A. Sheinerman, and L. A. Shmaenok, *Sov. Phys. JETP*, **49**, 439 (1979).
31. S. V. Bobashev, G. S. Volkov, A. V. Golubev, V. I. Zaitzev, V. Ya. Tsarfin, and L. A. Shmaenok, *Sov. Tech. Phys. Lett.*, **14**, 1283 (1979).
32. S. V. Bobashev, A. V. Golubev, Yu. Ya. Platonov, N. N. Salashenko, L. A. Shmaenok, G. S. Volkov, and V. I. Zaitsev, *Physica Scripta*, **43**, 356–367 (1991).
33. M. Ya. Amusia, *Atomic Photoeffect*. Plenum Press, New York and London, 1990.
34. Z. A. Albikov, V. P. Belik, S. V. Bobashev, G. S. Volkov, A. V. Golubev, N. G. Zabrodin, V. I. Zaitzev, M. I. Ivanov, O. V. Kozlov, A. K. Krasnov, A. N. Polyakov, N. N. Salashenko, V. P. Smirnov, N. I. Terentiev, V. Ya. Tsarfin, and L. A. Shmaenok, *Plasma Diagnostic* **6**, 48–52, Energoatomizdat, Moscow, 1989. [In Russian]
35. A. A.Sorokin, L. A. Shmaenok, S. V. Bobashev, B. Möbus, F. Scholze, and G. Ulm, *Abstracts of 17th International Conference X-ray and Inner Shell Processes*, Hamburg, Germany, 1996, p. 400 (1996).
36. H. Rabus, F. Scholze, R. Thornagel, and G. Ulm, *Nucl. Instrum. Meth.*, **A377**, 209–216 (1996).
37. A. Lau-Främbs, U. Kroth, H. Rabus, E. Tegeler, G. Ulm, and B. Wende, *Metrologia*, **32**, 571–574 (1995).
38. A. Lau-Främbs, U. Kroth, H. Rabus, E. Tegeler, and G. Ulm, *Rev. Sci. Instrum.*, **66**, 2324–2326 (1995).
39. F. Scholze, M. Krumrey, P. Müller, and D. Fuchs, *Rev. Sci. Instrum.*, **65**, 3229–3232 (1994).
40. T. Saito, I. Saito, T. Yamada, T. Zama, and H. Onuki, *J. Electron Spectr. Rel. Phenom.*, **80**, 397–400 (1996).
41. E. B. Saloman, S. C. Ebner, and L. R. Hughey, *SPIE Proceedings*, **279**, 76–83 (1981).
42. L. R. Canfield and N. Swanson, *J. Res. NBS*, **92**(2), 97–112 (1987).
43. R. Korde and J. Geist, *Appl. Opt.*, **26**, 5284 (1987).
44. E. M. Gullikson, R. Korde, L. R. Canfield, and R. E. Vest, *J. Electron Spectr. Rel. Phenom.*, **80**, 313–316 (1996).
45. M. Krumrey and E. Tegeler, *Rev. Sci. Instrum.*, **63**, 797 (1992).
46. F. Scholze, H. Rabus, and G. Ulm, *Appl. Phys. Lett.* **69**, 2974–2976 (1996).
47. S. V. Bobashev, L. A. Shmaenok, and V. P. Smirnov, *AIP Conference Proceedings* **215**, 242–258 (1990). *15 International Conference on X-ray and Inner Shell Processes (X-90)*, Knoxville, Tennessee, 1990, Invited Presentations, 242–258 (1990).
48. L. A. Shmaenok, Y. Platonov, N. N. Salaschenko, A. A. Sorokin, M. M. Simanovskii, A. V. Golubev, V. P. Belik, S. V. Bobashev, F. Bijkerk, E. Louis, F. G. Meijer, B. Etlisher, and A. Ya. Grudsky, *J. Electron Spectr. Rel. Phenom.*, **80**, 259–262 (1996).

9. VACUUM TECHNIQUES

Roger L. Stockbauer

Department of Physics and Astronomy
and
Center for Advanced Microstructures and Devices
Louisiana State University
Baton Rouge, Louisiana

9.1 Introduction

As the name vacuum ultraviolet (VUV) implies, this radiation does not propagate through air and must be transported and used in a vacuum environment. This chapter includes the information a researcher needs in order to house an experiment to take advantage of VUV radiation. We will not be concerned with the vacuum in the light source itself, since it is assumed that the source is either a synchrotron or a laboratory source that will have its own vacuum requirements.

There has been a large number of books published on vacuum technology with thorough discussions of pumps, pumping requirements, materials, valves, leak detection, and so forth. One of the more concise and useful books is by John O'Hanlon [1]. Additional books that will prove beneficial to the reader are handbooks of vacuum material properties, material preparation techniques, and fabrication methods by Rosebury [2] and Kohl [3]. The encyclopedic arrangement of Rosebury's book is especially convenient for ascertaining characteristics of laboratory and industrial materials.

These references are excellent resources for gaining a familiarity with the various types of vacuum equipment mentioned here. In order to proceed, the reader should have a working knowledge of fore, sorption, cryo, ion, titanium sublimation (TSP), and diffusion pumps; right angle, gate, and leak valves; thermocouple and ion gauges. In addition, familiarity with residual gas analyzers (RGA) and helium leak detectors will be helpful.

9.2 Design of the Vacuum Environment

Chambers to house VUV experiments fall into three main categories depending on their use: gas phase experiments including metal vapors, solid state experiments, and beamline optics. The latter two have very similar vacuum requirements and the need for vacuum cleanliness for optics in vacuum will be given special attention.

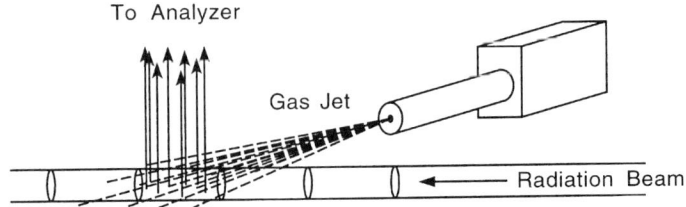

FIG. 1. Schematic of a generic gas phase experiment.

9.2.1 Gas Phase Experiments

Gas phase experiments below 1050 Å [4] will typically use a beam or jet of the gas under investigation to increase the number density of the atoms or molecules in the localized region where the interaction with the VUV light takes place [5]. The configuration consists of the gas jet along one axis with the light beam at right angles and the analyzer making the measurement along the third axis as shown in Fig. 1. The intersection of the photon beam, gas jet, and acceptance aperture of the analyzer defines the interaction region. The goal is to have as large a number density of the sample gas in the interaction region and as low a number density along the trajectory of the particles to be analyzed as possible to prevent self-absorption or collisions of the particle with the background gas. Such collisions will either disrupt the excitation of the particle to be analyzed or deflect it out of the acceptance cone of the analyzer.

This requirement of high gas pressure in the interaction region and low pressure elsewhere is met by using a small gas nozzle with high-speed pumps. Decreasing the nozzle diameter and increasing the pumping speed increases the pressure in the gas jet compared with the background pressure. The nozzle can be anything from a machine-drilled hole in a plate, a hypodermic needle, or a laser or e-beam drilled microaperture. Typical sizes are 1mm for mechanically drilled holes, 200 µm for hypodermic needles, and 10–100 µm for microapertures. The nozzle is placed as close to the interaction region as possible since the number density decreases away from the nozzle exit as the gas expands. Such expansion also rotationally cools the gas, which is beneficial in some experiments where thermal excitation would lead to spectral broadening.

The type of pump for the gas phase experiment must be selected with an eye toward the type of sample under investigation. For atmospheric and similar nonreactive gases, high-speed turbo, or cryopumps, are the most effective. Diffusion pumps, which were once the mainstay of gas phase experiments, have been banned at most synchrotron light sources because of the disastrous effects of a pump failure. The resultant clean-up of the entire light source and beamlines cannot be chanced. Preference is usually given to cryopumps in this application since they can economically be used to pump the sample during the experiment as well as keep the chamber under vacuum when the experiment is not being run. As an example, a cryopump with a speed of 2,500 liters per second for a particular sample gas will maintain a background pressure of 10^{-5} torr with a 50 µm diameter nozzle and a pressure behind the nozzle of 0.1–0.5 atm. The sample pressure in the interaction region is typically 1–2 orders of magnitude higher than the background pressure, resulting in a sample number density of 10^{13} cm^{-3} for the above example [6].

Although cryopumps are suitable for atmospheric and organic gases, special care must be taken with both poorly pumped gases and corrosive gases. Helium is a perfect example of the former. Although cryopumps do pump helium, they do so for only a limited time. As the absorber in the cryopump becomes saturated, the pumping speed gradually decreases and soon the pump must be recycled, which involves turning off the compressor, letting the pump warm up, and pumping away the released gases with a turbo or sorption pump. The frequency of this operation depends on the amount of gas pumped but typically a week's worth of running 16 hours of day can be accommodated. Although the loss of pumping speed as the absorber becomes saturated is gradual for most gases, this is not the case for helium. Here, the loss of pumping speed is catastrophic. The ability of the cryo-absorber to hold helium is a strong function of temperature since helium liquefies at 4 K and the absorber is typically at 10 K. Once the absorber becomes saturated, the background pressure in the chamber rises, increasing the heat load to the absorber, increasing its temperature, releasing more helium. In a very short time, helium fills the chamber to an atmosphere or more of pressure while the experiment is running. This occurs over a period of approximately 10 seconds with no warning and can have disastrous effects on channeltrons and multipliers if the high voltage cannot be turned off quickly enough.

Other types of pumps are not well suited for gas phase experiments. Ion and TSP pumps cannot handle the large, continuous gas loads over extended periods and are not usually used in these applications. However, very large turbopumps with speeds in excess of 2000 l/s are finding greater use in some of the more critical applications.

Since the gas phase experiments operate with a high number density of the sample gas in the interaction region, one is generally not concerned about contamination of the sample by the background gas in the experimental chamber. Base pressures on the order of 10^{-7}–10^{-8} are often considered adequate. Chambers are usually constructed of aluminum or stainless steel and flanges are sealed with O-rings. Likewise, valves need only be sealed with O-rings on both the gate and bonnet. The chambers usually are not baked (see Section 9.2.2) so the base pressure is on the order of 5×10^{-8} torr.

Special consideration must also be given to corrosive gases with regard to both the vacuum equipment and the experimenter's health. Halogens and their compounds are especially notorious for their deleterious effects on pumps. Few materials can survive even short exposures to F_2 and prolonged exposure to Cl_2 and Br_2. The only real recourse here is to clean and overhaul or replace the pump once the experiment is over.

Even if the sample is not directly harmful to the equipment, one must remember that the gas must eventually be removed from the cryopump. The gas should be vented to a hood that has been approved for the gas in question. Also, be careful of the gases (e.g., NO_2) that form acids when exposed to moisture in the air.

9.2.2 Solid State Experiments

Experiments on solids, especially probes of surfaces, have much more stringent vacuum requirements than gas phase experiments. In order to probe the properties of surfaces, they typically must be well ordered, well characterized and free of defects and contamination. Much effort has gone into developing methods and procedures for cleaning surfaces, and recipes exist for almost any metal or oxide surface [7]. Researchers spend considerable time preparing the sample surface before experiments are begun, and it is necessary that the surface remain clean for the typically 10-hour duration of an experiment. During that time, the surface becomes contaminated by the residual gas in the chamber. When one realizes that at a pressure of 10^{-6} torr, enough molecules strike the surface in one second to form a full monolayer, assuming that all of them stick, the importance of very high vacuum becomes apparent. In order to maintain a clean surface for the 10 hours of an experiment, a residual pressure five orders of magnitude lower (10^{-11} torr) is required. This vacuum regime, called ultra-high vacuum (UHV), presents special design challenges in the selection of pumps, valves, and materials. For the chamber itself, the material of choice is 304, 304L, 316, 316L stainless steel in order to provide the structural integrity and stability during bakeout cycles.

Obtaining UHV requires one additional step not encountered in the high vacuum system of gas phase experiments: The heating of the chamber to remove water vapor from the surfaces exposed to the vacuum. Enough water is adsorbed on the interior surfaces when they are exposed to air, that the vacuum will remain in the 10^{-9} torr range for days even with the fastest pumps. The water can, however, be quickly removed by heating the system to 200°C while under vacuum. This bakeout procedure entails keeping the entire system at the elevated temperature for 24 hours to ensure that all internal elements have reached this temperature. Any surfaces remaining near room temperature will readsorb the displaced water and continue to outgas after the bakeout is terminated. It is important that the heat be uniformly applied so the chamber is usually wrapped in aluminum foil or a fireproof insulated blanket. A 12-inch diameter chamber,

14 inches high with a 240 l/s ion pump typically requires approximately 5000 watts of power to maintain the 200°C. A typical bakeout usually takes three full days before an experiment can be run, one at the elevated temperature, the second to cool down, and the third to outgas the filaments and condition the various internal surfaces.

The high temperature during the bakeout constrains the types of materials that can be used in the chamber. Most plastics are not suitable and hydrocarbon-based lubricants cannot be used. Outgasing and gas permeation of elastomer O-rings used to seal vacuum flanges limits such chambers to an ultimate vacuum in the 10^{-9} torr range. Flanges must be metal sealed to withstand the thermal expansion and contraction during bakeout and to avoid the outgasing and gas permeation of elastomer O-rings. However, if two O-rings are used and the space between them pumped separately, UHV can be achieved. Such differentially pumped seals are used, for instance, in rotary stages.

UHV systems are almost always pumped with a combination of ion and TSP pumps. Although turbo- and cryopumps do perform well in this region, they are not practical for maintaining UHV conditions over sustained periods. This has more to do with economics than with performance, since the lifetime of both of these pumps is somewhat lower than ion pumps.

The selection of the type of flanges for a UHV chamber is rather straightforward. The scientific community and industry have settled on one standard, the copper sealed ConFlat® for any flange less than a 14-inch diameter. Since almost every vacuum instrument or every piece of vacuum hardware is supplied on a ConFlat® flange, one should have a very good reason to consider any other type of flange. For diameters over 14 inches, the ConFlat® is not always reliable. Critical applications in this size range require a heavier flange such as the wire-sealed Wheeler® flange, which has a good performance record. These larger flanges need to be treated with a bit more care and the manufacturer's recommendations concerning tightening procedures and torque sequences should be followed precisely. For unique applications, rectangular flanges are manufactured using either a ConFlat® or Helicoflex seals.

Valves for UHV chambers should have metal-sealed bonnets since the same restrictions apply here as in O-ring sealed flanges. The gate seals, however, can be elastomer O-ring if the valve is merely separating chambers or instruments both under vacuum. For instance, the valve separating the ion pump from the chamber in most systems is usually a viton sealed gate valve since the valve is always open while the chamber is at UHV. Gate valves that separate the UHV from atmosphere should be avoided since this requires a metal gate seal. Although these valves are available, they are expensive and have questionable reliability.

In designing instruments and devices to be used in UHV, one must again be aware of the high temperature properties of the materials being used since they will be subjected to the high temperature bakeout. Soft solder melts at typical bakeout temperatures. Teflon releases large amounts of water as it is heated and starts to decompose at 300°C. Copper becomes very soft after bakeout and can no longer be considered a good structural material. Metals or alloys containing Cd, Zn, As, P, Mg, Ca, Bi, or Se should be avoided since their vapor pressure at 300°C is too high. When in doubt, one should check the vapor pressure curve of the material and not use it if the vapor pressure is greater than 10^{-11} torr at 300°C. Vapor pressure charts and tables are available on pages 143–148 of Rosebury [2] and pages 448–450 of O'Hanlon [1].

The handling of parts and devices that are to be used in UHV require extra caution. Touching surfaces that will be exposed to the vacuum should be avoided. Even though the hydrocarbons in a fingerprint will be removed during bakeout, the residual carbon and salt deposits will not, and, while probably not harmful to the vacuum or experiment, their unsightly appearance indicates a lack of care and proper technique by the perpetrator.

Actual construction of UHV chambers is usually left to commercial companies with dedicated expertise in these methods. The typical cost of the chamber itself can be roughly estimated by adding the cost of all the flanges and multiplying by two.

9.2.3 Beamline Vacuum

Although the typical VUV experimenter will not be concerned with the design and construction of a synchrotron radiation beamline, one needs to be aware of its vacuum requirements since the experiment will usually be connected directly to the beamline with no intervening windows. Beamlines must typically maintain a vacuum of 10^{-10} torr or better to ensure that the vacuum integrity of the ring is maintained and to protect the optics from contamination. Since the optics are exposed to a large flux of high energy radiation, any hydrocarbons, CO or CO_2 on the surface of the optics, will be decomposed leaving behind a carbon film. There are comparably few beamlines that can be used to do an experiment in the 38–45 Å range since most of this light is absorbed by a carbon coating on the optics. Synchrotron managers typically forbid the use of any pump containing oil and usually demand an RGA analysis of the vacuum of a experimental chamber before the valve separating it from the beamline can be opened.

For the solid state experimenter, this is not a problem since the vacuum in the UHV chamber matches or exceeds the vacuum of the beamline. For gas phase experiments running at 10^{-5} torr background pressure, this, however, does present a problem. Some sort of isolation must be introduced between the chamber and the beamline. This can be achieved in several ways. A thin film window can be used if the material transmits enough radiation in the desired

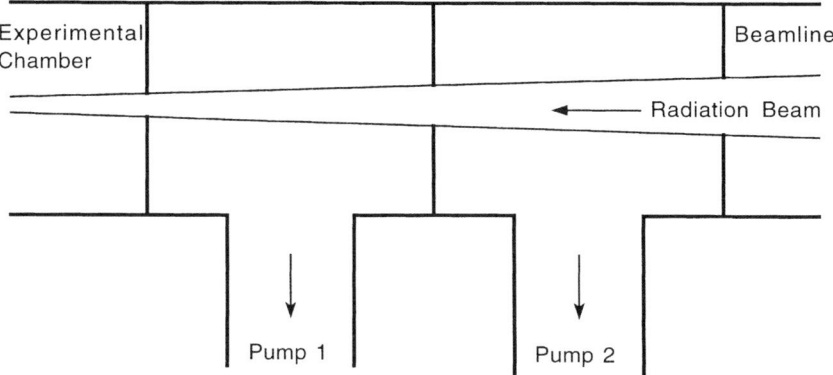

FIG. 2. Schematic of a differential pumping arrangement to isolate an experimental chamber from a beamline or lightsource.

energy range. Thin metal films (e.g., Al above 175 Å, Be above 120 Å, or Lexan above 45 Å) are typically used. These films must be thin enough to transmit the radiation, yet sturdy enough to withstand the handling during installation and the gas flow during pump down. They can be mounted in special window valves that are manufactured with viewports installed in the valve gate. These are typically used for alignment with visible light from the source while the beamline is at atmospheric pressure. The viewport can be replaced with a mount that holds the thin film window, allowing the window to be moved out of the way when the experiment is not running.

A more widely applicable solution to beamline-experiment isolation, is a differential pumping system illustrated in Fig. 2. This consists of a series of chambers between the experiment and beamline separated by baffles. Apertures in the baffles are made as small as possible to allow the synchrotron radiation to pass through yet minimizing the flow of gas from the experiment. Each chamber is individually pumped to remove as much gas as possible. Any number of stages of differential pumping can be operated in series to achieve the degree of isolation required. This method is limited, however, by the gas molecules whose initial velocity vector is directed straight down the centerline of the apertures.

An enhancement to this type of differential pumping uses an extended aperture in the form of a capillary tube [8, 9]. Typically, a glass capillary is used (1–2 mm ID) with the synchrotron light directed into one end of the tube as shown in Fig. 3. Because of the grazing angle of incidence, the radiation is almost totally reflected by the inside surface of the glass. The tube acts as a light pipe and can even sustain a small amount of bending to guide the light into the interaction region. Microchannel arrays can also be used if space is limited [10] Again, either of these can be combined in series to achieve very high isolation.

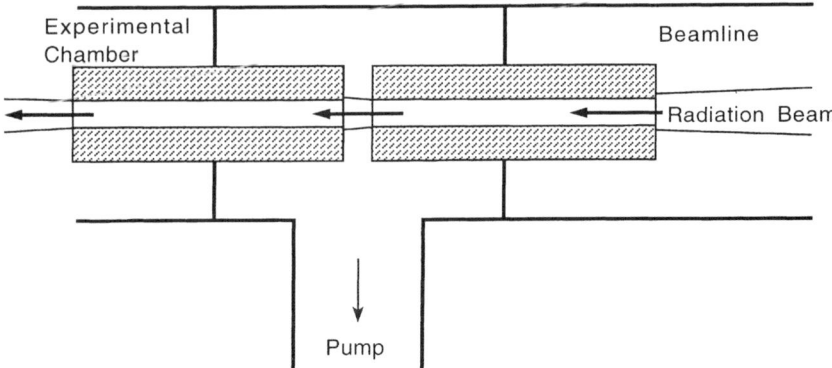

FIG. 3. Schematic of a differential pumping system using glass capillary light pipes.

The advantage of the capillary arrangement over the apertures lies in their low conductance. The conductance in l/s for air at 25°C for a capillary is much lower, $12d^3/l$ compared with $9d^2$ for the aperture, where d is the diameter and l is the length, both in centimeters.

The same type of differential pumping arrangement can be used for laboratory discharge lamp sources where the gas from the discharge must be isolated from the high vacuum of the experiment [8, 9].

The design of beamline vacuum chambers is identical to that of UHV chambers with one notable exception. Most optical elements cannot withstand the 200°C bakeout. For instance, gold is commonly used as an optical coating and, although it remains a film at room temperature, it starts to form micro-balls at temperatures above 100°C. Similarly, epoxies used to make replica gratings start to outgas and decompose around the same temperature. This then limits the bakeout temperature of these systems to 80°C. The procedure to obtain UHV in these systems is to bakeout at 200°C with the optical elements removed. When the system has been shown to achieve the desired ultimate vacuum, the optics are installed and an 80°C bakeout extended for several days is performed.

Vacuum integrity of the beamline and synchrotron storage ring dictates that only ion and TSP pumps be used. Hydrocarbon lubricated pumps are prohibited, and roughing is achieved with well-maintained turbopumps with trapped forepumps.

9.3 Leak Detection

One of the most time consuming chores in setting up a vacuum-based experiment is isolating vacuum problems. Typically, the system will not achieve its design's ultimate vacuum either when first commissioned or after changing instrumentation or making modifications. Marginally operating pumps, leaks, contamination, or improper bakeout are the common causes of vacuum problems with vacuum leaks being the most common. Although information from the pump manufacturer usually contains directions to verify pump performance and locate malfunctions, experimenters are mostly on their own in dealing with leaks, contamination, and bakeout problems. O'Hanlon [1] outlines several procedures for dealing with these difficulties. An RGA is the best tool for distinguishing between a leak, trapped gas (virtual leak), or improper bakeout. A helium leak detector can be used if the latter two possibilities have been eliminated.

Either of these two instruments represents a considerable investment and are not often readily available. There are several alternate techniques available that use the vacuum gauges that should already be installed on any vacuum chamber. Leaks in the 10^2 torr region are often large enough that they can be heard if the roughing pump and other equipment is silenced. Leaks in this range are usually caused by a missing or misaligned gasket or a partially opened vent valve. In the 10^{-3}–10^2 torr range, a thermal conductance (thermocouple) gauge can be used with acetone as a probe. The acetone is squirted onto the suspected areas and the pressure indication on the gauge monitored. When acetone enters the chamber through a leak, its expansion will cause it to freeze, restricting the flow rate through the leak and the indicated pressure will decrease. In the 10^{-8}–10^{-4} torr range, an ion gauge can be used with helium as a probe gas. An ion gauge is 5 times less sensitive to helium than to air, so helium flowing through a leak in place of air, will cause a drop in the indicated pressure. In a system that has achieved its expected base vacuum previously, one should suspect the flange seals, perhaps a scratched knife edge or a fiber or hair across a sealing surface.

Below 10^{-8} torr, one needs to rely on an RGA or helium leak detector. The RGA is preferred since it gives an indication of the composition of the residual gases thereby distinguishing between leaks (N_2, O_2, and a small amount of H_2O in the spectra) and inadequate bakeout (H_2O and hydrocarbon peaks above mass 44 in the spectra). In a well-baked ion TSP pumped system, the most intense mass peak should be CO at mass 28. Virtual leaks present their own problems. These are areas where gas or solvents can be trapped requiring very long times to pump out. Blind tapped holes where the cross sectional area between the screw thread and the tapped wall is small are the most common. This is prevented by filing a groove on the screw threads, drilling a pumpout hole through the screw, or drilling a hole through the tapped piece to the trapped volume. Gas can also be trapped between two surfaces bolted together.

Although flange seals should be the first culprit suspected as a leak source, others include electrical feedthroughs and brazed or welded joints. For a new system or new flange, the metal itself should be checked for porosity as a last resort. Experience and patience are the best resources in tracking down vacuum problems.

9.4 Optics Cleaning

With major investments in evermore complicated beamline optics, any procedures that prolong their useful life are most welcome. The major problem with the optics is their exposure to high intensity, high energy radiation for prolong periods of time. The result is the formation of a carbon coating that reduces the reflectivity and therefore the intensity delivered to an experiment in the 38–45 Å spectral region. Often, the intensity will be reduced by 5 orders of magnitude rendering the beamline useless in this range negating any investigations involving the carbon K-edge. The carbon coating also roughens the surface, increasing scattered light.

Several techniques have been developed to remove the carbon contamination either *in situ* or with the optics removed. The methods use some form of active oxygen to reduce the carbon to form CO or CO_2, which then desorbs from the surface and is removed from the system leaving behind a clean surface whose reflectivity sometimes exceeds the original. *In situ* cleaning techniques have the advantage that the optics do not have to be disturbed so no realignment of the beamline is necessary. If proper procedures are followed, it may not be necessary to bakeout the beamline afterward. Removing the optics, however, gives greater access to the optical surface, which can be easily inspected visually allowing for more control of the cleaning parameters.

Both methods described here produce clean surfaces as reflective as the original and minimize damage to the optical surface and metal reflective coating. The first [11] uses an rf glow discharge in a cavity attached to the optics vacuum chamber. The cavity is located with the shortest possible path of the activated oxygen to the optical surface without a direct line-of-sight. This maximizes the cleaning rate while minimizing sputtering of the surface or depositing sputtered material on the surface. A mixture of oxygen and 2% water vapor is introduced into the cavity at a pressure of 0.1–0.5 torr. This partial pressure of water is low enough that the chamber needs to have only a mild bakeout after the procedure and yet high enough to enhance the cleaning rate by a factor of 40. At a power of 100 W, 13.5 MHz rf is capacitively coupled to the cavity. The cleaning time, which depends on the geometry, is typically 16 hours. The same type of procedure can be performed with the optics removed to a special cleaning chamber [12].

The second method [13] uses an ultraviolet lamp in air to generate ozone and to promote the reaction of ozone with the carbon on the surface. The lamp, or lamp array, is placed within 1 centimeter of the surface to be cleaned and simply turned on. Clean-up occurs in 24 hours. The method is inherently simple and causes minimum damage, and there is no danger of sputtering the surface. The method can be used *in situ* if there is enough room in front of the optical surface to mount the lamps. Otherwise, the optics can be removed and the procedure performed on the bench. The reaction can be speeded up by exposing the surface to a mixture of 5% ozone in oxygen while the lamp is running [14]. The resulting clean-up time can be reduced to minutes. The drawback to any procedure using ozone, whether generated by the lamp or deliberately introduced, is its disposal. Ozone is an irritating gas and should be removed with an approved ventilation system.

These cleaning procedures should be used sparingly. Hydrocarbon material is 100 times more susceptible to damage from the discharge or ozone than the carbon film [15]. This is a problem for replica gratings where the surface is epoxy-coated with a thin reflective metal. As long as the metal film is intact, the epoxy will be protected but small pinholes might be enlarged [16].

Although there are dangers in reactive oxygen cleaning, the benefits far outweigh the drawbacks since the alternative is to replace or recoat the optics. The cleaning should be done at judicious intervals to minimize possible damage to the optics. Another option is to operate the beamline at the best possible vacuum with minimum hydrocarbon contamination. Flashing the TSP pumps on a daily basis after shutting off the light for the night will maintain the UHV and reduce the carbon buildup.

References

1. J. F. O'Hanlon, *A User's Guide to Vacuum Technology*. John Wiley & Sons, New York, 1989.
2. F. Rosebury, *Handbook of Electron Tube and Vacuum Techniques*. Addison-Wesley, Reading, MA, 1965. Reprinted by the American Institute of Physics, Woodbury, NY, 1995.
3. W. H. Kohl, *Handbook of Materials and Techniques for Vacuum Devices*. Van Nostrand Reinhold, New York, NY, 1967. Reprinted by the American Institute of Physics, Woodbury, NY, 1995.
4. Gas phase experiments using radiation between 1050 and 1800 Å typically use gas cells since VUV transmitting windows are available in this range.
5. E. Poliakoff, M.-H. Ho, G. E. Leroi, and M. G. White, *J. Chem. Phys.* **84**, 4779–4795 (1986).
6. E. Poliakoff, Private communication.
7. R. G. Musket, W. McLean, C. A. Colmenares, D. M. Makowiecki, and W. J. Siekhaus, *App. Surf. Sci.* **10**, 143–207 (1982).
8. J. L. Dehmer and J. Berkowitz, *Phys. Rev.* **A10**, 484–490 (1974).
9. J. L. Dehmer and D. Dill, *Phys. Rev.* **A18**, 164–171 (1978).
10. T. B. Lucatorto, T. J. McIlrath, and J. R. Roberts, *Appl. Opt.* **18**, 2505-2509 (1979).
11. E. D. Johnson, S. L. Hulbert, R. F. Garrett, G. P. Williams, and M. L. Knotek, *Rev. Sci. Instrum.* **58**, 1042–1045 (1987).
12. W. R. McKinney and P. Z. Takacs, *Nucl. Instrum. Meth.* **195**, 371–374 (1982).
13. R. W. C. Hansen, J. Wolske, D. Wallace, and M. Bissen, *Nucl. Instrum. Meth.* **A319**, 249–253 (1994).
14. T. Harada, S. Yamaguchi, M. Itou, S. Mitani, H. Maezawa, A. Mikuni, W. Okamoto, and H. Yamaoka, *Appl. Opt.* **30**, 1165–1168 (1991).

15. R. A. Rosenberg, J. A. Smith, and D. J. Wallace, *Rev. Sci. Instrum.* **63**, 1486–1489 (1992).
16. R. W. C. Hansen, J. Wolske, and P. Takacs, *Nucl. Instrum. Meth.* **A319**, 254–257 (1994).

10. LITHOGRAPHY

Yuli Vladimirsky

Center for X-ray Lithography
University of Wisconsin—Madison

The extraordinary progress and success of optical lithography using ultraviolet (UV) light in development and production of microelectronic devices and integrated circuits (IC) are truly phenomenal. Lithography, the printing of patterns on surfaces, is both a driving force and a bottleneck in semiconductor manufacturing. Optical lithography provided IC manufacturing with the unique combination of high-volume production, high precision, and low-cost fabrication. From 1960, when the monolithic IC was invented by Jack Kilby and Robert Noyce [1] and a computer chip contained only one memory bit, to 1997, the memory on a single chip increased by more than eight orders of magnitude [1–4], reaching 256 Mbit. Figure 1 presents the trend of memory chip nomination and minimum feature size (device gate length) from 1960 to 2012. The points shown for 1997 and the following years are a projection [4] made by the Semiconductor Industry Association (SIA). During those 35 years, the minimum feature size decreased by almost two orders of magnitude, from 24 μm to 0.35 μm. It will reach 50 nm by the year 2012, bringing the memory of a single chip to 256 Gbit. The realization of this prognostication depends on timely introduction of advanced lithography techniques, and particularly, vacuum ultraviolet (VUV) and soft x-ray light photolithography [5].

10.1 Integrated Circuit Fabrication and Lithographic Process

Semiconductor IC fabrication is based on the wafer (usually silicon wafer) fabrication process, involving hundreds of processing steps and including several photolithographic cycles. A simplified processing sequence [1, 2] consists of (1) deposition of some form of a thin film on a wafer (oxide, polysilicon, metal); (2) mask image transfer in the resist using photolithography; (3) etching the developed pattern into the deposited film, (4) some other form of wafer processing such as ion implantation, dopant diffusion, or substrate etching, and (5) removal of the previous resist layer and preparation for the next sequence. Depending on the complexity of the device being built, the entire processing sequence can be repeated 10 to 30 times to form all necessary layers [2]. The fabrication process sequence for a bipolar P-N diode junction requires a

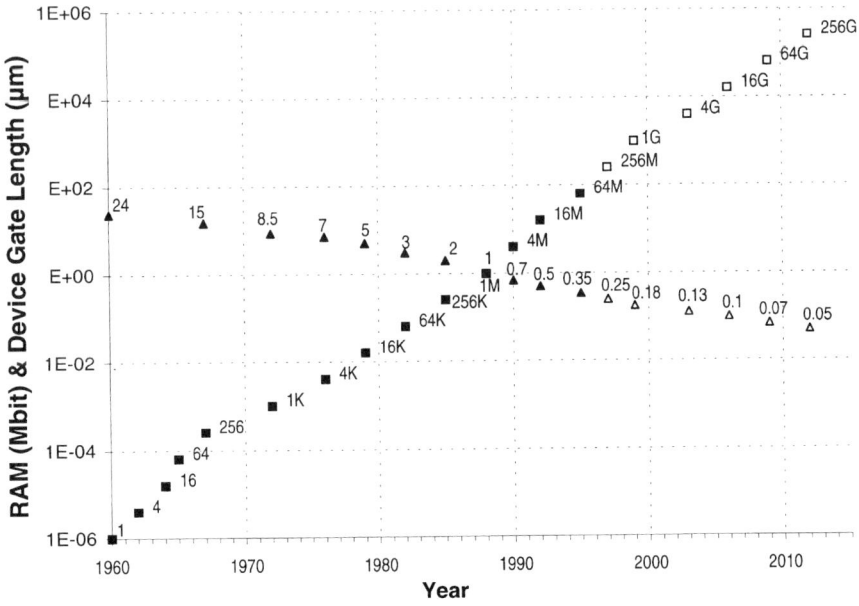

FIG. 1. Chronology of random access memory (RAM) nomination (■) and device gate length (▲) for computer chips. The "hollow" symbols (□, △) represent SIA prediction for the semiconductor industry.

minimum of two lithographic cycles, and a simple P-channel metal oxide semiconductor field effect transistor (MOSFET, PMOS) is based on four to five lithographic levels [6]. Today, fabrication of a computer chip requires more than 25 thoroughly aligned lithographic levels [1]. The purpose of the lithographic process is to pattern a thin film (e.g., oxide, polysilicon, metal, etc.) to be used as an insulator in a device, a stop layer for ion implantation, a conductor, or a wire. A three-step photolithographic process is schematically presented in Fig. 2. The first step is generation of a latent image in the resist by an exposure, the second step is relief image formation in the resist by development, and during the third step the resist image is transferred into the underlying film or semiconductor surface using a wet or dry etching process.

10.2 Photoresist in Lithography

The photolithography used in microcircuit fabrication sequences, also called lithography and microlithography, is an image formation and transfer process, which evolved from that used in the printing industry for about 200 years. The word *lithography* actually means "writing on or with a stone." In 1796, the

German playwright Alois Senefelder invented a planographic printing process later called lithography [7]. He found that an image created with a greasy substance on a surface of a water-absorbent stone, such as a limestone slab, could be printed onto paper using ink with high fidelity. In the early nineteenth century, a photoengraving printing technique was established. This technique uses a photosensitive material, called a resist, to form images on a metal plate. After exposure to light through a film negative, the resist becomes hard and insoluble where light hits it, while the area, protected from the light, can be washed away (negative resist), and exposed bare metal can be etched, creating a relief in the imaged area (see Fig. 2). The resist materials devised and developed for the printing industry have proven to be useful in manufacturing semiconductor devices, and of course, they have been undergoing serious development to meet the growing demands of the semiconductor industry in terms of resolution, sensitivity, adhesion to the semiconductor materials, and resistance to the aggressive processes [8]. Two types of photosensitive resist, negative and positive, are shown in Fig. 2. The resists used in photoengraving were negative, and the microlithography also started with negative resists. In contrast to a negative resist, a positive resist becomes soluble (soften) in the exposed areas

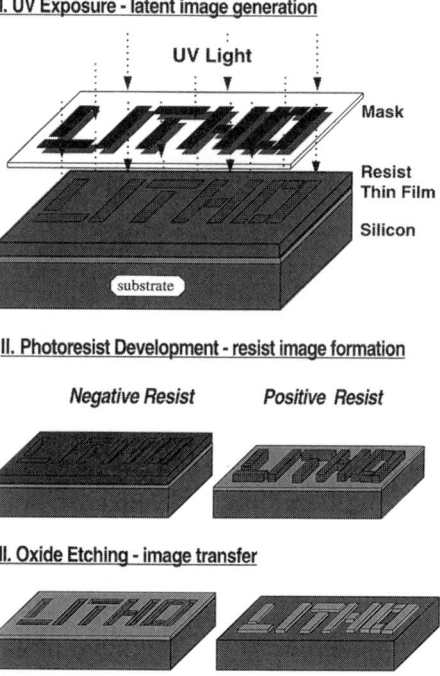

FIG. 2. Sketch of a lithographic process.

and is removed during development. Both negative and positive photoresists are photosensitive polymeric systems and can be composed of one or two major components [8]. The one-component resists have intrinsic light sensitivity. In a two-component resist, an inert matrix resin serving as a binder and film-forming material is sensitized by a compound, mostly monomeric in nature, which undergoes radiochemical reactions responsible for the change in solubility of the resin. The photosensitivity ranges from a few tens to several hundreds mJ/cm^2. The resist mixture is suspended in an appropriate solvent that keeps the resist liquid for ease of application to the semiconductor surfaces. To this solution, a few additives are added to control and modify the chemical and spectral characteristics of the photoresist. A classical one-component positive resist is Poly(methyl methacrylate) (PMMA). PMMA has been serving the lithographic community for more than for 30 years as UV, electron-beam, deep-UV, and x-ray resist [5, 6, 8, 9]. A well-known one-component negative resist is a copolymer of glycidyl methacrylate and ethyl acrylate (COP). The phenolic resin (novolak) is widely used as a matrix for many positive and negative two-component resists. The novolak readily dissolves in relatively low concentration alkaline solution. A diazonaphthoquinone photoactive compound (PAC), added to novolak, acts as an inhibitor. During irradiation this PAC is converted into a base soluble compound and is removed together with the novolak matrix during development. These are so-called AZ-type resists (AZ1300, AZ1400, etc.). The novolak-based resists have proven to be excellent materials and have been successfully used for manufacturing many generations of computer chips [6, 8, 9]. A new class of resists, resists with *chemical amplification*, was specifically developed for deep UV [10] and found application in VUV and soft x-rays [5]. These resists are based on acid-catalyzed chemical reactions [8, 10]. An acid labile group, attached to the main polymer, would react with the photogenerated acid, deprotect the polymer, and generate a new molecule of acid, beginning a catalytic process. In the course of this process, several hundreds of deprotected sites and molecules of acid will be produced by a single photon. The corresponding resist sensitivity could be many times higher compared with that of a conventional resist. An excellent novolak-based chemically amplified positive resist APEX-E was developed at IBM and is used for deep UV and x-ray lithography. The SAL-605 resist, produced by Shipley, is an example of a negative chemically amplified resist, suitable for x-ray lithography.

10.3 Optical Lithography

Two major photolithographic schemes, used for forming an image in a photoresist, are proximity printing and projection printing. Both techniques use a mask with a pattern consisting of transparent and opaque features. Originally, glass plates with photographic images were used. Today, as an optical mask substrate a high-quality fused quartz plate is used, and the opaque pattern is formed by a ~70 nm-thick chromium film, so called binary chrome-on-quartz mask [11].

The proximity printing is based on casting a shadow of the mask pattern, as depicted in the upper part of Fig. 2. As a light source for the optical lithography, a high-pressure mercury (Hg-arc) lamp is used, a high-intensity source with many bright spectral lines: $\lambda \approx 436$ nm (G-line), $\lambda \approx 404$ nm (H-line), $\lambda \approx 365$ nm (I-line) [11, 12]. The illumination system includes condenser optics for uniform exposure and interference filters to select desired wavelength. The mask and a wafer are positioned in close proximity with a small, accurately controlled gap (~25 µm or less), or even brought into direct contact, as it was used in the 1960s [1]. To decrease the number of defects, produced because of the mask/wafer contact, the proximity printing replaced the contact scheme in the late 1960s. It was popular in the 1970s, and still is used today in noncritical lithography levels and laboratory experiments.

The image formation in proximity printing is described by Fresnel diffraction [5, 12, 13]. The relation between object size w, wavelength λ, and distance G from the object to the "image" plane, where the shadow of the object is cast, can be formulated in terms of number of Fresnel zones [13] (FZ)

$$FZ = \frac{w^2}{\lambda G} \qquad (1)$$

when G is much longer than the wavelength λ ($G \gg \lambda$).

It requires at least two Fresnel zones ($FZ = 2$) to reproduce a simple structure (a thin line or a small square), and the resolution w_{min} of this technique is usually expressed as

$$w_{min} = k\sqrt{\lambda G} \qquad (2)$$

with $k = \sqrt{2}$ as a "theoretical" value. The actual resolution of the lithography depends also on the resist process and few other parameters, and factor k can be smaller [12, 14], especially, in the proximity x-ray lithography ($\lambda \approx 7$–12 Å)—the leading contender for sub-0.25 µm lithography [5]. An example of a pattern produced by two-level optical proximity printing and gold electroplating is presented in Fig. 3. The pattern of a gas discharge device [15] was printed with $\lambda \approx 365$ nm (I-line) and the gap of $G = 5$ µm in a 10-µm thick resist, what determines the "theoretical" resolution of $w_{min} \approx 3.5$ µm [Eq. (2)]. A close-up view of an electrode pin demonstrates a fine periodic structure. This periodic structure corresponds to a standing wave produced by the interference of the light reflected from the bottom and the top of the resist layer, revealed after development and faithfully reproduced in electroplated gold.

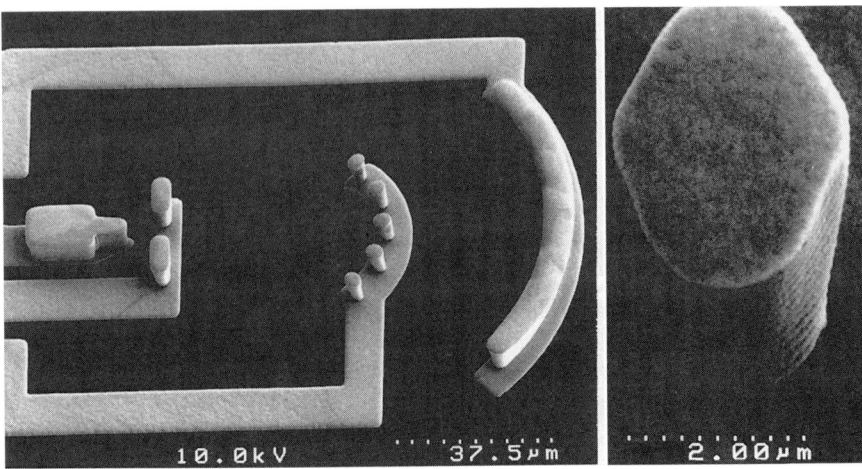

Fig. 3. A gas discharge device produced by two-level optical lithographic process.

In the late 1970s, the optical proximity printing was replaced by the projection printing technique [12]. The refractive elements of a lens for the projection printing system require a narrow bandwidth to avoid chromatic aberration, and selection of specific wavelength is important because of high optical dispersion of materials in this region. As in a proximity scheme, the light source for the projection optical lithography is a high-pressure Hg-arc lamp ($\lambda \approx 436$ nm, 365 nm). More recently, lines $\lambda \approx 266$ nm and $\lambda \approx 248$ nm of the lamp were used to improve the resolution of the imaging [16]. In the deep UV wavelength region, a new type of photon sources can be used. These are very efficient and powerful pulsed ultraviolet lasers—the excimer lasers [16, 17]. Experiments have been performed with the XeCl ($\lambda = 308$ nm) laser [17, 18]. Lithography exposure systems were built using the KrF ($\lambda = 248.4$ nm) laser [19, 20] and will be used for production of 256 Mbit dynamic random access memory (DRAM) chips with 0.25 μm minimum feature size [11]. To improve the resolution further, an ArF ($\lambda = 193$ nm) laser is proposed to replace the KrF laser [11]. An excimer laser produces a beam with relatively low spatial (few milliradians divergence) coherence and a wide bandwidth $\Delta \lambda \approx 1$–1.5 nm. These properties could be beneficial in reducing the speckle problem associated with conventional lasers. However, optical material limitations, such as strong chromaticity, could require significant narrowing [21] of the bandwidth to 0.004 nm, especially for the 193 nm ArF laser.

A lithographic projection system incorporates such subsystems as a light source, a condenser lens, a heat removing filter, beam-orienting mirrors, a shutter, a reduction projection lens between a mask and a wafer, an automatic mask, and a wafer handler, which includes a high-precision alignment system

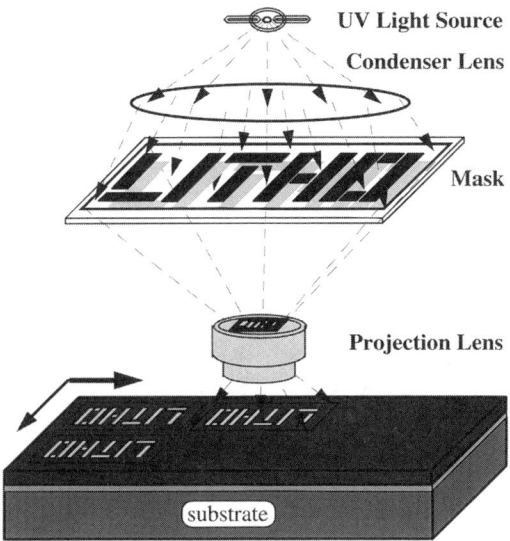

FIG. 4. An optical projection imaging scheme.

[11, 14]. The image formation in a projection printing system and its major elements are shown schematically in Fig. 4. The mask is illuminated from the back, or *transilluminated*, using a condenser lens, and the light that has passed through the mask is then focused onto the resist surface in the image plane of a microlithographic projection lens, positioned within a few millimeters above the wafer surface. The reduction factor of a projection lens used today is 5× and 4×, but from 1977 through 1984, different lithography tools had reduction factors of 1×, 10×, and even 20× [11, 12, 14]. To cover a whole wafer, several types of scanning and stepping schemes have been considered or used [1, 11, 12, 14].

The function of the projection lens is similar to that of a 35-mm photographic camera, yet it is much more complicated because it is providing a practically aberration-free diffraction-limited performance for a relatively large field size, significantly lower distortion of the image, and much smaller field curvature. The theoretical resolution of a projection system (aerial image) using transillumination is given by a relation [13]:

$$w_{\min} = \frac{k_1(\sigma)\lambda}{NA_{PL}} \qquad (3)$$

where $w_{\min}$ is the minimum feature size, NA_{PL} numerical aperture of a projection lens, and k_1 a factor depending on spatial coherence. The partial coherence σ is defined as a ratio of numerical aperture of a condenser lens NA_{CL} to that of the projection lens

$$\sigma = \frac{NA_{CL}}{NA_{PL}} \qquad (4)$$

Depending on the coherence of illumination the k_1 factor in Eq. (3) ranges from 0.82 for coherent illumination ($\sigma = 0$) to 0.61 for incoherent ($\sigma = \infty$). It has a minimum of $k_1 = 0.56$ for an optimal partial coherence ($\sigma = 1.5$) of a projector [13]. In practice, the partial coherence is $\sigma \approx 0.6$–0.9, and the actual resolution of the lithography depends also on the resist process and a few other lithographic conditions [11, 12, 22]. Consequently, values of k_1 typically range from 0.8 for conventional to 0.6 for complex resist processes [12]. In an IC fabrication process, the features are mostly of relatively simple shapes like thin lines, little squares, short gratings, and so on. To reconstruct such images, it could be sufficient to use the first diffraction orders together with the zeroth order, which requires an off-axis illumination. It is equivalent to annular illumination with a central obstructed area of a relative radius $\alpha = 0.5$, and in terms of resolution for the incoherent case, can be translated as $k_1 = 0.5$ [13]. In an asymptotic case representing $\alpha = 1.0$ and $\sigma = \infty$, the resolution is described by a corresponding factor $k_1 = 0.39$ [13]. In the combination of advanced off-axis illumination and resist processing [11] the best values of $k_1 = 0.45$–0.5 are expected to be achieved for 0.25 and 0.18 μm lithography (see Table I).

The depth of focus is another important characteristic of an imaging lens. A small amount of defect of focus introduced in excess of $2\lambda/\pi(NA)^2$ causes a very rapid deterioration of the image in terms of high spatial frequencies [13], and for incoherent illumination the total depth of focus is

$$\Delta z = \frac{k_2 \lambda}{(NA_{PL})^2} \qquad (5)$$

and the constant $k_2 = 4/\pi \approx 1.27$. The Rayleigh limit for depth of focus is $k_2 = 1$ [11, 12]. Similar to the resolution factor k_1, the depth-of-focus factor k_2 depends on the realistic lithographic condition, and in practice it is ~1.5–2 times smaller, but approaching the theoretical limit for the advanced lithographic lenses. [11]. Table I presents theoretical and demonstrated values of parameters for lithographic reduction lenses, including those under development [2, 3, 11, 12, 14]. To provide high resolution, the projection lens will have a large numerical aperture and be free of aberrations. These requirements are achieved by more and more complex lenses and increasing the number of elements: In 1965 a projection lens had 9 elements [11]; in 1978, the number of elements was 11 [14]; in 1985, 15 [12]. The modern microlithographic projection lens with an $NA \approx 0.4$–0.6 is a long (up to 1000 mm) cylindrical metal tube

TABLE I. Theoretical and Achieved Parameters of Reduction Lenses Developed for IC Fabrication

Numerical aperture	Resolution, μm		Depth of focus, μm		Year
	Theoretical $k_1 = 0.56$	Achieved	Theoretical $k_2 = 1$	Achieved	
$\lambda = 436$ nm (g-line*), reduction 5×					
0.20	1.22	1.74	11		1977
0.30	0.80	1.16	4.8		1982
0.35	0.70	1.00	3.6	1.5	1984
0.43	0.57	0.81	2.4	1.2	1988
0.54	0.45	0.60	1.5		1989
$\lambda = 365$ nm (i-line*), reduction 5×					
0.35	0.58	0.70	3.0		1986
0.40	0.51	0.65	2.3		1988
0.45	0.45	0.60	1.80		1989
0.48	0.43	0.46	1.58	1.0	1996
0.60	0.34	0.35	1.01		1990
$\lambda = 248$ nm (KrF excimer laser), reduction 4×					
0.50	0.28	0.30	1.0	0.8	1995
0.57	0.24	0.25	0.76		1996†
0.60	0.23	0.25	0.69		1997†
0.70	0.20		0.51		?†
$\lambda = 193$ nm (AeF excimer laser), reduction 4×					
0.50	0.22		0.77		?†
0.60	0.18		0.54		?†

* Hg-arc lamp; † under development.

containing ∼ 25–35 glass optical elements made of at least 10 different types of glass. It could weigh more than 300 kg [11] and be extremely expensive (≥$1 million) because of the limited availability of deep UV transparent materials and the high cost of optical-quality glass. A reduction lens using a combination of refractive and reflective optical elements (catadioptric optic) is finding its application in production in 0.35 μm lithography [11].

Today, when the field size is of the order of a few cm^2, it is not possible to project the entire image of an IC pattern on a 6–9-inch mask in a single exposure. The most successful approach, called step-scan, combines stepping from field to field and scanning a well-corrected narrow annular zone within the field [11, 12]. The stepper built for production of 64 Mbit DRAM chips uses i-line ($\lambda \approx 365$ nm) of the Hg-arc UV lamp and is equipped with a 5× reduction lens with $NA = 0.6$. This system complemented with advanced resist processing is able to provide 0.35 μm resolution [11]. The advanced resist processing includes tight temperature, humidity, environmental control, fully automated resist application, post-application and post-exposure bake, and use of anti-reflective films in the resist stack to suppress an undesirable standing wave (see Fig. 3). Modern state-of-the-art projection systems, being designed and built for the 0.25 μm resolution required for 256 Mbit, will use wavelength $\lambda \approx 248$ nm (KrF laser); advanced illumination techniques; enhanced, phase-shifting and attenuated (partially transmitting) optical masks designed to correct the wavefront and provide improved resolution and depth of focus [11]. This enhancement is achieved by adding subresolution phase-shifting features at edges of the main pattern, or/and making the opaque features partially transmitting. For the 1 Gbit memory chip generation with required resolution 0.18 μm (see Fig. 1) the ArF ($\lambda = 193$ nm) laser should be used [11]. Beyond this, the future of optical lithography is very uncertain, and shorter wavelength will have to be used to realize higher resolution. Some expectations have been raised with respect to the potential of UV lasers with $\lambda = 157$ nm (F_2 excimer) and $\lambda = 126$ nm. But optical properties of materials in this wavelength region are very marginal even for $\lambda = 157$ nm, and development of all-reflective optics would be necessary [11].

10.4 X-Ray Lithography

Moving to the shorter wavelengths brings us to the VUV region, or more specifically—in the soft x-ray region. The application of soft x-ray lithography for microcircuit fabrication was proposed at IBM in 1969 by R. Feder to overcome diffraction limitations of the proximity optical lithography, and early work was performed together with E. Spiller [23, 24]. The replication of a high-resolution pattern using soft x-rays was independently suggested and demonstrated by D. Spears and H. Smith in 1972 using AlK_α radiation ($\lambda = 8.3$ Å) and a 3-μm thick Si membrane as x-ray mask substrate with gold absorber [25]. The major efforts have been directed to the proximity scheme (see Fig. 5).

The wavelengths suitable for x-ray lithography (XRL) are quite soft, so an appropriate combination of a relatively transmissive substrate and a reasonably thin, but sufficiently opaque absorber has to be achieved. This compromise is found by using 1–3 μm thick uniform silicon, silicon nitride, or silicon carbide membranes as a mask carrier and a 0.4–0.7 μm thick gold, tungsten, or tantalum absorber. The silicon wafer with the membrane is mounted on a rigid glass ring [5]. The contrast of x-ray masks is usually lower than that of binary optical masks. The partial transmission of a mask absorber (attenuated mask) is a

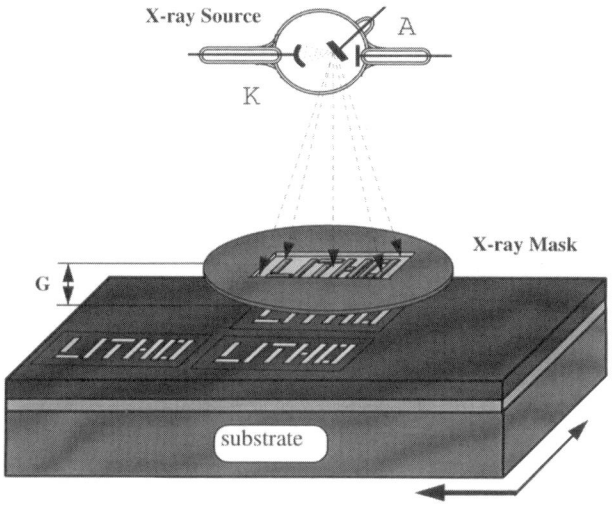

FIG. 5. An x-ray proximity imaging scheme.

desirable feature [5], and this feature was recently introduced in the optical mask design [11].

The resolution [Eq. (2)] of proximity lithography, optical and x-ray, is described by Fresnel diffraction. However, it is controlled not only by the diffraction phenomena but also by the lithographic conditions. One of those factors is penumbra blur associated with the finite size of source translated into a local divergence of the incident beam [5]. Similar to projection printing, the k-factor can vary (1.22–1.5) depending on a type of the feature to be imaged [5]. Limiting the penumbra blur to $\sim$20–40% of the feature size is actually beneficial [26], and imaging in these conditions corresponds to $k = 1$. Advanced resist processes and mask patterning techniques make it feasible to reduce this factor even further and reach values of $k = 0.7$–0.6 [5, 14]. Other important factors influencing the image formation and limiting the resolution are the photoelectron and Auger effects, which constitute the process of x-ray absorption. The electrons with energies of several hundred eV or a few keV are generated as a result. Those electrons are responsible for distributing the x-ray energy in the resist [27]. The optimal wavelength can be found by analyzing diffraction blur, which decreases with the wavelength, together with the photoelectron blur [5], which is more pronounced for shorter wavelength. As follows from this analysis, the optimal wavelength range, or x-ray "lithographic exposure window" (XRL window) is $\lambda = 7$–12 Å, which corresponds to 1.0–1.8 keV photon energies [5]. The patterning capabilities of XRL are sufficient to image features far below 0.1 μm and at a mask/wafer gap of $\sim$10 μm can facilitate line width of $w = 0.060$ μm for $k = 0.7$ and $w = 0.050$ μm for $k = 0.6$. The mask/wafer gap of 10 μm can be provided by advanced x-ray steppers.

An important requirement of an XRL system is a bright x-ray source. The early studies were performed in a relatively wide range of wavelength using conventional x-ray sources based on electron beam bombardment of some suitable target producing bremsstrahlung radiation with characteristic lines: $\lambda = 4.4$ Å (Pd L_α) [28, 29], $\lambda = 7.1$ Å (Si K_α) [28, 29], $\lambda = 8.3$ Å (Al K_α) [23, 25, 28, 29], $\lambda = 13.3$ Å (Cu L) [1, 28, 29], and $\lambda = 44.6$ Å (C K_α) [23, 24, 28, 29]. The conversion efficiency of the e-beam energy into x-rays is very small, and a 10–20 kW x-ray tube would deliver $\sim$1–2 W/srd or a few tenths of a mW/cm^2 to the exposure plane at 50–100 cm away from the source. This corresponds to a 15–30 minute long exposure for a single field [29], which is too long for microcircuit production. Point x-ray sources, based on confined hot plasma, offer higher x-ray conversion efficiency compared with that of x-ray tubes. A few mW/cm^2 power density can be delivered to the exposure plane, and the exposure time of a single field can be on the order of a few minutes. The hot plasma can be produced by focusing a high-energy short laser pulse on a target such as thin metal film or solid gas pellets (laser plasma), or by the collapse of an electrical discharge (dense plasma) [5]. The temperature of the plasma is high enough to ionize the material and produce characteristic and continuous radiation in the 8–20 Å wavelength region centered at 13–14 Å. These sources are still under development, and are not available commercially. Although the plasma sources are more efficient than the x-ray tubes, they do not deliver x-ray flux sufficient to provide acceptable throughput of the lithographic system [5].

Viable candidates as x-ray sources for XRL are synchrotrons and electron-storage rings [5, 29]. An electron beam with energies of several hundreds MeV to a few GeV is confined on an orbit of a "circular" accelerator. A highly collimated, in the vertical direction, electromagnetic radiation is emitted by the relativistic electrons tangentially when they are turning in the magnetic field of the "bending" magnets, keeping the electrons on an orbit. The spectrum of the synchrotron radiation is similar to that of black body radiation with corresponding temperature of several degrees MK. These machines were developed for high-energy and nuclear physics research, but found an extensive use in spectroscopy as a source of continuous radiation. The parameters of a synchrotron or a storage ring constructed for XRL are chosen to provide the spectrum with median energy (critical energy) of 1–1.8 keV to match the optimal wavelength range $\lambda = 7$–12 Å, or the XRL exposure window. There are companies manufacturing synchrotron rings suitable for XRL: Sumitomo Heavy Industries, Japan; and Oxford Instruments, England [5].

Both companies offer superconducting storage rings. In 1991, a compact ring Helios 1, manufactured by Oxford Instruments, was bought and installed at IBM to facilitate the XRL program. The compact ring Helios 2 was purchased by the National University of Singapore (NUS). A photograph of a Helios compact synchrotron ring is presented in Fig. 6, and its parameters in Table II. The power density of 500–800 mW/cm^2 in the exposure field for a synchrotron-based x-ray source is providing an exposure time in fractions of a second. This is expected from a lithographic exposure station for IC fabrication in a production environment in order to fulfill high throughput requirements. An ability to accommodate up to 20 exposure stations combined with high reliability makes this machine a truly practical x-ray source for the semiconductor industry.

An XRL proximity system based on synchrotron x-ray source is schematically shown in Fig. 7. A typical beamline includes a collimating grazing incidence mirror (1.5–2°), which collimates a horizontally diverging x-ray beam and delivers it through a ~25 μm thick Be window to the mask/wafer plane with only a few mrad of local divergence. A combination of a grazing incident mirror and a Be window is very suitable for forming the XRL exposure window with the optimal wavelength range of $\lambda = 7$–12 Å [5]. Utilizing an advanced process based on phase mask approach [30], very high resolution patterns can be produced. Figure 8 illustrates the use of a phase mask capable to image 65 nm lines in a 2.5 μm thick positive resist at a 30 μm mask/wafer gap. Penetrating ability of x-rays depends on the wavelength. This can be used to produce patterns with high aspect ratio (height-to-width) features, limited mostly by the

FIG. 6. A Helios superconducting synchrotron ring. (Courtesy of Oxford Instruments.)

TABLE II. Parameters of Helios Synchrotron (Oxford Instruments)

Beam energy	0.7 GeV
Magnetic field	4.5 T
Radius	0.519 m
Average current	300–500 mA
Beam lifetime (400 mA)	>28 h
Injection energy	100–200 MeV
Source size	
Radial	0.7 mm
Vertical	0.5–0.7 mm
Source divergence	
Radial	3.0 mrad
Vertical	0.8 mrad
Total power	14–18 kW
Power in XRL window	
Total	5–8 kW
Beamline (30 mrad)	14–18 W
Exposure	0.5–0.7 W/cm^2
Number of port	16–22
Uptime	>95%

mechanical stability of the resist material. A pattern of 0.75 μm lines formed in 15 μm thick PMMA resist is shown in Fig. 9. This pattern with an aspect ratio of ~20:1 was obtained using $\lambda \approx$ 6–10 Å wavelength range from a synchrotron storage ring. Another example of a high aspect ratio pattern [15] produced in

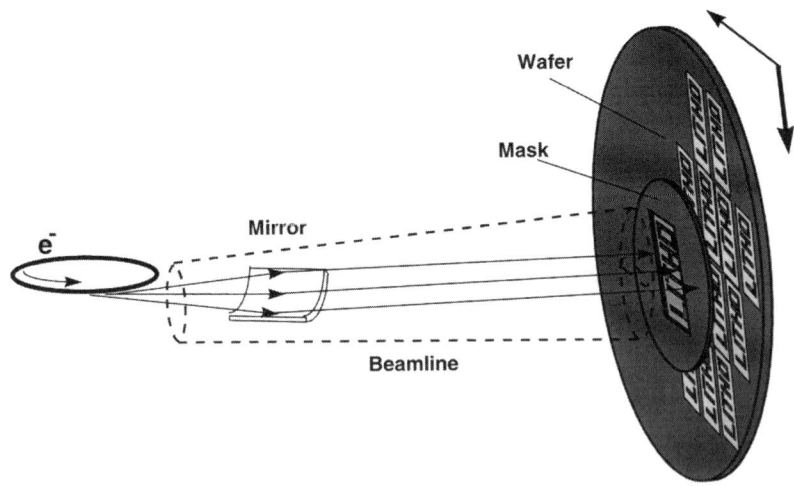

FIG. 7. An x-ray lithography system with the synchrotron x-ray source.

FIG. 8. A high-resolution (65 nm lines) resist pattern obtained by using x-ray phase mask. (Courtesy of Zheng Chen, CXrL, University of Wisconsin—Madison.)

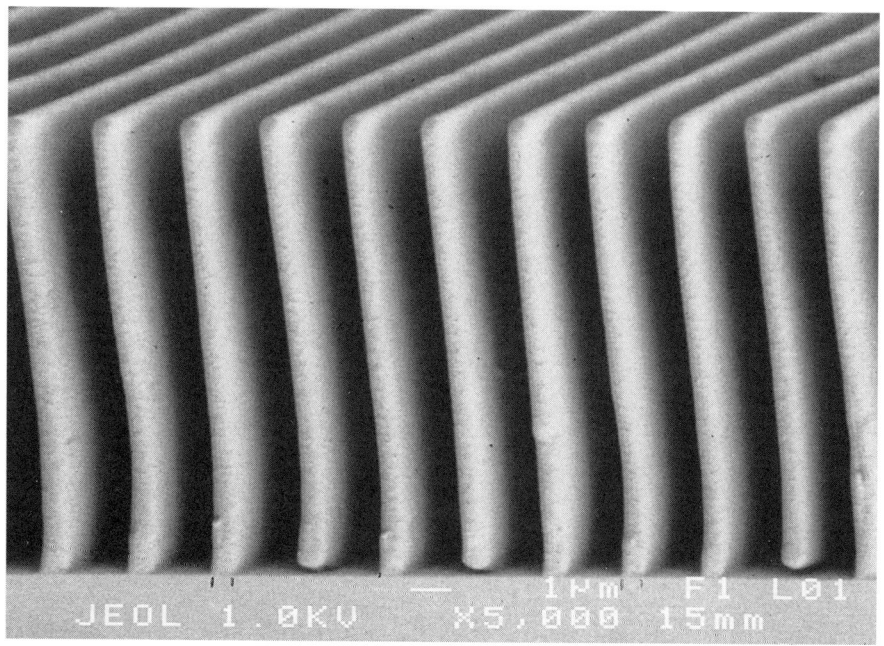

FIG. 9. A 20:1 aspect ratio pattern (1.5 μm period) formed in a 15 μm thick PMMA. (Courtesy of Zheng Chen, CXrL, University of Wisconsin—Madison.)

500 μm thick PMMA with 10 μm wide lines, revealing an aspect ratio of ~50:1 is presented in Fig. 10. The pattern was exposed in a $\lambda \approx 3.5$–7 Å wavelength range. Both patterns presented in Figs. 9 and 10 have been obtained in a contact printing scheme.

Following the path of optical lithography, one can propose a reduction projection scheme [12, 31]. This scheme is based on a soft x-ray microscope system with multilayer coated Schwarzschild optics [31] (see Fig. 11). The wavelength for soft x-ray (also called extreme UV) projection lithography was chosen in the region of $\lambda \approx 130$–140 Å, where reliable reflective multilayer optical elements with good efficiency can be produced [31]. In a simplistic scheme presented in Fig. 11, only four symmetrical reflective surfaces are shown. However, to meet requirements of high resolution and a relatively large projection field, seven or more curved reflective surfaces and off-axis imaging could be necessary [12, 31]. These optical elements would have to be produced with extremely small figure error, which is very difficult to achieve in this wavelength region. Another serious problem to consider is patterning a mask with a curved surface. High absorption in the resist will limit its thickness below

FIG. 10. "Plexiglas boxes"—a 50:1 aspect ratio pattern imaged in 500 μm thick PMMA.

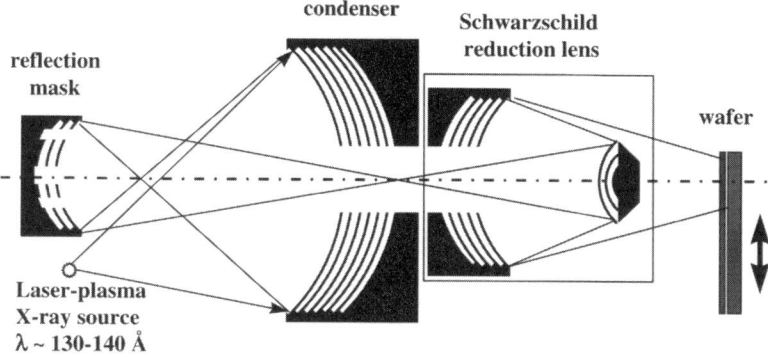

FIG. 11. The principle of soft x-ray projection lithography.

100 nm and additional processing steps would have to be used to produce thicker patterns [31]. The development of soft x-ray projection lithography is a technically exciting and extremely challenging program. As it stands now, the proximity x-ray lithography is the main contender for the sub-0.25 µm lithographies. The proximity XRL offers a combination of characteristics satisfying the manufacturing requirements: high resolution, increased process latitude, large "depth-of-focus," and high throughput [5].

References

1. L. F. Thompson, "An Introduction to Lithography," in *Introduction to Microlithography*, Second Edition, L. F. Thompson, C. G. Wilson, and M. J. Bowden (eds.), pp. 1–17, American Chemical Society, Washington, DC (1994).
2. F. W. Voltmer, "Manufacturing Process Technology for MOS VLSI," in *VLSI Electronics, Microstructure Science*, Vol. 1, N. G. Einspruch (ed.), pp. 2–40, Academic Press (1981).
3. J. L. Prince, "VLSI Device Fundamentals," in *Very Large Scale Integration (VLSI) Fundamentals and Applications*, D. F. Barbe (ed.), pp. 4–41, Springer-Verlag (1982).
4. *The National Technology Roadmap for Semiconductors*, Semiconductor Industry Association (SIA) publication, p. 14 (1997).
5. F. Cerrina, "X-Ray Lithography," in *Handbook of Microlithography, Micromachining, and Microfabrication*, Vol. 1, P. Rai-Choudhury (ed.), pp. 253–319, SPIE Press, Bellingham, Washington (1997).
6. A. N. Broers and T. H. P. Chang, "High Resolution Lithography for Microcircuits," in *Microcircuit Engineering*, H. Ahmed and W. C. Nixon (eds.), pp. 1–74, Cambridge University Press (1980).
7. *The Encyclopedia Americana*, Vol. 17, p. 585, Groiler Incorporated, Danbury, Connecticut (1985).

8. C. G. Wilson, "Organic Resist Materials," in *Introduction to Microlithography*, Second Edition, L. F. Thompson, C. G. Wilson, and M. J. Bowden (eds.), pp. 139–267, American Chemical Society, Washington, DC (1994).
9. J. G. Lane, "Resists for storage ring x-ray lithography," *Procc. SPIE*, **448**, 119–129 (1983).
10. R. D. Allen, W. E. Conley, and R. R. Kunz, "Deep-UV Resist Technology: The Evolution of Materials and Processes for 250-nm Lithography and Beyond," in *Handbook of Microlithography, Micromachining, and Microfabrication*, Vol. 1, P. Rai-Choudhury (ed.), pp. 312–375, SPIE Press, Bellingham, Washington, USA (1997).
11. H. J. Levinson and W. H. Arnold, "Optical Lithography," in *Handbook of Microlithography, Micromachining, and Microfabrication*, Vol. 1, P. Rai-Choudhury (ed.), pp. 11–141, SPIE Press, Bellingham, Washington (1997).
12. M. J. Bouden, "The Lithographic Process: The Physics," in *Introduction to Microlithography*, Second Edition, L. F. Thompson, C. G. Wilson, and M. J. Bowden (eds.), pp. 20–138, American Chemical Society, Washington, DC (1994).
13. M. Born and E. Wolf, *Principles of Optics*, Pergamon Press (1980).
14. B. J. Lin, "Optical Methods for Fine Line Lithography," in *Fine Line Lithography*, R. Newman (ed.), pp. 105–232, North-Holland Publishing Company (1980).
15. Y. Vladimirsky, V. Saile, J. D. Scott, O. Vladimirsky, K. Morris, J. M. Klopf, G. Calderon, H. P. Bluem, and B. C. Craft, "X-Ray Lithography Program at CAMD," *The Electrochemical Society Proceedings*, **95–18**, 366–375 (1996).
16. B. Ruff, E. Tai, and R. Brown, "Broadband deep-UV high NA photolithography system," *Proc. SPIE*, **1088**, 441–446 (1989).
17. K. Jain, C. G. Wilson, and B. J. Lin, "Ultrafast high resolution contact lithography using excimer lasers," *Proc. SPIE*, **334**, 259–262 (1982).
18. K. Jain and R. T. Kerth, "Excimer laser projection lithography," *Appl. Opt.*, **23**(5), 648–650 (1984).
19. V. Pol, J. H. Bennewitz, G. C. Escher, M. Feldman, V. A. Firtion, T. E. Jewel, B. E. Wilcomb, and J. T. Clemens, "Excimer laser-based lithography: a deep ultraviolet wafer stepper," *Proc. SPIE*, **633**, 6–16 (1986).
20. R. F. Hollman, F. Cleveland, E. M. Da Silvera, R. W. McCleary, and R. W. Strauten, "Design and performance of a production-oriented deep UV stepper," *Proc. SPIE*, **1264**, 548–555 (1990).
21. W. McClearly, P. J. Tompkims, M. D. Dunn, K. F. Walsh, J. F. Conway, and R. P. Mueller, "Performance of a KrF laser stepper," *Proc. SPIE*, **922**, 396–399 (1988).
22. B. J. Lin, "Methods to print optical images at low-k_1 factors," *Proc. SPIE*, **1264**, 2–13 (1990).
23. E. Spiller, "Early history of x-ray lithography at IBM," *IBM Journ. Res. Devel.*, **37**, 291–297 (1993).
24. E. Spiller and R. Feder, "X-ray lithography," in *X-ray Optics—Application to Solids, Topics in Applied Physics*, **22**, H. J. Queisser (ed), pp. 36–92, Springer-Verlag (1977).
25. D. Spears and H. I. Smith, "High-resolution pattern replication using soft x-rays," *Electr. Lett.*, **8**, 102–104 (1972).
26. Y. Vladimirsky and J. Maldonado, "Illumination Effects of Image Formation in X-ray Proximity Printing," *Microcircuit Engineering* **90**, pp. 343–346, North-Holland (1991).
27. Y. Vladimirsky, J. D. Scott, and P. K. Bhattacharia, "Ionization Radiation Effects and Mechanisms," *Mat. Res. Soc. Symp. Proc.*, **306**, 21–27 (1993).

28. M. P. Lepselter and W. T. Lynch, "Resolution Limitations for Submicron Lithography, in *VLSI Electronics, Microstructure Science*, Vol. 1, N. G. Einspruch (ed.), pp. 83–127, Academic Press (1981).
29. R. K. Watts, "Advanced Lithography," in *Very Large Scale Integration (VLSI) Fundamentals and Applications*, D. F. Barbe (ed.), pp. 42–88, Springer-Verlag (1982).
30. Z. Chen, Q. Leonard, and F. Cerrina, "X-ray Phase-Shifting Mask: Nanostructures," *Proc. SPIE*, pp. 183–192, **3048** (1997).
31. E. Spiller, "Soft X-ray Optics," pp. 256–258, SPIE Press, Bellingham, Washington (1994).

11. X-RAY SPECTROMICROSCOPY

Harald Ade

Department of Physics
North Carolina State University
Raleigh, North Carolina

11.1 Introduction

11.1.1 Background and Motivation for Spectromicroscopy

X-ray photoemission, absorption, and fluorescent spectroscopies with synchrotron radiation have been enormously successful and have seen rapid growth during the last two decades. The electronic and geometric structure information obtained with these techniques is unique, and a variety of characterization methods have been developed. Advances in these spectroscopies typically require new forms of instrumentation, either better beamlines or better or different spectrometers. These efforts have recently been extended and complemented by efforts to provide high spatial resolution spectroscopy to investigate heterogeneous materials; a marriage between spectroscopy and microscopy. In principle, one can distinguish spectromicroscopy, the generation of images with spectral content, from microspectroscopy, the generation of detailed spectra from small areas. Occasionally, we will use this fine distinction, but most of the time we will use spectromicroscopy in a more generic sense that encompasses microspectroscopy.

Among the novel characterization techniques in the soft x-ray energy range that were most explicitly spawned or greatly enhanced by synchrotron radiation are Near Edge X-ray Absorption Fine Structure (NEXAFS) [1] or X-ray Absorption Near Edge Structure (XANES) spectroscopy, X-ray Magnetic Circular Dichroism (XMCD) spectroscopy [2], fluorescence spectroscopy, and x-ray photoelectron spectroscopy (XPS). All these techniques have benefited directly from the high intensity. high brightness and tunability of synchrotron radiation, even though some NEXAFS experiments, for example, have been and still are performed with bremsstrahlung radiation. In order to fully appreciate the potential of spectromicroscopy, one has to first appreciate the power and advantages of these spectroscopy techniques.

As an example, we illustrate the chemical specificity of NEXAFS and its implication as an analytical tool for polymer applications by presenting several polymer NEXAFS spectra in Fig. 1 [3]. Even without going into detail about the meaning and assignment of the various spectral features observed in these

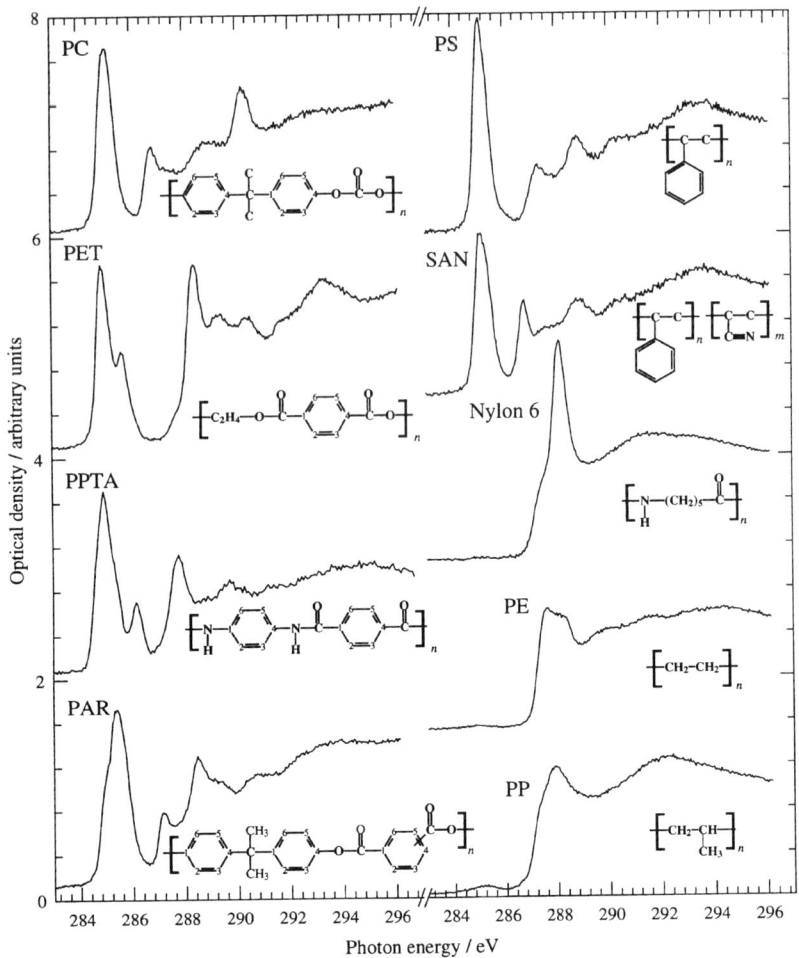

FIG. 1. Polymer NEXAFS spectra of a variety of saturated and unsaturated polymers with their chemical structure as indicated. All spectra show differences in intensity for the various peaks. Even polypropylene and polyethylene have subtle different yet distinguishable NEXAFS spectra. PC, polycarbonate; PET, poly(ethylene terephthalate); PPTA, poly(p-phenylene terephthalamide); PAR, polyarylate; PS, polystyrene; SAN, styrene-acrylonitrile; Nylon-6, poly(ε-caprolactam); PP, polypropylene; PE, polyethylene. (Figure adapted from [3] with permission of Elsevier Science.)

spectra, it is obvious that each polymer has a unique spectral signature that can be used as a fingerprint for analysis. Although all major changes in the spectra occur in a narrow energy range between 284.5 eV and about 295 eV, the differences both in intensity and in lineshapes are quite significant. The spectral sensitivity of NEXAFS even extends to isomeric substitution in some conjugated aromatic components and polymers [4]. The overlap of contributions from different chemical moieties in the same energy range will, however, limit the applicability of NEXAFS and the overall chemical sensitivity might not be as high as the one with Infra Red (IR), Raman or Nuclear Magnetic Resonance (NMR) spectroscopy. Similar spectroscopic richness is found in XPS, XMCD [2], and fluorescence spectroscopy. However, we will leave it up to others, including authors of other chapters in this book, to further elaborate on this and to elucidate the reader in detail on the subject of spectroscopy.

11.1.2 Organization of This Overview and General Comments

We will not be able to present an exhaustive and detailed review of the whole field of x-ray microscopy, neither in terms of all the various techniques developed nor the numerous applications performed. We intend to primarily provide a flavor of the different approaches that are pursued in spectromicroscopy and will emphasize applications, particularly in materials science and related fields, rather than biology. The latter and the development of x-ray microscopy in general has been recently reviewed by Kirz *et al.* [5], and we have deliberately chosen a different focus. The development of the field of x-ray microscopy over the last 15 years is also documented in several conference proceedings [6–10], popular articles [11, 12], and review articles [13–17]. We will restrict ourselves to the 50–1000 eV energy range, which will exclude commercial instruments [18–20] based on either Al and Mg Kα radiation, with a photon energy of 1486.7 eV and 1253.6 eV, respectively, from a detailed discussion. This restriction will also exclude hard x-ray microprobe experiments at synchrotron facilities, which typically use photon energies well above 1 keV. We will focus on microscopy developments that have generated the most applications and will use a selected number of applications as illustrations. We will also be biased toward microscopy efforts that have achieved the highest spatial resolution and that show the promise to achieve 10 nm or better in the future. Details about various soft x-ray spectromicroscopy efforts and applications not provided by this overview might be found in a recent special issue of *J. Electron. Spectrosc.* [21] and the most recent x-ray microscopy conference proceeding [10].

After a general comparison of microscopy techniques and some historical comments in the final sections of the introduction, we will present the various approaches to spectromicroscopy in Section 11.2 and provide numerous applications in Section 11.3 as examples. We will conclude with a discussion and a future outlook.

11.1.3 General Comparison of X-Ray Spectromicroscopy to Other Microscopy Techniques

Although the value of x-ray spectromicroscopy might be obvious to some researchers, its value may not be obvious to others. Detailed comparisons to other techniques to assess this value can be made and have been partially made [22] and have to include important parameters such as spatial resolution, spectral resolution (i.e., chemical sensitivity), energy range, data acquisition rates, surface/bulk sensitivity, sample damage, sample preparation requirements, availability, costs, and so forth. These parameters span a multidimensional parameter space, and the various microscopy techniques typically occupy a separate, non-overlapping volume. If indeed there is little overlap, this indicates that these microscopy techniques are complementary, rather than competing techniques.

We will restrict our short discussion on this issue to three of the most important parameters: spatial resolution, spectral information, and radiation damage. There are indeed numerous techniques that can provide much higher spatial resolution than x-ray microscopy. Among them are Electron Microscopy (EM), Scanning Tunneling Microscopy (STM), Atomic Force Microscopy (AFM), all of which have atomic resolution. However, these techniques typically provide only limited spectral information. Other techniques, such as IR spectroscopy and Raman spectroscopy, have quite high spectral content, but only very modest spatial resolution (>1 μm). It is generally the combination of relatively high spatial resolution (<1 μm) coupled with high and unique spectral information at modest radiation damage and the variable surface sensitivity that provides a niche for x-ray microscopy. To short-circuit what could be a lengthy discussion, one might simply ask: Why are so many x-ray photoemission and x-ray absorption spectroscopy experiments performed even without spatial resolution? It seems without question that high spatial resolution would add something useful to spectroscopy methods and materials analysis. This is exemplified by a recent specific and direct comparison between Electron Energy Loss Spectroscopy (EELS) in a Scanning Transmission Electron Microscope (STEM) and NEXAFS microscopy [23]. It monitored the spectral changes in the polymer poly(ethylene terephthalate) (PET) as an indication of radiation damage as a function of dose. This study found that the potentially much higher spatial resolution in an EELS-STEM of <1 nm [24] cannot be used with polymers and that NEXAFS microscopy can analyze a sample area about 500 times smaller than EELS given the same radiation damage. Hence, with low damage and excellent energy resolution and spectral richness (see Fig 1) numerous NEXAFS spectromicroscopy applications in polymer science and related fields have already been performed, some of which will be detailed in Section 11.3. Similarly, there is a distinct advantage to investigate magnetic materials with spectromicroscopy, as recognized and evidenced by the increasing number of groups that perform magnetic x-ray microscopy. We think we can let the applications and examples shown in Section 11.3 speak for themselves.

11.1.4 Historical Remarks

The first spectromicroscopy experiment with photoelectrons at high spatial resolution was performed by D. W. Turner's group in Oxford [25] in 1981, who achieved a spatial resolution of about 2 μm with a magnetic projection microscope. Energy resolution of about 300 meV, and hence spectroscopic information, was obtained with a retarding field analyzer after the image had been already magnified by about 100×. The general advantages of photon spectromicroscopy have been pointed out and were further expounded on by others [26]. To some extent one can also classify all the x-ray microscopy developments driven by biological applications that started in the late 1970s and early 1980s as spectromicroscopy, as the difference in absorption cross section between water and carbon based matter was to be used as contrast mechanism for hydrated biological samples. This strategy exploits differences in the electronic structure of constituent atoms, as photon energies in the so-called water window between 290 and 540 eV cannot excite the K-shell electrons of oxygen but can excite the K-shell electrons of carbon. However, this strategy relied solely on elemental contrast and did not get associated with spectromicroscopy, which was pursuing true chemical sensitivity, that is, valence/oxidation states, and other more sophisticated aspects associated with photon spectroscopy techniques.

The first spectromicroscopy effort involving the author was started in the second half of the 1980s with the development of a Scanning Photoemission Microscope (SPEM) [27] at beamline X1A at the NSLS. It was based on zone plate optics and used a Cylindrical Mirror Analyzer (CMA) as electron spectrometer. Although it had a rather modest energy resolution of about 4–7 eV, this SPEM was the first spectromicroscope to achieve sub-micron XPS microscopy, mapping oxidation states of silicon with a spatial resolution of about 500 nm [28]. The spatial resolution was subsequently improved to about 150 nm [29]. At about the same time of breaking the micron barrier in XPS, Tonner's group had achieved the first sub-micron XANES microscopy from surfaces with electrostatic imaging techniques [30, 31]. The older "sibling" microscope to the X1A-SPEM, the X1A-Scanning Transmission X-ray Microscope (STXM) developed by Kirz and Jacobsen *et al.* [32], was later used to achieve NEXAFS/XANES imaging in transmission at 30–50 nm spatial resolution [33]. The project MAXIMUM subsequently pushed the spatial resolution limit of XPS spectromicroscopy to about 100 nm [34], and recently, Bauer's group reported having achieved XPS imaging with a spatial resolution of about 40 nm [35]. The X1A STXM still provides the highest spatial resolution spectromicroscopy to date, with an energy resolution of about 300 meV. However, one should keep in mind that it is generally more difficult to achieve high spatial resolution XPS spectromicroscopy than it is to achieve NEXAFS microscopy in transmission.

There are, of course, other important spectromicroscopy efforts not mentioned so far that have evolved concurrently. In particular, Kunz's group at the Hasylab has built a microscope based on a gracing incidence ellipsoidal mirror. Although this effort never held a record in spatial resolution it has been quite successful and innovative, particularly in implementing the widest range of detection methods such as detection of photoelectrons, photoluminescence, fluorescence, desorbed ions, scattered and transmitted photons.

11.2 X-Ray Spectromicroscopy Approaches

11.2.1 Background

The developments since Beamson *et al.* [25] started the field can be grouped into two categories: (i) microprobe instruments and (ii) imaging instruments. Microprobe instruments generate a finely focused photon beam, whose size and shape determines the spatial resolution and image qualities. The sample or the optics has to be scanned and an image is acquired pixel by pixel in a serial fashion. In contrast to this approach, imaging instruments generate full images at any given time and all image pixels are recorded in parallel. In most cases, electrostatic or magnetic fields or lenses are used to magnify photoelectrons emitted from a surface directly.

11.2.2 Microprobe Optics

High spatial resolution microprobes in the soft x-ray range (50–1000 eV) can be achieved by three different physical phenomena and technological approaches: (i) diffraction, (ii) gracing incidence reflection, and (iii) multilayer enhanced normal incidence reflection. Each of these approaches has its own strengths and weaknesses. A fourth approach might be to just use small apertures, either directly by themselves or at the end of a tapered capillary. We will discuss these possibilities in turn.

11.2.2.1 Diffraction: Zone Plate Optics.
Instruments that use diffraction are typically based on zone plate optics, which are circular, variable line spacing, transmission gratings. The placement of alternating transparent and opaque or phase shifting zones are arranged in such a way as to diffract the beam to the same spot on the optical axis in a given diffraction order, and hence focus photons. The spot size achievable depends on the accuracy of the placement of the zones and the size of the outermost zone width. Zone plates with outermost zone widths of 20–30 nm have been fabricated [36–38], and marks

and spaces in test pattern as small as 25 nm have been observed [39]. The probe size achieved with zone plates in the soft x-ray range is unparalleled over the entire electromagnetic spectrum with any focusing device. It is thus not surprising that zone plates are the most widely used optics for microprobe formation in the soft x-ray range. The primary shortcoming of zone plates in a very short focal length and hence working distance. Like all microprobe optics, zone plates have to be coherently illuminated to provide the highest spatial resolution possible. Given this constraint, generally only undulator sources at high brightness storage rings and their associated monochromator beamlines result in small microprobes of sufficient intensity to perform experiments. Since zone plates are based on diffractive optical principles, only a small fraction (about 10–15%) of the coherent flux is diffracted into the positive first order that is used for micro probe formation. All other orders have to be eliminated with a strategically placed small pinhole, referred to as Order Selecting Aperture (OSA), to achieve high signal-to-background (S/B) ratios (see Fig. 2). This further reduces the usable working distance.

There are presently three zone-plate based STXMs in operation worldwide: The X1-STXM at the NSLS, the BL7.0 STXM at the ALS, and the King's College STXM at beamline 5U2 in Daresbury. These microscopes are optimized and dedicated to transmission microscopy of samples 50–1000 nm in thickness.

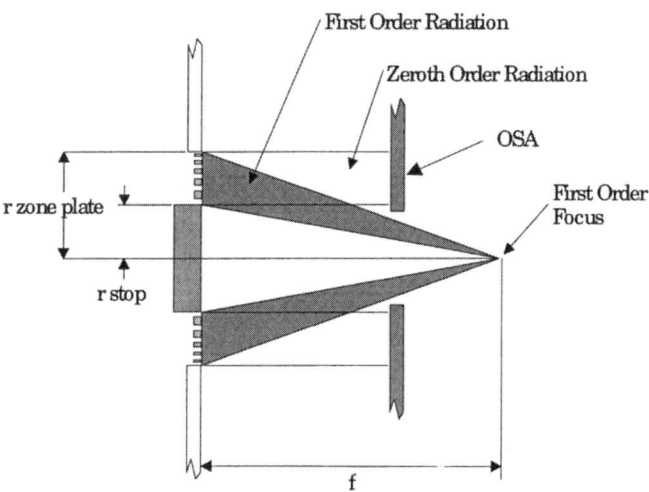

FIG. 2. Side view of zone plate (ZP) and two of its diffraction orders. All orders other than the positive first order are stopped by the Order Selecting Aperture (OSA) placed in the shadow of a central ZP stop. The first order focal spot is used as the microprobe in a STXM, and its size determines the spatial resolution achievable. (Figure adapted from [85].)

The precise thickness depends primarily on the photon energy used. Typically, sections 100–300 nm in thickness are used for carbon K-edge NEXAFS.

We describe briefly the operation of a STXM. Typically, the sample is located in a helium purged-atmospheric pressure environment, and is raster scanned under computer control with capacitance controlled piezo transducers to acquire images. Alternatively, the X, Y position is held fixed while the photon energy and the Z position of the zone plate are scanned in synchrony to acquire an energy scan from a small spot. The transmitted flux is detected with a gas flow counter or a solid-state detector. To collect absorption spectra, an energy scan (I) from the sample is recorded, and subsequently or just prior to this another energy scan (I_0) is recorded without a sample or through an open area of the sample. The negative log ratio of these energy scans ($-\ln(I/I_0)$) is an optical density spectrum in units of absorption lengths. It takes a few minutes to acquire an image with compositional information (spectromicroscopy) and about the same time to record several energy scans from small sample areas (microspectroscopy) and normalization scans from open areas.

Presently, the highest resolution zone plates at X1A have a diameter of 80–90 µm, an outermost zone width of 20–30 nm, and hence a focal length of less than 1 mm at 280 eV. Because of the OSA, the usable working distance is reduced to less than 200 µm. The situation is similar for the King's College STXM. This small working distance, the width of a hair, provides one of the main challenges in constructing and in operating a STXM. Since the focal length f of a zone plate is proportional to the photon energy E ($f \propto E$), the instrument has to be refocused if absorption spectra are to be acquired from the smallest possible spot. This requires that there be no significant runout, that is, transverse wobble, during the refocusing motion. Otherwise, the area spectroscopically analyzed is not as small as theoretically possible. This significantly complicates the microscope instrumentation. Runout during spectroscopy mode is generally a problem and is controlled only after extraordinary alignment and dynamic correction efforts. In routine operation, the corresponding spatial resolution during spectrum acquisition mode is slightly degraded. The BL7.0 STXM presently uses a ZP with an outermost zone width of 80 nm and has a spatial resolution of about 150 nm [40]. Additional STXMs are planned at the ALS and at BESSY II as well as other sources [5].

The basic optical concepts of a zone-plate based SPEM are the same as those of a STXM, except that photoelectrons are energy analyzed with conventional spectrometers in a "back-reflection" geometry to provide spectroscopic information. This makes it even more difficult to operate with the small working distances provided by zone plates, particularly since the spectrometer has to be shielded from the photoelectrons emerging from the zone plate/OSA assembly. Typically, the zone plate parameters in a SPEM are thus somewhat different than those used in a STXM and zone plates with larger outermost zone widths

Δd, and therefore worse spatial resolution, are used in a SPEM. Larger Δd puts less stringent requirements on the zone plate absolute placement and allows larger zone plates to be fabricated. Both aspects, larger diameter d and larger Δd provide longer focal lengths f, since $f = d*\Delta d/\lambda$, where λ is the photon wavelength. Even then, the small working distance puts constraints on the design and operation of an XPS microscope, whereas the necessity for refocusing in a XANES/NFXAFS mode would put extreme demands on the mechanical stages. So far, only XPS mode is used in existing SPEM instruments.

Presently, there are two operating SPEMs based on zone plates. One at beamline X1A at the NSLS (X1-SPEM) [41, 42], and one at ELETTRA in Trieste [43, 44]. At the ALS, a SPEM is at an advanced stage of construction and commissioning [40]. Zone plates used for the X1-SPEM-II and the ALS SPEM have a diameter of 140–200 μm and an outermost zone width of 80 or 100 nm. The SPEM at ELETTRA is using zone plates with an outermost zone width of 100 nm as well, and all instruments have or should have similar spatial resolution. Various schemes have been devised to handle the change in focal length as the photon energy is changed. At the NSLS and ELETTRA, two independent manipulators allow independent placement of the zone plate and the OSA. At the ALS SPEM, five zone plates/OSA pairs are prealigned at different distances, with each pair covering a separate photon energy range.

A zone plate–based SPEM is in the advanced construction phase at the SRRC in Taiwan [45], while an additional SPEM is planned at the Korean synchrotron radiation source in Pohang. A zone plate–based XMCD microscope detecting total electron yield has been constructed by Kagoshima *et al.* [46, 47].

11.2.2.2 Grazing Incidence Optics.
Metal coated grazing incidence optics are used essentially in all soft x-ray beamlines at synchrotron radiation facilities. The primary advantage of adopting grazing incidence optics for use as a high-resolution microscope objective is tunability over a large energy range (visible to 1200 eV) without having to adjust the optics or the sample position. In addition, typically the optics used has a long working distance and it is easy to implement several detection schemes simultaneously. Presently, however, these optics suffer from figure errors and the spatial resolution is somewhat limited. Various geometries are possible and have been explored. Kirkpatrick–Baez (KB) systems, consisting of two crossed, single focusing mirrors, either spheres or ellipsoids, have a small numerical aperture and are primarily used for harder x-rays. A system optimized for throughput on an ALS bending magnetic source at modest spatial resolution, uses two bend elliptical mirrors in the KB geometry [40] for soft x-ray XPS microscopy. A spatial resolution of 1 μm is anticipated. This requires exceptional control of figure errors, and the slope deviation from the perfect ellipsoidal surface has to be less than 1 μrad rms along the illuminated length.

An elliptical ring mirror, with symmetry of revolution around the optical axis, is used in the Hamburg microscope [48–50]. It is the most successful spectromicroscope based on grazing incident optics [49, 51]. It achieved the highest spatial resolution (0.4 μm) with usable countrates and has performed the most applications. A similar microscope is also operated at MAXlab in Lund, Sweden [52]. One of the main advantages of this microscope, particularly when compared to zone plate microscopes, is the relatively long working distance offered by the ellipsoidal geometry used, and the fact that this optics is completely achromatic. The sample to optics distance is 32 mm, which provides relatively easy access for various detectors. This has been fully exploited by the Hamburg group, which has installed a visible and UV luminescence spectrometer; a time of flight (TOF) electron analyzer; a hemispherical electron analyzer; multichannel plate detector for desorbed ions and scattered, emitted, and reflected soft x-rays; a transmission detector; and a TOF detector for desorbed ions. Not all detectors can operate at the same time, but several of them can and they are easily interchanged [49].

The ellipsoidal ring mirror offers a larger numerical aperture (NA) than KB objectives and hence would possess a better diffraction-limited spatial resolution compared with that of KB optics. However, the throughput is greatly reduced because of the necessary central obstruction. The contrast is also much reduced in diffraction-limited operation compared with an unobstructed optics with the same NA. Wolter mirrors, a combination of an elliptical and a hyperbolic surface, have similar characteristics to the ellipsoidal ring mirror used in the Hamburg microscope. They are rotationally symmetric, offer large numerical apertures, but also need a central obstruction. The figure errors are also quite difficult to control [53], but resolutions as low as 100 nm have been reported [54]. Additional details about grazing incidence optics can be found in an article by Kunz and Voss [48] who assert that spot sizes smaller than 100 nm should be possible with grazing incidence optics if fabrication technology advances sufficiently. Recent results at BESSY have achieved a spatial resolution of better than 1 μm with a single off axis ellipsoidal mirror [55]. The mirror used has not achieved yet the required surface quality and the unexpectedly good results are attributed to the use of only a small part of the mirror over which the figure errors are presumably smaller.

11.2.2.3 Schwarzschield Objectives Microscopes. In contrast to grazing incidence optics, aberrations are not as problematic for near-normal incidence. This is fully exploited in Schwarzschield objectives (SO), which are stigmatic and aplanatic to third order. Most present day SO typically use two spherical surfaces, for which it is a lot easier to control the figure errors than for aspherical surfaces. The normal incidence reflectivity is enhanced with multilayer coatings, and reflectivity in excess of 60% have been achieved with energies near 100 eV [56]. Because of interface roughness and interdiffusion, it is particularly difficult to make effective normal incidence multilayers at higher energies, that is, above the carbon K-edge. This can be appreciated if one considers that individual layers are only $\lambda/4$ thick, which for 500 eV photons would result in a thickness of about 0.6 nm, or a few atomic layers. Schwarzschield microscopes typically offer a spatial resolution intermediate between that of zone plates and grazing incidence optics. Objectives with a NA of up to 0.4 are readily available without coatings. Within the bandwidth of the multilayer, these objectives have a fixed focal length, but the bandwidth is limited to about 5%. If larger photon energy changes are required, the SO has to be exchanged with one that has a different coating. Most likely, it will at least be necessary to refocus the sample and possibly to realign the SOs.

The first x-ray microscopy efforts with SO were undertaken by Haelbich, Spiller, and Kunz [57]. More recent deployment of multilayer coated SOs are in the scanning photoemission microscope MAXIMUM [34, 58] and in the AT&T projection lithography project at the NSLS [59]. MAXIMUM has a similar general setup to other scanning photoemission microscopes, and shares a relatively large working distance with the grazing incidence instruments. Images in transmission have resolved features 90 nm in size. The main applications of MAXIMUM have focused on semiconducting surfaces and electronic materials processing issues. Presently, there is only one SO-based spectromicroscope in operation, with a second, SuperMAXIMUM, in commission at ELETTRA [60]." SuperMAXIMUM will have five SOs to cover different energy ranges.

11.2.2.4 Apertures. A spectromicroscope originally exclusively based on apertures is the Photon Induced Scanning Auger Microscope (PISAM) at BESSY [55, 61]. Adjustable apertures on the order of a few µm are directly placed into the unmonochromatized beam of an undulator. This provides exceptionally high photon flux, but the spatial resolution is only a modest 3–4 µm. A second-generation instrument is presently being commissioned, which will use an ellipsoidal mirror to achieve higher spatial resolution [55]. Apertures could also be used at the end of capillary concentrators to provide high spatial resolution [62]. Although impressive results with a spatial resolution of better than 100 nm have been achieved with tapered capillaries for hard x-rays [63], no working instrument based on capillary optics has so far been produced in the softer energy range.

11.2.3 Imaging Microscopes

11.2.3.1 Magnetic Projection Microscope. The magnetic projection microscope (MPM) played a special role in the development of the field of spectromicroscopy and was, used by Beamson *et al.* [25] in 1981 in their pioneering experiment. The sample is placed in a high field region of a strong magnet, typically produced by a superconducting solenoid. The magnetic field is made to diverge, and photoelectrons emitted from the sample are trapped by the magnetic field into spiral paths around the magnetic field lines. The image is magnified as the magnetic field lines separate. The magnification is given by

$$M = (B_{sample}/B_{detector})^{1/2}$$

where $B_{sample}/B_{detector}$ is the ratio of the magnetic field at the detector to that at the sample. The spatial resolution is related to the electron kinetic energy and magnetic field at the sample, as the radius of the spiral paths of the photoelectrons is proportional to $(K_{sample})^{1/2}/B_{sample}$. A spatial resolution on the order of 1 µm has been achieved.

Although only few groups have further pursued the development and use of a MPM [64–68], it has several key advantages. For samples with negligible permeability, the magnetic field lines penetrate into materials without distortions. Hence, the effective depth of focus is very large, and electrons can be extracted from small holes, or from any area of all irregular surface. This is particularly useful when investigating fractured surfaces [69, 70] or micropatterned structures that have deep trenches. In addition, all electrons emitted with an initial angle up to 90° with respect to the magnetic field lines will be detected. This unparalleled collection efficiency would provide low dose microscopy of radiation-sensitive materials. At the same time, the high collection efficiency is coupled to one of the biggest disadvantages of a MPM. The transverse velocity of the electrons is what is setting the spatial resolution, which is typically in the µm range. In practice, it might thus be desirable to control the collection angle with "skimmers" which might improve the spatial resolution to a few tens of micrometers [65]. However, the MPM is not forming conjugate image and diffraction planes and it is difficult to control the collection angle, and hence, spatial resolution. It is difficult to foresee much improvement in spatial resolution in the future, and when compared with other instruments further instrument developments of MPM seem to be somewhat limited.

11.2.3.2 Electron Optics Microscopy: Photoemission Electron Microscope.
Rather than projecting photoelectrons with magnetic fields, it is possible to image photoelectrons from flat and reasonably conducting samples directly with electrostatic and magnetic lenses. This possibility goes back to the earliest days of electron microscopy, and the concepts used in a Photoemission Electron Microscope (PEEM) have been developed over many years. A historical perspective of these developments can be found in a journal article by Griffith and Engel [71].

While differences in design and technical detail exist, the overall basic concept of all PEEMs is rather similar. An immersion objective lens, in which the sample is an integral part of the electron optical system, both accelerates the low kinetic energy electrons to 10–20 kV and forms a magnified image. This image is further magnified, transferred and projected onto a detector. Depending on the design, one or several lenses are used for projection and additional magnification. Typically, the detector is an image intensifier, consisting of a microchannel plate, a phosphor screen and a phase plate, and a camera. A pinhole in the backfocal plane of the objective is used as a contrast aperture, which plays an important role in high-resolution imaging. We show as an example, the schematic of Tonner's PEEM in Fig. 3.

Initially, PEEM was exclusively used with near photothreshold excitation. In this operating mode, a variety of experiments have been performed and a spatial resolution of 10–15 nm has been achieved [72, 73]. Tonner was first to explore PEEM with soft x-ray excitation and recorded the first XANES spectra from microscopic areas on surfaces [74]. The main disadvantage of high photon energy excitation is a reduction in spatial resolution because of chromatic aberrations. In order to eliminate these aberrations, and also to be able to switch between XPS and XANES/NEXAFS microscopy, both Bauer and Tonner have designed and commissioned an energy-filtered PEEM [35, 75].

PEEM microscopy and its applications have grown quite rapidly in the recent past. There are at least three commercial PEEM microscopes available at this time (Staib Instrumente, Elmitec, and Omicron) and the number of instruments in use and soon to be in use is growing rapidly. Part of this popularity is most likely based on the less stringent source requirements. A PEEM can essentially be bolted onto any synchrotron radiation beamline with quite respectable performance. The PEEM is technologically somewhat simpler than the scanning instruments and doesn't involve precision motion. It is based on a much longer history of instrument development: the development of electron microscopy.

11.2.3.3 Transmission X-Ray Microscopes. The Göttingen x-ray microscopy group has been operating full field, transmission x-ray microscopes based on zone plate optics for more than two decades [76]. The sample is immediately downstream of a pinhole, which is illuminated by a zone plate condenser that in conjunction with the pinhole also serves as a monochromator (see Fig. 4). A micro zone plate magnifies the x-rays transmitted by the object by about 1000× and the image is recorded in the latest generation of these instruments with a backside-thinned CCD camera with a detection efficiency as high as 73% [77]. In a test pattern, lines and spaces as small as 25 nm have been observed [39]. Initial applications have focused on biological samples. Starting a few years ago, the Göttingen microscope has also been used to study clay minerals and soil colloids [78, 79], and various other samples. Microscopes similar to the Göttingen microscopes have been built at the Aarhus University Storage Ring in Denmark [80], the ALS [81], as well as at the Photon Factory in Tsukuba, Japan. The Göttingen microscope has recently been modified to allow XMCD imaging [82]. A schematic of the modified microscope is shown in

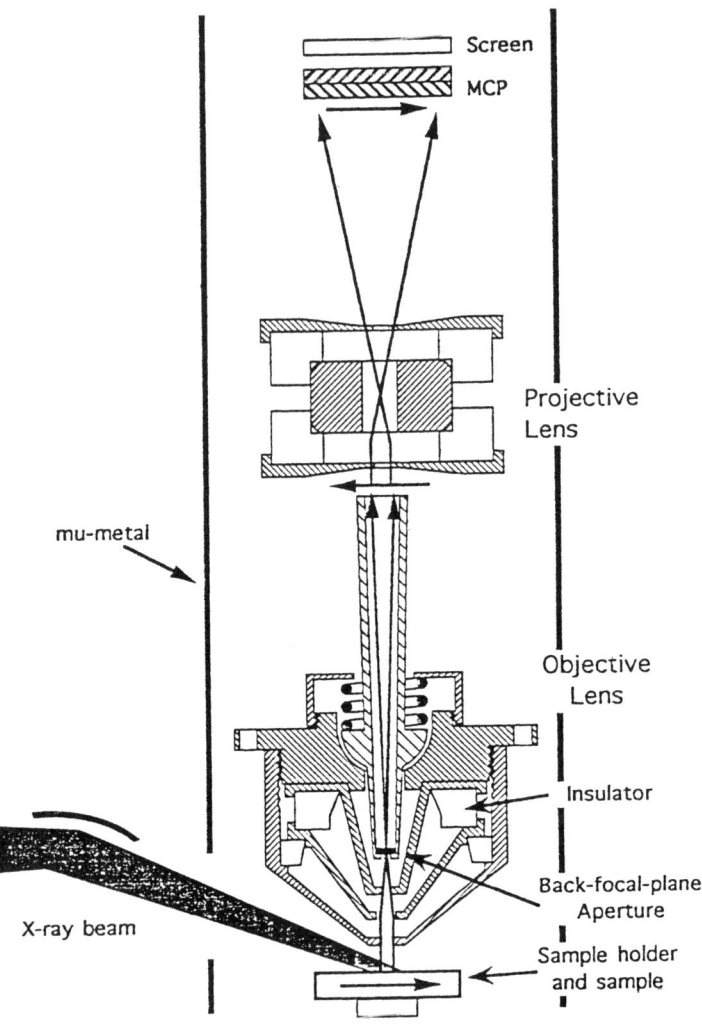

FIG. 3. Schematic of Tonner's PEEM. X-rays illuminate the field of view on the sample, which emits secondary electrons in proportion to the absorption of x-rays. The secondary electrons are accelerated and focused with electrostatic lenses to form a magnified image of the sample surface on the microchannelplate intensifier. (Figure reprinted from [122] with permission from Elsevier Science.)

Fig. 4. A mask can be used to select only out-of-plane radiation and the degree of circular polarization achieved is 60%. An external magnet can be used to adjust the magnetic field at the sample locations.

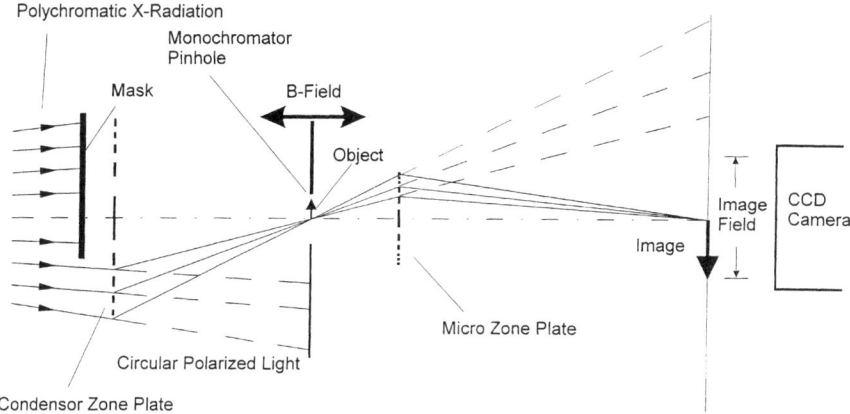

FIG. 4. Schematic of the Göttingen microscope at BESSY I. Circularly polarized light was produced by masking the condenser zone plate in the upper half. A small solenoid produces a magnetic field parallel/antiparallel to the beam direction. (Figure courtesy of G. Schmahl [82].)

11.3 Applications

We will describe several selected applications of x-ray microscopy in some detail, while only alluding to others. Most of these projects are continuing research projects, while a few of the results presented are demonstration experiments or preliminary. Nevertheless, they should illustrate the range of problems for which researchers from different communities have found x-ray microscopy to be a useful or promising analytical tool.

11.3.1 Polymers

In many applications, the physiomechanical properties of a polymer system can be tailored by altering the chemical and morphological structure. In most cases, the polymer system of interest is not a homogeneous or single-component system, but is a blend, composite, or copolymer. With these heterogeneous molecular systems, many physical processes characteristic of interfacial boundaries are present, such as phase separation, diffusion, adsorption, adhesion, and intermolecular binding. The role of these and other processes is important in a wide array of commercial applications, including biomedical devices. To understand the nature of the molecular and physical processes in these heterogeneous systems, it is important to simultaneously understand both the spatial distribution and the chemical behavior of the various components within the polymer matrix. Although one can use a variety of analytical tools that combine chemical and morphological information, NEXAFS STXM has become a primary technique for these applications. Most samples can be microtomed without problems or fabricated as a thin film and subsequently examined in transmission. With only a few microscopes worldwide, the primary limitation regarding the number of polymer applications with STXM is a lack of sufficient experimental capacity.

11.3.1.1 Polymer Blends.

Traditionally, conventional microscopies, particularly electron microscopy in conjunction with heavy metal staining methods, are used for the characterization of binary, ternary, or quaternary polymer blends. However, absorption rates for staining agents are very similar for many polymer components of interest. It is then difficult to delineate these components in an electron microscope. Because of the non-linearity of the staining process, it is furthermore very difficult to acquire quantitative information about composition. There is also the danger of inducing artifacts associated with the staining process and the possibility of misinterpretation.

Given the technological prevalence of polymer blends, it is not surprising that the initial demonstration of NEXAFS microscopy in transmission [33] involved a blend of a random copolymer of polystyrene and styrene-acrylonitrile (PS-r-SAN) and polypropylene (PP). Several researchers have since investigated a variety of other multicomponent polymer blends [83–85]. As an example, we show the investigation of the morphology of an elastic ternary blend of poly(ethylene terephthalate), PET; low-density polyethylene, LDPE; and Maleated Kraton with the X1A-STXM. Of particular interest in this material, is the distribution of the Kraton, a rubbery component, and more specifically whether the Kraton is also located at the PET/LDPE interface or only inside the LDPE domains. The interpretation of electron micrographs of stained samples of this material yielded ambiguous results. NEXAFS micrographs showed that Kraton is present at the PET/LDPE interface, as well as in the LDPE phase. The features in the lower right-hand corner of Fig. 5 make this particularly obvious: dark domains are touching each other (Kraton around the LDPE domain) in Fig. 5(a), whereas in comparison the LDPE domains are sharply delineated and separated in Fig. 5(b) [85].

Other multicomponent polymer systems investigated are polycarbonate/acrylonitrile-butadiene-styrene blends. These are complex mixtures consisting of three polymeric components, polycarbonate (PC), styrene-acrylonitrile copolymer (SAN), and SAN copolymer grafted onto polybutadiene in latex form (SAN-g-PB). These systems also contain small amounts of titanium oxide. Of particular interest in these materials is the characterization of morphological and compositional changes during repeated recycling. In initial experiments, NEXAFS microscopy delineated the distribution of all constituents in an un-recycled material [84]. The SAN was shown to accumulate at the interface between the continuous PC phase and the dispersed SAN-g-PB particles. Some

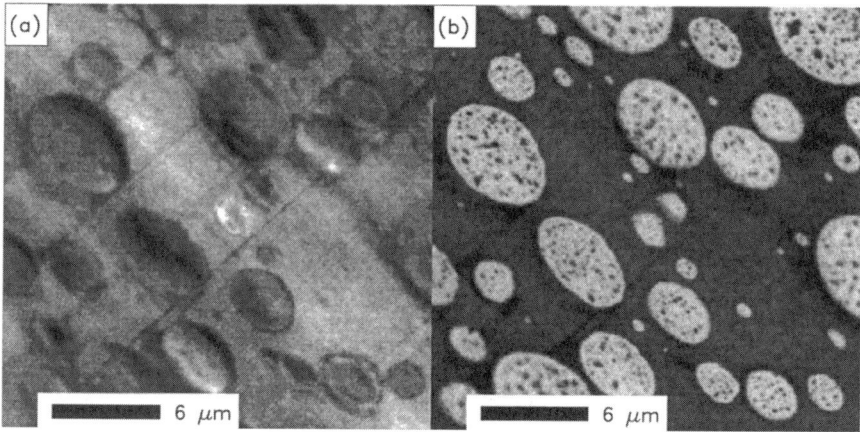

FIG. 5. (a) Micrograph of LDPE, PET, and Kraton ternary blend acquired at a photon energy of 299 eV. Both LDPE and Kraton appear relatively dark because of their high number of single C–H bonds. (b) Micrograph acquired near 285 eV. LDPE is very transparent and appears bright while PET and Kraton are dark. Comparing these images, we find that Kraton also accumulates at the LDPE–PET interface, rather than just inside the LDPE domains as was expected. (Figure adapted from [84].)

free SAN also appeared to be in the PC matrix, not associated with the SAN-g-PB. The titanium oxide was typically inside or in close proximity to the SAN/SAN-g-PB agglomerates.

11.3.1.2 Multiphase Polymers: Phase Separation During Processing

The materials discussed previously contained relatively well defined polymer components and the objective was to delineate their spatial distribution. A complementary problem where x-ray microspectroscopy can make meaningful contributions is when various phases segregate during synthesis and processing of polymers (or other materials) and the chemical composition of the phases has to be ascertained with microspectroscopy. These phases are not just mixtures of two or more polymers as complicated reactions can occur during processing.

A class of materials where one can easily view the morphology created by phase separation with traditional techniques, but where more specific chemical information is needed at high spatial resolution, is polyurethanes. Methylene-diphenyl-diisocyanate (MDI) and toluene-diisocyanate (TDI) based polyurethanes are produced in large quantities, with a world-wide production of about five million metric tons a year. Both MDI- and TDI-based materials can form precipitates during processing. The relative concentration of polyol, urea, or urethane functionality in the precipitates might differ significantly from that of the matrix, and the question of whether the precipitates are more urea or more urethane in character is of central importance in order to tailor the material properties. Rightor, Hitchcock, and Ade *et al.* [40, 85, 86] are thus in the process of systematically determining the sensitivity of NEXAFS to urea and urethane functionalities by acquiring NEXAFS spectra of specially prepared model polyurethane polymers, polyols, and polyurea, and comparing these against model polymers with known urea and urethane fractional composition. Although the chemical difference, and hence spectral difference, between a urea and a urethane functionality in a polymer is subtle, quantitative analysis with an accuracy of a few percent for the urea and urethane components seems possible, even if the polyol is the predominant constituent and interferes with the urea/urethane analysis [87].

NEXAFS microscopy has also been used to view for the first time a complex morphology created by phase separation during processing. For example, the morphology of a liquid crystalline polyester (LCP) based on several aromatic monomers has been successfully determined with NEXAFS microscopy. In a particular material based on this LCP, NEXAFS microscopy revealed an unusual morphology consisting of four chemically distinct phases. Micrographs in Fig. 6 provide an illustration of the texture observed in this material. Figure 6(a) emphasizes aromatic functionalities in the sample and a discontinuous phase with domains smaller then 100 nm, as well as a continuous phase enclosing elliptical bodies with dimensions of a few microns are clearly discernible. Figure 6(b) reverses the contrast between the larger features, while there is virtually no contrast between the small features at this energy because of the particular NEXAFS cross section for this chemical composition. Fig. 6(c) is acquired at a photon energy that has only residual chemical sensitivity and is a mixed "density/thickness/chemistry map". Figure 6(d) is acquired below the carbon edge, and emphasizes elements other than carbon. The dark features in Fig. 6(d) are regions interpreted as rich in oxygen, presumably from carbonyl functionalities. This complex morphology based on different chemical compositions was observed only in this material and entirely absent in materials with nominally the same composition, but that were processed differently.

11.3.1.3 Additional Examples of Polymer Applications.

The dewetting and spinodal decomposition in polymer thin films has also been studied with NEXAFS STXM. Experiments and theoretical work on spinodal decomposition in binary polymer blends indicated large changes in phase behavior and dynamics relative to bulk behavior. Currently used techniques, such as AFM, laser light scattering, and neutron scattering are able to sense growing domain sizes but are unable to sense local two- and three-dimensional variations within domains. Studies have been performed on an annealed bilayer of the immiscible polymers polystyrene (PS) and poly(methyl methacrylate) PMMA. Previous work with AFM has shown segregation of one polymer into spheroids within the other polymer upon annealing but was unable to chemically differentiate the species [88]. NEXAFS micrographs acquired at an energy most specific to the

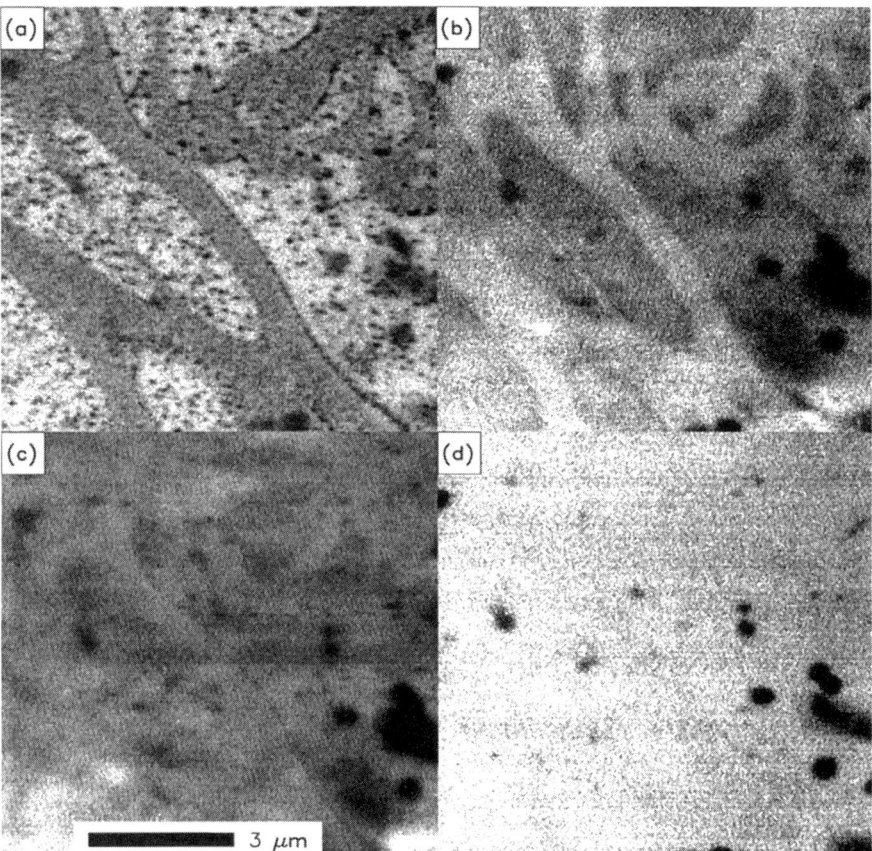

FIG. 6. Micrographs of the same region of a thin section of an aromatic liquid crystalline polyester imaged at a photon energy of (a) 285.0 eV, (b) 286.8 eV, (c) 296.2 eV, and (d) 281.1 eV, revealing four chemically distinct phases. In contrast to this material, a polyester of the same nominal composition but different processing route did not exhibit any detectable phase separation. Scale bar applicable to all figures. (Figure adapted from [84].)

PMMA (288.4 eV) show that the PS segregates into thick ellipsoid-like regions with the PMMA forming a shell around these ellipsoids and a thin layer of PMMA between the PS dispersions [89]. Brominated PS/PS systems have also been investigated. A similar study has been performed with a PEEM by Cossy–Favre et al. [90]. The homogeneity of thin films of polymer blends were studied for various film thicknesses. A 19.4-nm thick PS/polyvinylmethylether film exhibited protrusions 1–2 μm in size, which were enriched in PS. A thicker (50.4 nm) film of PS/SAN film showed 5–6 μm segregated regions without any topological structure.

Kikuma et al. [75, 91] used the chemical sensitivity of NEXAFS microscopy to investigate the chemical and structural changes in a polyacrylonitrile (PAN) fiber induced by heat treatment that makes the fiber more fire resistant. Kikuma could clearly observe a core-rim structure that developed because of the heat treatment. However, the work surprisingly indicates that the chemical conversion is higher in the core than in the rim, in apparent contradiction, to the prevailing model in which the conversion proceeds from the outside in. This model is based on the assumption that the fiber-air interface is the site of reaction initiation. It was suggested that the NEXAFS observations could be explained by the fact that the conversion reaction is exothermic and the core could be at higher temperatures, either leading to a higher conversion rate in the core or to an increase in disorder of oriented moieties of the fiber because of melting.

NEXAFS microscopy in a STXM has also been quite useful in the characterization of multilayer polymer films. Of interest in this work, is the influence of coating conditions and thermal history, the level of diffusion of small molecules and the level of interpenetration of adjacent polymer layers. In one structure of four layers, 0.7–3 µm in thickness, coated on a base polymer layer, the images acquired clearly showed each of the layers and NEXAFS spectra of closely spaced areas indicated that there was not a significant interpenetration between the two layers of interest, poly(styrene acrylonitrile) and a porous carbon black [92]. We believe that this is the first quantitative measurement of the interpenetration of adjacent polymer layers with x-ray microscopy. A second laminate consisted of nine layers on a base layer [93]. A microstructure of undetermined origin was found dispersed through the fourth layer, parallel to the layer boundary. This microstructure was completely absent in the transmission electron microscope, and is probably caused by microphase separation or a preferential orientation of aromatic groups.

Polymer applications such as the chemical mapping of microstructured thin films have also been pursued with the PISAM. A structured perylene tetracarboxylic dianhydride (PTCDA) film on a Si substrate was exposed to air in order to obtain two different species of carbon and oxygen. Subsequently, differences in the oxygen Auger fine structures were detected and used to provide line scans with chemical information from the patterned sample. The PISAM has also been used to write organic "microwires" from thiophenes [55]. Similar experiments were performed in which metal structures have been written by the photolytic decomposition of organometallic compounds condensed onto a surface [94].

11.3.1.4 Linear Dichroism Microscopy of Kevlar® Fibers. The polarization dependence of NEXAFS spectra from oriented materials is a well-documented and understood phenomenon [1]. The resulting linear dichroism in transmission microscopy, in which the absorbance depends on the orientation of the sample, can be utilized to assess orientation and the degree of orientational order in materials down to the size scale of the spatial resolution of the respective microscope [95]. Dichroism is most straightforward in materials that have uniform composition, but could prove to introduce unwanted complexities in materials that exhibit heterogeneous chemical composition and anisotropy.

Smith and Ade, for example, have determined the relative lateral orientational order of various poly(p-phenylene terephthalamide) Kevlar® fiber grades [96], a chemically uniform material, with x-ray linear dichroism microscopy. The internal structure of these technologically important, high crystallinity fibers is highly complex and certain aspects are still subject to debate. In particular, an accurate measure of the lateral orientational order of these fibers is still elusive. Micrographs of thin sections of these fibers exhibit a "butterfly" pattern when imaged at photon energies specific to certain chemical functionalities. This pattern reflects the average lateral orientation of these groups and shows, for example, that the average aromatic ring planes and carbonyl groups are pointing radially outwards.

The difference in the degree of radial order between different fiber grades is most readily observed in comparing spectra acquired from locations within these fibers where the radial position vector for that location was either perpendicular or parallel to the polarization vector of the photons (see Fig. 7 and inset). Larger spectral differences are observed for Kevlar® 149 than for Kevlar® 49, which makes Kevlar® 149 the more oriented fiber. It is also interesting to note that the spectral differences observed have opposite signs in the spectral regions dominated by π^* transitions (<293 eV) and σ^* transitions (>293 eV). Smith and Ade found that on average Kevlar® 149 is 1.6 more radially ordered than Kevlar® 49 and 2.3 times more radially ordered than Kevlar® 29. This has to be compared to the degree of crystallinity which is >90% for Kevlar® 149, 90% for Kevlar® 49, and 85% for Kevlar® 29 [97].

11.3.2 Magnetic Materials

XMCD spectroscopy has become an active field of research during the last few years because of its scientific and technological significance. Since many technological applications of magnetic materials and phenomena, particularly those related to information storage, are dependent on small structures, activity and interest in XMCD spectromicroscopy has also greatly increased. The first demonstration that element-specific magnetic information could be obtained from small sample areas about 1 μm in size has been accomplished by Stöhr et al. [98], using Tonner's PEEM. An important aspect of this demonstration was that the magnetic bits that had been imaged were buried under a surface layer of 13 nm of carbon and 4 nm of an organic fluorocarbon lubricant. Because of the long mean free path of the photons and the secondary electrons

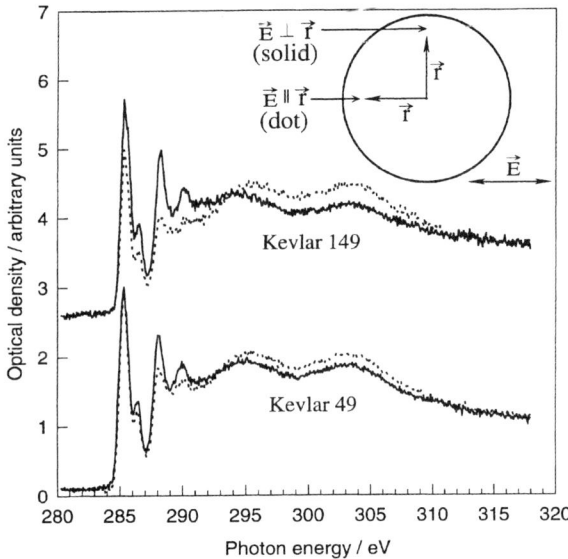

FIG. 7. Spectra of Kevlar® 149 and Kevlar® 49 fibers obtained from locations within the fiber as indicated by the inset. The differences in the peak intensities are caused by differences in the degree of radial orientational order in this fiber. Note that the peak intensity differences are much smaller in Kevlar® 49 than in Kevlar® 149. By fitting the intensities of these spectral differences, quantitative information of the relative degree of radial orientational order between fiber grades has been obtained. Kevlar® 149 is about 1.6 times as radially oriented as Kevlar® 49. [Reprinted with permission from A. P. Smith and H. Ade, *Appl. Phys. Lett.* **69**, 3833 (1996). Copyright 1996 American Institute of Physics.]

used for recording the absorption spectra and images, a remarkably large magnetic contrast was observed. Figure 8 shows an example of these results. Contrast reversal between images at the Co L_3 and Co L_2 edges was observed. The main advantages of XMCD over other techniques, such as Bitter microscopy, Kerr microscopy, Lorentz microscopy, scanning electron microscopy with polarization analysis (SEMPA), spin-polarized low-energy electron microscopy (SPLEEM), or magnetic force microscopy, is elemental and chemical state specificity as well as variable probing depth. All these aspects arise from the fact that NEXAFS spectra can be obtained at each element-specific absorption edge, and that NEXAFS spectra can be recorded with partial or total electron yield, and in the future with scanning instruments with fluorescence yield. These advantages mirror precisely the advantages of the spectroscopy method by itself. A somewhat detailed comparison with other techniques can be found by Tonner *et al.* [75]. Tonner and colleagues also imaged the bit pattern in an Fe-Th-Co alloy magnetic optical disk medium in which the Fe and Tb are antiferromagnetically coupled. They also demonstrated that XMCD microscopy in a PEEM can

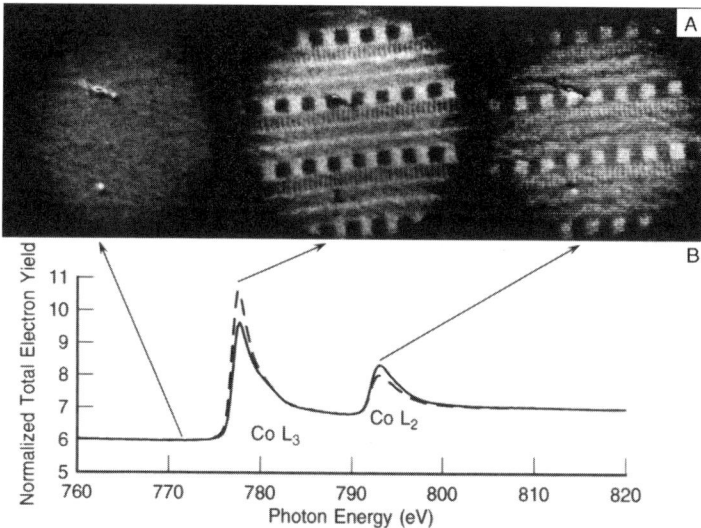

FIG. 8. (a) XMCD micrographs of magnetic domains of a magnetic recording disk recorded with right circularly polarized light with energies as indicated in the Co L-edge spectra in (b). The rows of the magnetic domains have bit patterns 10 × 10 μm, 10 × 2 μm, 10 × 1 μm, and 10 × 0.5 μm in size. The magnetization direction of the domains lies along the rows but alternatively point to the right and left in the figure. This alternation is responsible for the contrast reversal observed in the two right images. [Reprinted with permission from J. Stöhr et al., Science **259**, 658–661 (1993). Copyright 1993 American Association for the Advancement of Science.]

be used to inspect a damaged area of a Co-Pt magnetic recording disk. The topographical image of parts of this disk exhibited clear marks and signs of relatively severe damage in the coating overlayer, while the XMCD images specific to the magnetic pattern beneath the overlayer showed that it is intact.

In an application and demonstration complementary to Stöhr et al. and Tonner et al., Hillebrecht et al. [99] used a commercial PEEM from Staib and made use of the element-specific information in XMCD to distinguish the magnetization of substrate domains [15 ML of Fe grown on Ag(110)] from those of a Mn overlayer of 0.3 monolayer equivalent thickness. They found that the Mn magnetic moments are antiparallel to that of the Fe substrate, and that the Mn possibly forms a surface alloy. Hence, electron yield XMCD can both probe fairly deep into the sample, as Stöhr and Tonner showed, and can also be very surface sensitive if the thin layer can be distinguished spectroscopically. In different experiments, Hillebrecht et al. also demonstrated that a form of linear dichroism can be used as contrast mechanism in magnetic imaging [100].

In another interesting demonstration of the complementary nature of the chemical and magnetic information that can be obtained with XMCD in a

PEEM, Swiech et al. [101] imaged permalloy squares micropatterned on a Si surface with circularly polarized light. The squares are clearly distinct from the substrate and appear as four triangles with three contrast levels reflecting the magnetization direction. Typically, the magnetization vector observed in each triangle is oriented in such a way as to provide flux closure. This indicates that the system tends to minimize the magnetic stray field. Kagoshima et al. [47] have also achieved XMCD imaging with a spatial resolution of about 1 µm. They used a scanning microscope based on zone plate optics, with the circularly polarized light produced by an insertion device, and imaged the magnetic domains of a commercial 8-mm videotape.

Dynamic measurements in XMCD imaging have also been explored. Hillebrecht reported the time evolution of a thin Fe film that had been magnetized to saturation and subsequently subjected to a 10 mGauss field in the opposite direction to the original transverse magnetic field [99]. Because of the external magnetic field and the use of electron optics in these experiments, image distortions occur that have to be accounted for. Axial magnetic fields will cause a much smaller distortion of the PEEM image, so that experiments could be carried out at higher magnetic fields. Although these and most XMCD imaging experiments have been accomplished with electron optical instruments, the same magnetic contrast mechanism can be explored in any x-ray optical and scanning instruments if circularly polarized light is used. Recently, Schmahl et al. have indeed successfully pursued this path and adopted their transmission x-ray microscope for XMCD imaging. In addition to demonstrating the expected contrast mechanism in transmission, they have been able to change external magnetic fields *in situ* without penalty to the image quality and without distortions, as neither the photons nor the microscope components are influenced by the external magnetic fields. We show, as example, images obtained from a $Gd_{72.3}Fe_{27.7}$ thin film acquired at the Fe L_3 edge at 706 eV with a modest energy resolution of $E/\Delta E = 225$ (Fig. 9). Images with the sample close to magnetic saturation and for a state where the magnetization is close to zero have been obtained and exhibit quite different magnetic morphologies. Schmahl et al. report the spatial resolution in these images to be about 50 nm and that they observed domain walls 50–100 nm in thickness. This thickness was slightly larger than previously estimated.

For surface and magnetic thin film analysis it would often be advantageous to combine techniques that provide chemical, structural, and magnetic information in one instrument. Stöhr et al. [98] had suggested that XMCD might be combined with EXAFS in order to provide structural information beyond what might be extracted from NEXAFS, while Bauer et al. [35] argued that it is tempting to combine XMCD imaging with Low Energy Electron Microscopy (LEEM), as the former gives only limited structural information, whereas LEEM has excellent structural sensitivity. Hence, if performed in the same instrument

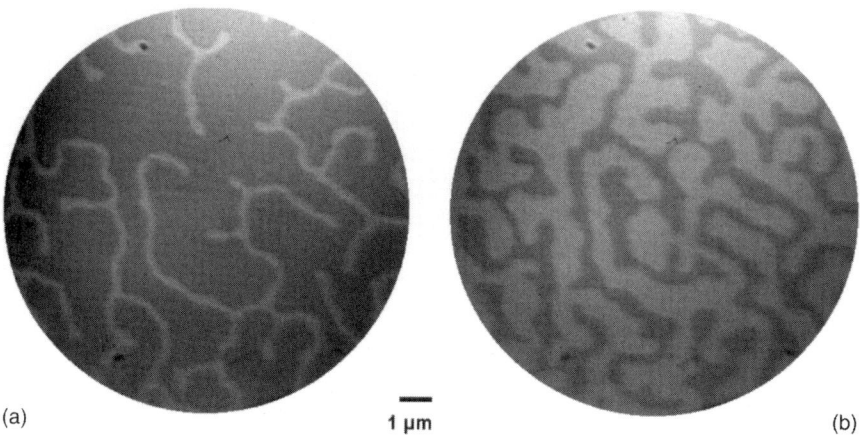

FIG. 9. Images acquired at 706 eV with the Gottingen microscope. (a) Magnetic field close to the saturation magnetization and (b) magnetic state where the magnetization is close to zero. (Figure courtesy of G. Schmahl [82].)

on the same sample, these techniques would be nicely complementary. In a first demonstration of recording XMCD data in a LEEM, Bauer's group has generated XMCD images of magnetic regions of a "nonmagnetic" steel with his LEEM/XPEEM microscope that was installed at the SX700-3 beamline at BESSY I [35]. His group has also directly combined LEEM and PEEM [102].

11.3.3 Semiconducting and Metal Surfaces and Overlayers

Tonner's initial demonstration of surface XANES microscopy with an XPEEM involved assessing the oxidation states of Si in native oxide and suboxide layers. When recording the desorption of a thin SiO_2 layer on Si, two different chemical desorption regimes were observed as a function of temperature; void nucleation and uniform, diffusion-rate limited desorption [30, 31]. A related and complementary study was recently performed with the MAXlab microscope, using XPS techniques [103]. This effort too found different desorption patterns of both native and thick Si oxide layers depending on temperature. Surface-shifted components in the Si $2p$ core level spectra furthermore indicated that at least part of the surface is reconstructed, both during and after desorption of the oxide layer.

During wafer processing, surface contaminants are of major importance for the microelectronic industry. Laser scattering methods can locate particulates, and electron microscopy can provide some compositional information on them. There is a big need, however, to rapidly locate and chemically characterize particulates that might be much smaller than 1 µm. The chemical composition, that is, knowledge about the oxidation state and possibly crystallographic phase, is needed in order to identify the contamination source during processing. Both XPEEM and micro-XPS have characteristics that appear promising as a solution to this problem. In a demonstration experiment, Brundle et al. [75, 104] used Tonner's microscope to show that 0.5 µm Alumina particulates can be readily located with high contrast in the PEEM (see Fig. 10) and that the Al XANES spectra reveal that the particulates are not just any aluminum oxide, but are indeed Alumina. The alternative approach, micro-XPS, is also pursued for this application. An instrument based on ellipsoidal KB mirrors is in the final commissioning stages at the ALS [40].

Various semiconducting surfaces have also been investigated with MAXIMUM. One of the major findings was the existence of spatial inhomogeneity, attributed to high defect densities, on cleaved surfaces that cause spectral differences either by pinning the fermi level or enhanced surface recombination [58]. Even if the fraction of the cleaved surface that is affected by high detect densities is relatively small, it could potentially substantially influence subsequent interface formation processes. Using spatially averaged spectra alone as a judgment if a cleaved surface is detect free is therefore a questionable strategy, as the signal from the defect-free regions can completely dominate these spectra.

The formation of $TiSi_2$ in confined spaces was recently studied by Singh et al. [105]. $TiSi_2$ has the lowest resistivity of all the refractory metal silicides and excellent compatibility with Al metallization. However, as dimensions in electronic circuits shrink, it has been observed that the $TiSi_2$ transformation from the high-resistivity C49 phase to a low-resistivity phase C54 is inhibited. Spectromicroscopy might be an excellent tool to investigate the evolution of these and related processes. Indeed, in the first spectromicroscopy studies of this system, the lateral variations of the local structure of the $TiSi_2$ could be directly imaged and the observations were attributed to the formation of the C54 phase in large areas and to the C49 phase at edges and in narrow spaces (see Fig. 11).

The ELETTRA SPEM has been used to study the interface of Au-Ag binary metal layers on Si. It revealed the existence of distinguished surface phases, attributed to surface alloying and growth and nucleation, and has illuminated the differences in the local electronic structure of these phases [44]. Clear spectral differences between reacted and metallic Au have been observed that vary across the sample. We show some of the images and spectra obtained in Fig. 12. In addition, the composition and electronic structure of 3D islands could not be revealed or detected from area-averaged spectra. In conceptually similar studies of the formation of complex interfaces, Bauer's group has started to investigate substrate-metal and metal-metal interactions [35, 106]. Energy filtered XPEEM was used to chemically distinguish crystals of different shapes as observed with

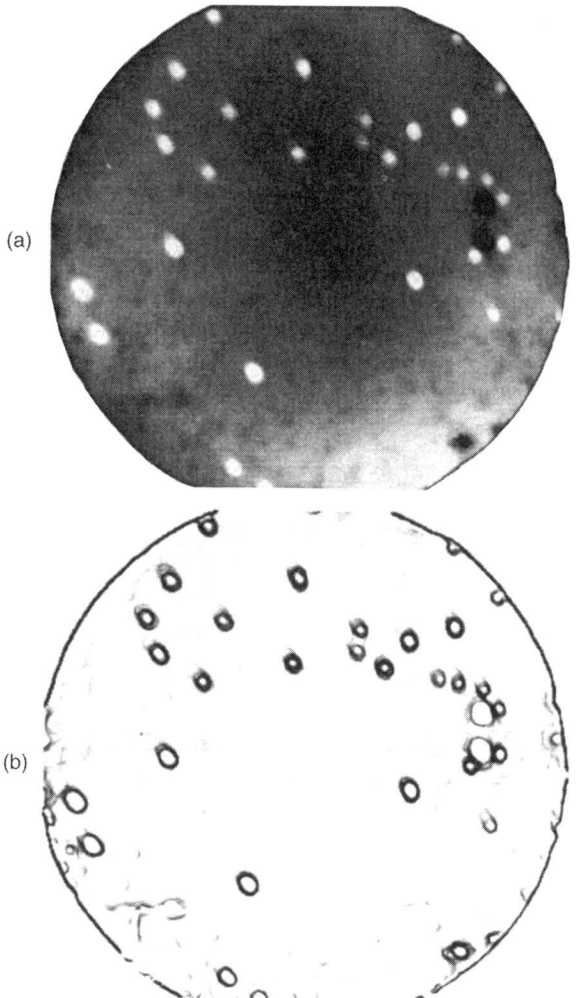

FIG. 10. (a) XPEEM images of 0.5 μm sized Alumina particulates on a Si wafer, acquired with a photon energy near the 72 eV Al 2p resonance. (b) The high chemical contrast makes it easy to automate finding the particulates with even simple image processing. (Figure reprinted from [75] with permission from Elsevier Science.)

the LEEM in Pd-Ag codeposited films on W(110) in order to correlate the composition to the observed morphology. They also determined the composition of differently shaped Pb crystals grown on Mo(110) in the presence of contamination and confirmed that both crystal types are Pb.

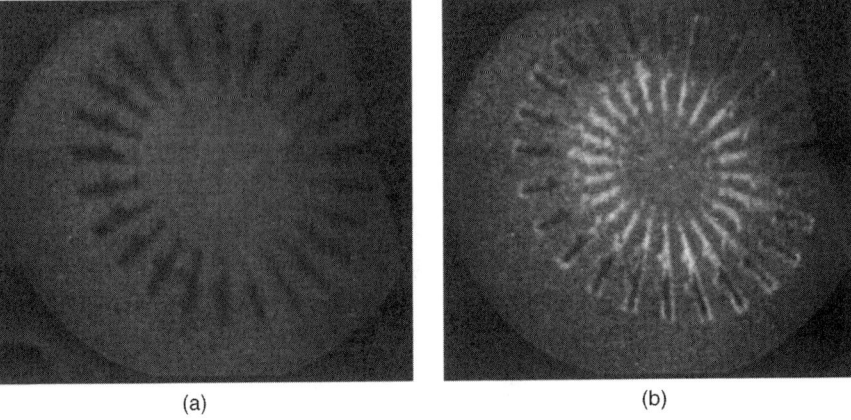

FIG. 11. XANES images of a part of a titanium silicide star test pattern. (a) Image acquired below the Ti $L_{2,3}$ edge at 445.5 eV. Only the centers of the spokes can been seen. (b) Image acquired at 455 eV, the shoulder of the absorption edge. The spoke tips and edges emit much more strongly than the spoke centers, indicating a change in the bonding in the silicide. [Reprinted with permission from S. Singh et al., Appl. Phys. Lett. **71**, 55 (1997). Copyright 1997 American Institute of Physics.]

11.3.4 Tribology and Corrosion

Tribology, that is, the study of wear and tear on the surface caused by rubbing, as well as corrosion are two promising areas where spectromicroscopy might make important contributions. The study of rubbed and worn surfaces, particularly when generated in a lubricant with additives, almost always demands the use of microcharacterization methods. Traditionally, SEM, Auger, and standard XPS methods have been used to characterize the surface morphologies and compositions, that are created in tribo-contacts. More recently, Atomic Force Microscopy/Lateral Force Microscopy (AFM/LFM) has been used to view surface textures, relative frictional properties, and adsorbate molecular structures on the wear surfaces [107] and their spatial distribution. The chemical specificity of spectromicroscopy can potentially address several longstanding issues in tribology. Ade et al. [85] have thus started to use the X1-SPEM to investigate wear scars on steel produced under controlled conditions. Tonner has also used his PEEM for tribological experiments [108]. Similar to tribology, corrosion proceeds as a very nonuniform process, and better understanding of the local chemical processes and composition involved might be obtained from laterally resolved spectroscopic data provided by spectromicroscopy. This has been first explored with MAXIMUM, which has been used to investigate the corrosion of metal Al-Cu-Si alloy films on Si [34].

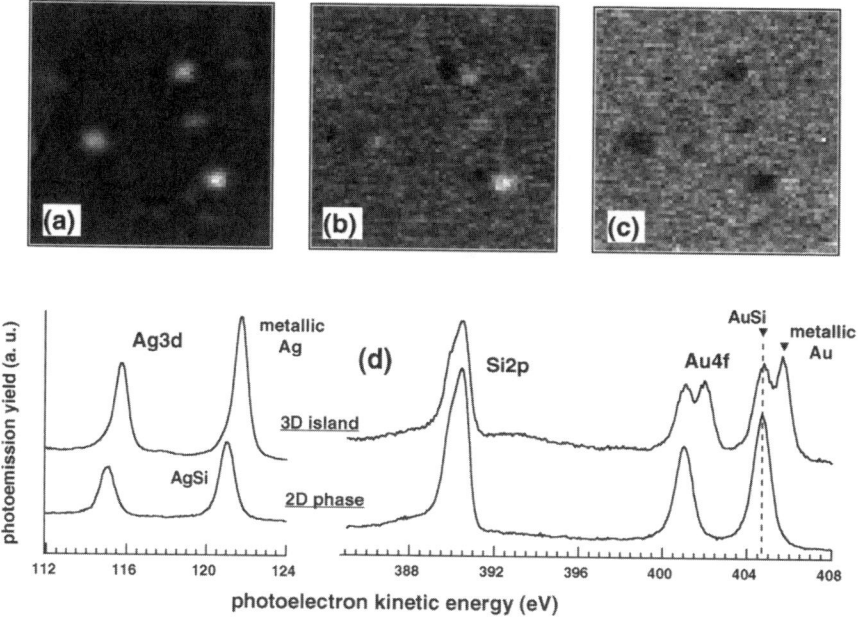

FIG. 12. Au chemical map acquired with an electron spectrometer tuned to metallic (a) and reacted (b) Au 4f spectral components. (c) Ag 3d, Si 2p, and Au 4f XPS spectra that show the different chemical state of the Ag and Au on the flat 2D areas and the 3D submicron metal islands of Au + Ag/Si(111). (Figure reprinted from [44] with permission from Elsevier Science.)

11.3.5 Composites

Ma et al. [69, 70] have investigated metal matrix composites with both XPEEM and MPM and compared their results with those of Auger microscopy. Of interest is an understanding of the interfacial interaction that occurs between the metal matrix and the ceramic reinforcement. Hence, detailed chemical information is highly desirable. The samples examined were SiC-coated carbon fibers in a titanium metal matrix. Samples imaged with XPEEM were polished and sputtered, whereas the samples studied with the MPM were fractured without further preparation. The fractured samples showed that the carbon-core is graphitic with a dominant sp^2 electronic structure, and the SiC coating is similar to that of crystalline, rather than amorphous, SiC with a dominant sp^3 electronic structure. The Auger microscopy did not have the chemical sensitivity to provide these details. It was also found that sample preparation methods play a crucial role in understanding the results of the measurements. While XPEEM yielded excellent contrast and resolution, the data were compromised by artifacts introduced by both the polishing and the sputtering necessary for sample preparation.

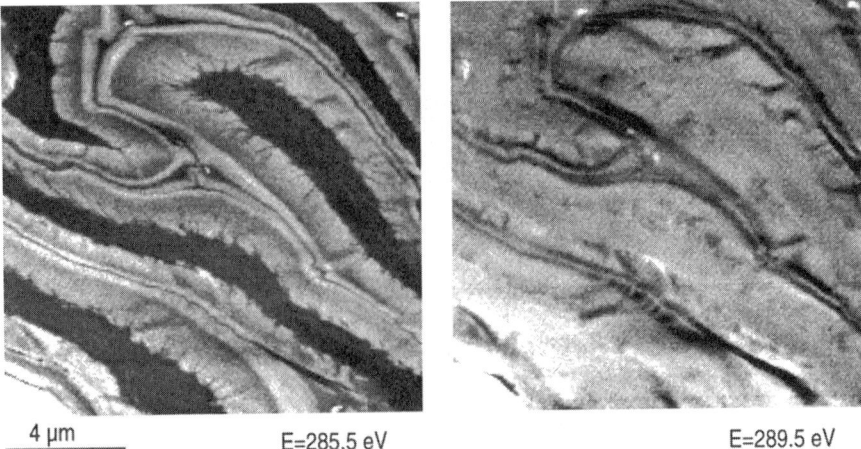

4 µm   E=285.5 eV   E=289.5 eV

FIG. 13. NEXAFS images of 40 million-year-old wood at the photon energies indicated. The variation of aromatic-olefinic concentration (a) as well as the carbohydrate concentration (b) can be mapped. (Figure courtesy of G. Cody.)

11.3.6 Geological Samples

Most of the applied and fundamental problems in fuel chemistry and organic geochemistry are related to determining the molecular structure of solid phase organics and relating this information to the time-dependent response of such systems when subjected to environmental stresses, such as temperature, pressure, and time. The difficulty in achieving characterization lies in the intrinsic microheterogeneous nature of such organic solids, such as coals and kerogens. Using NEXAFS microscopy as a new tool in this field, Cody et al. [109–112] have initiated a research program to address longstanding problems. They have, for example, followed the evolution of the molecular structure of sporopollenin, an important microscopic constituent in coals that can be identified by its shape. NEXAFS images of sporopollenin surrounded by a matrix of coal were acquired from a number of samples. Point NEXAFS spectra of the sporopollenin in these samples allowed Cody et al. to track the chemical structural evolution across a range of samples that had been subjected to progressively higher degrees of thermal metamorphism [109]. Determining molecular structure of sporopollenin and how it evolves when subjected to temperature pressure and time, holds the promise of addressing fundamental questions related to the mechanism and timing of oil generation.

NEXAFS microscopy can also unravel the complexity of solid-phase biomolecular materials. Figure 13(a) is a high-resolution image of 40 million-year-old wood acquired at 285.1 eV, where contrast is based on variations in the concentration of C=C bonded carbon, for example, aromatic or olefinic carbon.

Figure 13(b) is acquired at 289.1 eV, and much of the contrast is based on the concentration of carbon σ bonded with oxygen. The STXM reveals thus in enormous detail the high degree of chemical differentiation within the cell wall of wood and the carbohydrate distribution in these samples. Future work might address fundamental questions related to the fate of carbohydrates in organic-rich sediments over geological time.

As part of their studies on organic geochemical samples, Cody et al. also used linear dichroism microscopy to characterize the physical and chemical transformations that occur within coal during coking. It is important to study this process because high-quality metallurgical coke is crucial to the fabrication of steel. The physical and chemical changes that occur during coking are difficult to characterize because the critical stages involve nucleation at a very fine scale. Figure 14 shows the *in situ* NEXAFS analysis of the chemistry of nematic phases in a quenched coke. In Fig. 14(a), a pair of C-NEXAFS spectra highlight the high degree of molecular orientation in nematic phases of the same sample. Only orientation and not chemical differences are responsible for the observed intensity changes. The linear dichroism image of one of these samples in Fig. 14(b) reveals a spectacular tapestry, where the contrast is based entirely on molecular orientation relative to the polarization of the x-ray beam.

Inorganic geological materials have been studied with spectromicroscopy by Drouhay et al. [113]. They imaged the chemical composition of Ilminite in a PEEM and showed that samples that had been thought to be too insulating to be investigated in a PEEM still yield excellent images. The x-ray micrographs show the well-known lamellar domain structure (as observed in reflectance

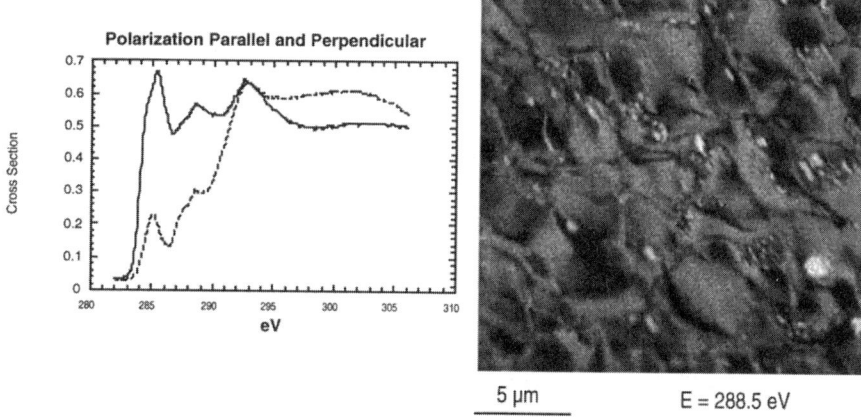

FIG. 14. NEXAFS spectroscopy (a) and dichroism microscopy (b) of a coke. Spectral difference and image contrast are solely due to variations in orientation and not chemical variations. (Figure courtesy of G. Cody.)

optical microscopy) and exhibit very strong contrast changes when imaged at the iron and titanium L edges, respectively, because of compositional variations. High-resolution spectroscopy from the two different regions show that iron is in two different charge states in the different domains, whereas titanium is in the same chemical state in both domains.

11.4 Discussion

We have presented an overview of the various approaches to spectromicroscopy and several classes of material applications as examples. In several cases we provided more details than in others in order to convey at least in a few instances the strengths of spectromicroscopy in depth. Some of the most popular applications involve carbonaceous materials that show rich spectroscopic variations, magnetic materials, as well as mesoscopic systems as they occur during thin film deposition and microstructuring. We have omitted a variety of spectromicroscopy applications involving biological samples [114, 115]. Numerous other interesting x-ray microscopy applications remained unmentioned so far, such as the study of high T_c materials [116], environmental studies, the aggregation in silica gels and zeolite precursors [117, 118], characterization of soot in lubricating oils [119], the curing of cement, characterization of defects in multilayer coatings [120], as well as others. This is partly because of a lack of space, but also because some of these applications do not explicitly make use of spectroscopic information.

In a series of interesting experiments, for example, Niemeyer and colleagues used the Göttingen microscope to image clay and similar minerals in suspension [78, 79]. In particular, they studied montmorillinite and found evidence for both the "cardhouse" and "brookhouse" structures previously postulated. Studying the samples in "wet cells" allowed the change of various aspects of the suspension medium, such as the pH and ionic concentration, resulting in dramatic morphological changes. The effects of surface surfactants, potentially used in soil decontamination processes, have also been investigated. We believe that these studies would greatly benefit if spectroscopic information could be added to the present studies that rely on morphological observations only. We are certain that spectromicroscopy experiments of these samples and problems will occur in the future, as more spectromicroscopy capacity will become available. Tonner, for example, started to use spectromicroscopy for environmental studies of hydrated samples with the ALS STXM.

Many of the projects and efforts discussed have been made possible by advances in source brightness with the advent of second- and third-generation synchrotron facilities and the concurrent development of high-resolution optics. We anticipate these advances to continue for both microprobe and imaging microscopes. Several efforts are under way to further improve zone plate optics, and it now seems possible to reach 10 nm spatial resolution with zone plates in the near future. Other possibilities for improvements exist. A particularly ambitious spectromicroscopy effort, for example, is planning to achieve energy-filtered PEEM with an aberration-corrected electron optical system [121]. The chromatic and spherical aberrations of the objective lens are corrected by means of an electrostatic mirror in combination with a corrected magnetic beam separator. The proposed instrument has a calculated theoretical spatial resolution of better than 1 nm. Energy resolution is provided by a corrected omega filter and should be as good as 100 meV. Although the realization of this instrument will take several years, it does provide a glimpse of the future and represents the ambitions of the field of spectromicroscopy.

11.5 Conclusions

The capabilities of x-ray microscopy have been greatly enhanced with the addition of spectroscopic capabilities. Increasingly varied applications are performed with x-ray spectromicroscopy, which is evolving into an ever more efficient general purpose analytical tool. Although traditional x-ray microscopy as a whole might be starting to mature, we are clearly only at the beginning of spectromicroscopy. Almost every new microscopy project at synchrotron facilities around the world is aimed at combining high-spatial resolution with spectroscopic information, and new and different technological approaches are pursued. Spectromicroscopy instrumentation is complex and new solutions have to be found in order to increase the productivity and ease of use of these instruments. Given the advances that have been made recently, one can look forward to an exciting period in the future.

Acknowledgments

I am grateful to R. Brundle, F. Cerrina, G. Cody, M. Kiskinova, J. Stöhr, G. Schmahl, and B. Tonner for sharing their results, for permission to publish their figures, and for providing various figures.

Of course, much of the work presented is based on the efforts of a rather large group of people affiliated with the various spectromicroscopy efforts. There are too many to list them here, but I would like to thank them all for making spectromicroscopy an exciting and growing field. Many of my own personal results would have been impossible without the efforts of J. Kirz and C. Jacobsen and their research groups who build and maintain the Stony Brook STXM at the NSLS. I am grateful to everybody involved.

This work is supported in part by a National Science Foundation Young Investigator Award (DMR-9458060).

References

1. J. Stöhr, *NEXAFS Spectroscopy*. Springer-Verlag, Berlin, 1992.
2. J. Stöhr, *J. Electron Spectros. Relat. Phenom.* **75**, 253–272 (1995).
3. H. Ade, *Trends Polym. Sci.* **5**, 58–66 (1997).
4. S. G. Urquhart, A. P. Hitchcock, A. P. Smith, H. Ade, and E. G. Rightor, *J. Phys. Chem.* B **101**, 2267–2276 (1997).
5. J. Kirz, C. Jacobsen, and M. Howells, *Q. Rev. Biophys.* **28**, 33–130 (1995).
6. G. Schmahl and D. Rudolf, eds., *X-ray Microscopy*. Springer-Verlag, Berlin, 1984.
7. D. Sayre, M. Howells, J. Kirz, and H. Rarback, eds., *X-ray Microscopy II*. Springer-Verlag, Berlin, 1988.
8. A. G. Michette, G. R. Morrison, and C. J. Buckley, eds., *X-ray Microscopy III*. Springer-Verlag, Berlin, 1990.
9. V. V. Aristov and A. I. Erko, eds., *X-ray Microscopy IV*. Bogorodski Pechatnik, Chernogolovka, Moscow Region, 1994.
10. J. Thieme, G. Schmahl, E. Umbach, and D. Rudolph, *X-ray Microscopy and Spectromicroscopy*. Springer-Verlag, Berlin, 1998.
11. M. Howells, J. Kirz, D. Sayre, and G. Schmahl, *Physics Today* **38**, 22–32 (1985).
12. M. R. Howells, J. Kirz, and D. Sayre, *Scient. Am.* **264**, 88–94 (1991).
13. J. Kirz and D. Sayre, in *Synchrotron Radiation Research*, H. Winick and S. Doniach, eds. pp. 277–322. Plenum Press, New York, 1980.
14. J. Kirz and H. Rarback, *Rev. Sci. Instrum.* **56**, 1–13 (1985).
15. G. Schmahl, D. Rudolph, B. Niemann, and O. Christ, *Q. Rev. Biophys.* **13**, 297–315 (1980).
16. G. Schmahl and P.-C. Cheng, in *Handbook on Synchrotron Radiation*, S. Ebashi, M. Koch, and E. Rubenstein, eds. Vol. 4, pp. 483–536. Elsevier, Amsterdam, 1991.
17. D. Sayre and H. N. Chapman, *Acta Crystallographica* **A51** (1995).
18. P. Coxon, J. Krizek, M. Humpherson, and I. K. M. Wardell, *J. Electron Spectros. Relat. Phenom.* **51–52**, 821–836 (1990).
19. U. Gelius, B. Wannberg, P. Baltzer *et al.*, *J. Electron. Spectrosc. Relat. Phenom.* **52**, 747–785 (1990).
20. C. Coluzza and R. Moberg, *J. Electron Spectros. Relat. Phenom.* **84**, 109 (1997).
21. H. Ade, ed., *J. Electron Spectrosc. Relat. Phenom.* **84** (1997).
22. B. Tonner, *Syn. Rad. News* **4**, 27–32 (1991).
23. E. G. RightorA. P. Hitchcock, H. Ade *et al.*, *J. Phys. Chem.* **B101**, 1950–1961 (1997).
24. D. A. Muller, Y. Tzou, R. Raj, and J. Silcox, *Nature* **366**, 725–727 (1993).
25. G. Beamson, H. Q. Porter, and D. W. Turner, *Nature* **290**, 556–561 (1981).
26. J. Kirschner, in *X-Ray Microscopy*, C. Schmahl and D. Rudolph, eds. pp. 308–313. Springer-Verlag, Berlin, 1984.
27. H. Ade, J. Kirz, H. Rarback, S. Hulbert, E. Johnson, D. Kern, P. Chang, and Y. Vladimirsky, in *X-Ray Microscopy II*, D. Sayre, M. Howells, J. Kirz, and H. Rarback, eds. pp. 280–283. Springer-Verlag, Berlin, 1988.
28. H. Ade, J. Kirz, S. Hulbert, E. Johnson, E. Anderson, and D. Kern, *Appl. Phys. Lett.* **56**, 1841–1843 (1990).
29. H. Ade, C. H. Ko, and E. Anderson *Appl. Phys. Lett.* **60**, 1040 (1992).
30. G. R. Harp, Z. L. Han, and B. P Tonner, *Physica Scripta* **T31**, 23–27 (1990).
31. G. R. Harp, Z. L. Han, and B. P. Tonner, *J. Vac. Sci. Technol.* **8**, 2566 (1990).

32. C. Jacobsen, S. Williams, E. Anderson et al., *Opt. Commun.* **86**, 351 (1991).
33. H. Ade, X. Zhang, S. Cameron, C. Costello, J. Kirz, and S. Williams, *Science* **258**, 972 (1992).
34. W. Ng, A. K. Ray-Chaudhuri, S. Liang et al., *Nucl. Instrum. Meth. in Phys. Res.* **A347**, 422–430 (1994).
35. E. Bauer, C. Koziol, G. Lilienkamp, and T. Schmidt, *J. Electron Spectrosc. Relat. Phenom.* **84**, 201 (1997).
36. S. Spector, C. Jacobsen, and D. Tennant, in *X-ray Microscopy and Spectromicroscopy*, J. Thieme, G. Schmahl, E. Umbach, and D. Rudolph, eds. Springer-Verlag, Berlin, 1997.
37. J. Thieme, C. David, N. Fay et al., in *X-ray Microscopy IV*, V. V. Aristov and A. I. Erko, eds. pp. 487–493. Bogorodski Pechatnik, Chernogolovka, Moscow Region, 1994.
38. D. Attwood, in *X-ray Microscopy, IV*, V. V. Aristov and A. I. Erko, eds. pp. 20–34. Bogorodski Pechatnik, Chernogolovka, Moscow Region, 1994.
39. H. Aschoff, *Diplomarbeit*. Universität Göttingen, 1994.
40. T. Warwick, H. Ade, A. P. Hitchcock, H. Padmore, B. Tonner, and E. Rightor, *J. Electron Spectrosc. Relat. Phenom.* **84**, 85 (1997).
41. C.-H. Ko, J. Kirz, H. Ade, E. Johnson, S. Hulbert, and E. Anderson, *Rev. Sci. Instrum.* **66** (1995).
42. C. R. Zhuang, H. Zhang, H. Ade et al., in *X-ray Microscopy and Spectromicrosopy*, J. Thieme, G. Schmahl, E. Umbach, and D. Rudolph, eds. Springer-Verlag, Berlin, 1997.
43. M. Marsi, L. Casalis, L. Gregoratti, S. Günther, A. Kolmakov, J. Kovac, D. Lonza, and M. Kiskinova, in *X-ray Microscopy and Spectromicroscopy*, J. Thieme, G. Schmahl, E. Umbach, and D. Rudolph, eds. Springer-Verlag, Berlin, 1997.
44. M. Marsi, L. Casalis, L. Gregoratti, S. Günther, A. Kolmakov, J. Kovac, D. Lonza, and M. Kiskinova, *J. Electron Spectrosc. Relat. Phenom.* **84**, 73 (1997).
45. C.-H. Ko, R. Klauser, and T. J. Chuang, in *X-ray Microscopy and Spectromicroscopy*, J. Thieme, G. Schmahl, E. Umbach, and D. Rudolph, eds. Springer-Verlag, Berlin, 1997.
46. J.-D. Wang, Y. Kagoshina, T. Miyahara, M. Ando, S. Aoki, E. Anderson, D. Attwood, and D. Kern, *Rev. Sci. Instrum.* **66**, 1401–1403 (1995).
47. Y. Kagoshima, J. Wang, T. Miyahara, M. Ando, and S. Aoki, in *X-ray Microscopy and Spectromicroscopy*, J. Thieme, C. Schmahl, E. Umbach, and D. Rudolph, eds. Springer-Verlag, Berlin, 1997.
48. C. Kunz and J. Voss, *Rev. Sci. Instrum.* **66**, 2021–2029 (1995).
49. J. Voss, *J. Electron Spectrosc. Relat. Phenom.* **84**, 29 (1997).
50. J. Voss, L. Storjohann, C. Kunz et al., in *X-ray Microscopy IV*, A. I. Erko and V. V. Aristov, eds. Bogorodski Pechatnik, Chernogolovka, Moscow Region, 1994.
51. J. Voss, H. Dadras, C. Kunz, A. Moewes, G. Roy, H. Sievers, I. Storjohann, and H. Wongel, *J. X-ray Sci. Technol.* **3**, 85–108 (1992).
52. J. Johansson, R. Nyholm, C. Törnevik, and A. Flodström, *Rev. Sci. Instrum.* **66**, 1398 (1995).
53. S. Aoki, T. Ogato, S. Sudo, and T. Onuky, *Jpn. J. Appl. Phys.* **31**, 3477 (1992).
54. S. Aoki, T. Ogata, K. Iimura, N. Watanabe, T. Yoshidomi, K. Shinada, and T. Kato, in *X-Ray Microscopy and Spectromicroscopy*, J. Thieme, G. Schmahl, E. Umbach, and D. Rudolph, eds. Springer-Verlag, Berlin, 1997.
55. M. R. Weiss, V. Wüstenhagen, R. Fink, and E. Umbach, *J. Electron Spectrosc. Relat. Phenom.* **84**, 9, (1997).

56. E. Spiller, in *Methods of Vacuum Ultraviolet Spectroscopy*, J. A. R. Samson and E. I. Ederer, eds. Academic Press, Chestnut Hill, MA, 1997.
57. R.-P. Haelbich, in *Scanned Image Microscopy*, E. A. Ash, eds. Academic Press, London, 1980.
58. F. Cerrina, A. K. Ray-Chaudhuri, W. Ng et al., *Appl. Phys. Lett.* **63**, 63–65 (1993).
59. J. E. Bjorkholm, J. Bokor, L. Eichner et al., *J. Vac. Sci. Tech.* **B8**, 1509–1513 (1990).
60. F. Barbo, M. Bertolo, A. Bianco et al., in *X-ray Microscopy and Spectromicroscopy*, J. Thieme, G. Schmahl, E. Umbach, and D. Rudolph, eds. Springer-Verlag, Berlin, 1997.
61. V. Wustenhagen, M. Schneider, J. Taborsi, W. Weiss, and E. Umbach, *Vacuum* **41**, 1577–1580 (1990).
62. N. V. Smith, W. A. Royer, and J. E. Rowe, *Rev. Sci. Instrum.* **65**, 1954–1958 (1994).
63. D. H. Bilderback and D. J. Thiel, *Rev. Sci. Instrum.* **66**, 2059–2063 (1995).
64. D. W. Turner, I. R. Plummer, and H. W. Porter, *Rev. Sci. Instrum.* **59**, 797 (1988).
65. P. Pianetta, P. L. King, A. Borg, C. Kim, I. Lindau, G. Knapp, M. Keenlyside, and R. Browning, *J. Electron Spectrosc. Relat. Phenom.* **52**, 797–811 (1990).
66. P. L. King, A. Borg, C. Kim, S. A. Yoshikawa, P. Pianetta, and I. Lindau, *Ultramicroscopy* **36**, 117–129 (1991).
67. P. L. King, A. Borg, C. Kim, P. Pianetta, I. Lindau, G. S. Knapp, M. Keenlyside, and R. Browning, *Nucl. Instr. Meth. in Phys. Res.* **A 291**, 19–25 (1990).
68. G. D. Waddill, T. Komeda, P. J. Benning, J. H. Weaver, and G. S. Knapp, *J. Vac. Sci. Technol.* **A9**, 1634–1639 (1991).
69. Q. Ma, R. A. Rosenberg, C. Kim, J. Grepstad, and P. Pianetta, *Appl. Phys. Lett.* **70**, 2389 (1997).
70. Q. Ma, R. A. Rosenberg, C. Kim, J. Grepstad, P. Pianetta, T. Droubay, D. Dunham, and B. Tonner, *J. Electron Spectrosc. Relat. Phenom.* **84**, 99–107 (1997).
71. O. H. Griffith and W. Engel, *Ultramicroscopy* **36**, 1–28 (1991).
72. M. Mundschau, E. Bauer, and W. Swiech, *Surf. Sci.* **203**, 412–422 (1988).
73. O. H. Griffith and G. Rempfer, *Advances in Optical and Electron Microscopy* **10** (1987).
74. B. P. Tonner and G. R. Harp, *J. Vac. Sci. Technol.* **7**, 1–4 (1989).
75. B. P. Tonner, D. Dunham, T. Droubay, J. Kikuma, J. Denlinger, E. Rotenberg, and A. Warwick, *J. Electron Spectros. Relat. Phenom.* **75**, 309–332 (1995).
76. B. Nieman, D. Rudolph, and G. Schmahl, *Appl Optics* **15**, 1883–1884 (1976).
77. W. Meyer-Ilse, T. Wilhein, and P. Guttmann, in *In Charge-Coupled Devices and Solid State Optical Sensors III*, M. M. Blouke, eds. Vol. 1900. SPIE, 1993.
78. J. Niemeyer, J. Thieme, and P. Guttmann, in *X-ray Microscopy IV*, V. V. Aristov and A. I. Erko, eds. Pp. 164–170. Bogorodski Pechatnik, Chernogolovka, Moscow Region, 1994.
79. J. Niemeyer, J. Thieme, P. Guttmann, D. Rudolph, and G. Schmahl, in *X-ray Microscopy and Spectromicroscopy*, J. Thieme, G. Schmahl, E. Umbach, and D. Rudolph, eds. Springer-Verlag, Berlin, 1997.
80. J. Abraham, R. Medenwaldt, E. Uggerhoj et al., in *X-ray Microscopy and Spectromicroscopy*, J. Thieme, G. Schmahl, E. Umbach, and D. Rudolph, eds. Springer-Verlag, Berlin, 1997.
81. W. Meyer-Ilse, D. Attwood; and M. Koike, in *Synchrotron Radiation in Biosciences*, B. Chance et al., eds. Pp. 624–636. Clarendon Press, Oxford, 1994.
82. G. Schmahl, P. Guttmann, D. Raasch, P. Fischer, and G. Schütz, *Synchr. Rad. News* **9**, 35–39 (1996).

83. H. Adc, A. Smith, S. Cameron, R. Cieslinski, C. Costello, B. Hsiao, G. Mitchell, and E. Rightor, *Polymer* **36**, 1843–1848 (1995).
84. H. Ade, A. P. Smith, G. R. Zhuang et al., *Mater. Res. Soc. Symp. Proc.* **437**, 99 (1996).
85. H. Ade, A. P. Smith, H. Zhuang, J. Kirz, E. Rightor, and A. Hitchcock, *J. Electron Spectrosc. Relat. Phenom.* **84**, 53 (1997).
86. E. G. Rightor, A. P. Hitchcock, S. G. Urquhart et al., *ALS compendium for 1993–96*, LBNL 39981 (1997).
87. A. P. Hitchcock, S. G. Urquhart, E. G. Rightor, W. Lidy, H. Ade, A. P. Smith, and T. Warwick, in *Microscopy and Microanalysis*, **3**, (1997).
88. S. Qu, et al., (1997).
89. A. P. Smith, H. Ade, M. Rafailovich, and J. Sokolov, (unpublished) (1997).
90. A. Cossy-Favre, J. Diaz, S. Anders et al., *Acta. Phys. Polonica* **A91** (1997).
91. J. Kikuma, T. Warwick, J. Zhang, and B. P. Tonner (to be published).
92. C. Zimba, A. P. Smith, and H. Ade, *Macromolecules* (to be submitted).
93. C. Zimba, A. P. Smith, and H. Ade, *Analytical Chemistry* (to be submitted).
94. P. Vaterlein, V. Wüstenhagen, and E. Umbach, *Appl. Phys. Lctt.* **66**, 2200–2202 (1995).
95. H. Ade and B. Hsiao, *Science* **262**, 1427 (1993).
96. A. P. Smith and H. Ade, *Appl. Phys. Lett.* **69**, 3833 (1996).
97. H. H. Yang, *Aromatic High Strength Fibers*. Wiley-Interscience, New York, 1989.
98. J. Stöhr, M. G. Samant, Y. Wu, B. D. Hermsmeier, G. R. Harp, S. Koranda, D. Dunham, and B. P. Tonner, *Science* **259**, 658–661 (1993).
99. F. U. Hillebrecht, D. Spanke, J. Dresselhaus, and V. Solinus, *J. Electron Spectrosc. Relat. Phenom.* **85**, 189 (1997).
100. F. U. Hillebrecht, T. Kinoshita, D. Spanke, J. Dresselhaus, Ch. Roth, H. B. Rose, and E. Kisker, *Phys. Rev. Lett.* **75**, 2224–2227 (1995).
101. W. Swiech, G. H. Fecher, Ch. Ziethen et al., *J. Electron Spectrosc. Relat. Phenom.* **85** (1997).
102. C. Koziol, T. Schmidt, M. Altman, T. Kachel, G. Lilienkamp, E. Bauer, and W. Gudat (unpublished).
103. U. Johansson, H. Zhang, and R. Nyholm, *J. Electron Spectrosc. Relat. Phenom.* **85**, 45 (1997).
104. C. R. Brundle, A. Warwick, R. Hockett et al. (unpublished).
105. S. Singh, H. Solak, N. Krasnoperov, S. Cerrina, A. Cossy, J. Diaz, J. Stohr, and M. Samant, *Appl. Phys. Lett.* **71**, 55 (1997).
106. G. Lilienkamp, C. Koziol, Th. Schmidt, and E. Bauer, in *X-ray Microscopy and Spectromicroscopy*, J. Thieme, G. Schmahl, E. Umbach, and D. Rudolph, eds. Springer-Verlag, Berlin, 1997.
107. R. Overney and E. Meyer, *MRS Bulletin* **18**, 26–34 (1993).
108. B. P. Tonner, *APS Bulletin* **42**, 142 (1997).
109. G. D. Cody, R. E. Botto, H. Ade, and S. Wirick, *Int. J. Coal Geol.* **32**, 69–86 (1996).
110. G. D. Cody, R. E. Botto, H. Ade, S. Behal, M. Disko, and S. Wirick, *Energy & Fuels* **9**, 525–533 (1995).
111. G. D. Cody, R. E. Botto, H. Ade, S. Behal, M. Disko, and S. Wirick, *Energy & Fuels* **153** (1995).
112. R. E. Botto, G. D. Cody, J. Kirz, H. Ade, S. Behal, and M. Disko, *Energy & Fuels* **8**, 151–154 (1994).

113. T. Droubay, G. Mursky, and B. P. Tonner, *J. Electron Spectrosc. Relat. Phenom.* **85** (1997).
114. G. De Stasio and G. Margaritondo, *J. Electron Spectros. Relat. Phenom.* **85** (1997).
115. X. Zhang, R. Balhorn, J. Mazrimas, and J. Kirz, *J. Struc. Biol.* **116**, 335–344 (1996).
116. Y. Hwu, *J. Electron Spectrosc. Relat. Phenom.* **85** (1997).
117. G. R. Morrison, M. T. Browne, T. P. M. Beelan, H. F. van Garderen, and P. A. F. Anastasi, presented at the X-ray Microscopy '93, Chernogolovka, Moscow, 20–24 Sept., 1993 (1993).
118. G. R. Morrison, M. T. Browne, T. P. M. Beelen, and H. F. van Garderen, *SPIE 1741 Soft X-ray Microscopy*, 312–315 (1992).
119. C. R. Morrison, A. D. H. Clague, J. T. Gauntlett, and P. J. Shuff, in *X-ray Microscopy and Spectromicroscopy*, J. Thieme, G. Schmahl, E. Umbach, and D. Rudolph, eds. Springer-Verlag, Berlin, 1997.
120. J. Friedrich, K. Behrens, V. Rautenfeldt, I. Diel *et al.*, in *X-ray Microscopy and Spectromicroscopy*, J. Thieme, G. Schmahl, E. Umbach, and D. Rudolph, eds. Springer-Verlag, Berlin, 1997.
121. R. Fink, M. R. Weiss, E. Umbach *et al.*, *J. Electron Spectrosc. Relat. Phenom.* **84**, 231 (1997).
122. B. P. Tonner, D. Dunham, J. Zhang, W. L. O'Brien, M. Samant, D. Weller, B. D. Hermsmeier, and J. Stöhr, *Nucl. Instr. Meth. in Phys. Res.* **A 347**, 142–147 (1994).

12. OPTICAL SPECTROSCOPY IN THE VUV REGION

Marshall L. Ginter
Institute for Physical Science and Technology
University of Maryland
College Park, Maryland

Kouichi Yoshino
Harvard-Smithsonian Center for Astrophysics
Cambridge, Massachusetts

12.1 Introduction

Atomic and molecular spectra in the VUV region are major contributors to the experimental characterizations of the structures of excited electronic states. Such spectra also are important to the identification of species participating in physical and chemical processes in environments ranging from terrestrial to astrophysical. In this chapter, VUV will imply wavelengths from ~2000 Å, (200 nm) into the edge of the soft x-ray region (~150 Å).

Traditionally, VUV spectroscopy has relied heavily on diffraction grating-based instrumentation, for which spectral resolution is strongly dependent on such factors as instrumental dispersion, slit widths, and grating quality. Because there are practical limits to minimum slit widths (~10 μ) for such systems, instrumental resolving power (RP) usually is limited by focal-plane dispersion [1]. Thus, the high RP (~80,000 or better) necessary to many VUV applications [1] usually is obtained from grating-based instrumentation by various combinations of long focal length, high spectral order, and grating ruling densities [1–3].

In addition to refinement of traditional spectrometers, a number of new spectroscopic tools have been added recently. These include laser-based single and multiphoton spectrometers [4, 5], pump-probe systems that access states traditionally observed by VUV spectroscopies [6, 7], and interferometers that operate efficiently at wavelengths shorter than 2000 Å. Instrumental RP limitations for these systems depend strongly on laser line widths in the first two examples and on VUV reflectivities and optical surface quality in the third.

In some VUV spectroscopic observations, RP is limited by line broadening (lifetime and/or Doppler) rather than by instrumental effects. Usually, lifetime broadening is of greater importance because the upper state of a VUV spectroscopic transition often lies energetically above ionization and/or dissociation limits.

Figure 1 provides an example of how resolution affects both the separability of different spectral features lying close together and the line shape and apparent

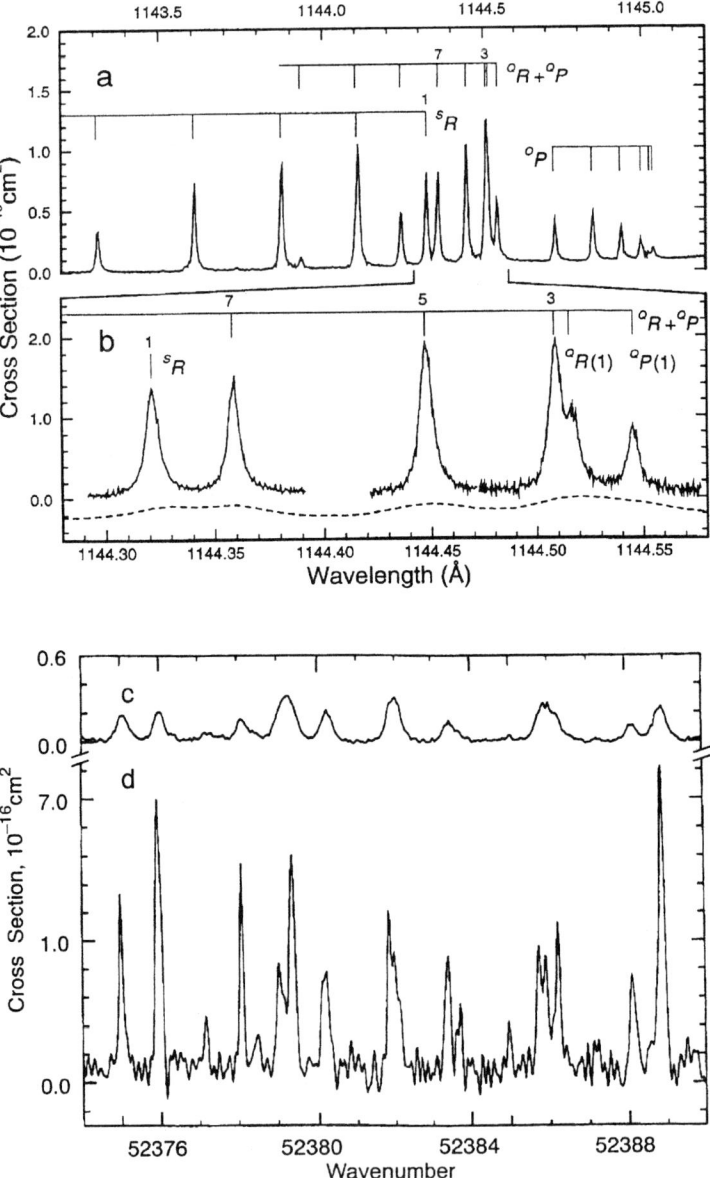

FIG. 1. Photoabsorption spectrum of O_2 (a and b) and NO (c and d) at various instrumental resolving powers. The resolving powers are: ~90,000 in a; ~300,000 and 20,000 for the solid and dashed spectra, respectively, in b; ~130,000 in c; and ~830,000 in d. See text for discussion. The spectra of O_2 in a and b were provided by Dr. B. R. Lewis.

intensities of such features. Figures 1a and 1b show photoabsorption spectra of the (0, 0) band of the $4p\pi_u j\,^1\Sigma_u^+ \leftarrow X\,^3\Sigma_g^-$ transition in $^{16}O_2$ measured at 79 K. The scan in Fig. 1a was obtained using nonlinear third harmonic generation in Kr with a scanning band width of ~0.5 cm^{-1} full-width half maximum (FWHM). The solid curve in Fig. 1b utilized a similar generating system with a scanning band width of ~0.15 cm^{-1} FWHM, and the dashed curve (displaced downward by ~0.2 for improved clarity) utilized a scanning VUV monochromer (2.2 m focal length) with a scanning band width of ~2.2 cm^{-1} FWHM. Figures 1c and 1d show photoabsorption spectra of the NO δ (0, 0) band. Figure 1c was taken using a 6.65 m VUV focal plane scanning spectrometer with a resolution of ~0.4 cm^{-1}, while Fig. 1d was obtained using a VUV Fourier transform spectrometer [8] with a resolution of ~0.06 cm^{-1}. Notice that increasing resolution not only improves line separation but also increases apparent line intensities and shapes (especially in the Lorentzian wings). In terms of $\lambda/\Delta\lambda$ or $\nu/\Delta\nu$, and assuming $\Delta\lambda$ or $\Delta\nu$ represent FWHM line widths, the instrumental resolving powers are: ~90,000 in Fig. 1a; ~300,000 and 20,000 for the solid and dashed spectra, respectively, in Fig. 1b; ~130,000 in Fig. 1c; and ~830,000 in Fig. 1d. Also notice that when spectral line widths are small compared with instrumental resolution (such as for the dashed curve in Fig. 1b), the lines disappear into the background, a well-known phenomenon that can be seen in other regions [9] in O_2 as well as many other spectra.

12.2 Wavelength Measurements and Energy Levels

Because wavelength is inversely proportional to energy, relative and absolute uncertainties in wavelength determinations translate directly to analogous uncertainties in transition energies and empirical energy level structures. Traditional high RP photographic plate detection has been especially important in the area of precise VUV wavelengths because hundreds of Ångstroms of a spectrum can be measured against standard reference lines on a single photographic plate, a process that ensures high relative precision and simplifies absolute wavelength determinations [10, 11]. On the other hand, laser-based instrumentation typically takes data in sub- or few-Ångstrom increments mechanically disjoint from one another. These data segments must be pieced together smoothly, often with few VUV reference lines available in many of the increments. Although interferometry, which at present is limited to the longest wavelength segment of the VUV, may eventually become a player in the shorter wavelength VUV, large array and/or plate detection will remain important for precise determination of the wavelengths of sharp spectral features in the VUV region.

Empirical determinations of atomic and molecular structures from the analysis of VUV spectra utilize pattern recognition to establish starting points for an analysis and the combination principle (CP) to verify and quantify level energies. Energy structures themselves can become very complex, with Rydberg series and their associated continua forming channels [12, 13] built on ionic cores. Since each energy level in the parent ion becomes a separate core with its own large manifold of channel structures, the potential for interactions between these many different Rydberg structures and between Rydberg and valence states is very large. Strong interactions affect both spectral energies and intensities [14], as can be seen from Fig. 2, which shows an example of channel-coupling effects in Ge I

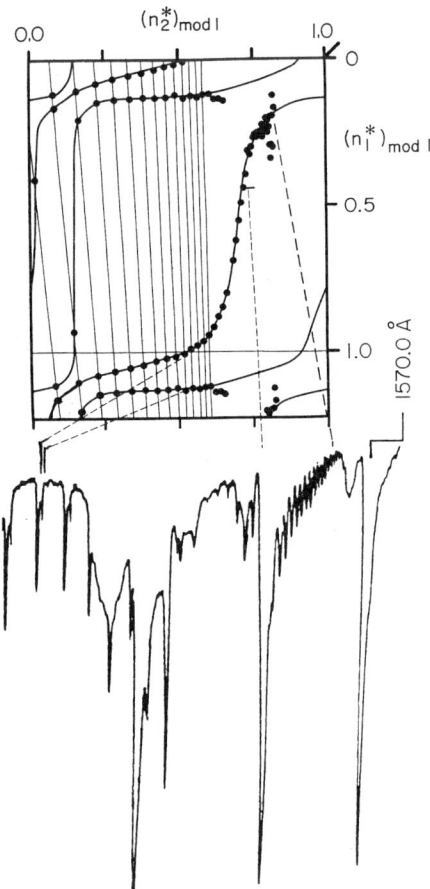

FIG. 2. Photoabsorption spectrum of Ge I near 1570 Å (bottom) compared with the associated multichannel quantum defect theory (MQDT) (Lu-Fano) diagram correlating a portion of the $J = 1°$ channel couplings (see [15] for details). The bottom panel is a densitometer trace of the photographically recorded spectrogram (top) taken using a grating spectrograph at a RP > 150,000.

[15] typical in the Group IVB elements [16, 17]. Additional atomic [12, 13, 18, 19] and comparable molecular examples abound [12, 13, 20, 21].

In addition to illustrating the effects of resolution on spectral line separabilities, intensities and shapes, Fig. 1 provides examples of one-photon absorption spectra obtained by harmonic-generated wavelength scanning (Figs. 1a, 1b) and focal plane scanning grating spectroscopy (Fig. 1c) with photoelectric detection and their use in the characterization of highly excited energy levels. Similarly, the analysis and identifications for Ge I illustrated in Fig. 2 (top panel) provide an example of a one-photon high-resolution (HR) spectrum obtained using a long focal length spectrograph with photographic plate detection [15] and a rare gas continuum background light source [22]. Additional examples of the analysis of one-photon HR grating spectra appear later in the section on intensity measurements.

Another technique for obtaining HR spectral data in the VUV region is illustrated in Fig. 3, which shows a resonance-enhanced multiphoton ionization (REMPI) spectrum of a $2^\pm \leftarrow X\,^1\Sigma^+$ (0-0) transition in HI observed near 1390 Å (72,000 cm^{-1}). The (2 + 1) REMPI spectrum in Fig. 3 was obtained using a frequency doubled Nd:YAG pumped tunable dye laser system [23], with the "+1" photon providing charged particles for signal detection. Other detection schemes are often used in multiphoton spectroscopies, such as fluorescence from excited or from probed dissociation products. Two-photon spectra usually provide information complementary to that obtained from one-photon spectra

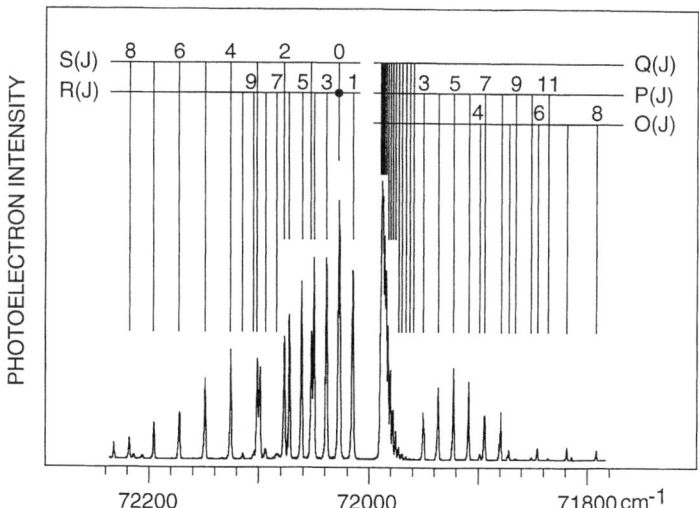

FIG. 3. (2 + 1) REMPI spectrum of the $2^\pm \leftarrow X\,^1\Sigma^+$ (0-0) band in HI observed near 1390 Å (72,000 cm^{-1}).

because of different selection rules. For example, for molecules with centers of symmetry, one- and two-photon spectra involve totally different manifolds of excited states (i.e., electronic states with the opposite [5, 9] and the same [24] parity as the initial state, respectively). Even in the case of molecules without centers of symmetry, the differences in one- and two-photon selection rules tend to produce much complementary information (i.e., the very intense $2^{\pm} \leftarrow X^1\Sigma^+$ spectrum in Fig. 3 is not seen in one-photon observations, whereas many transitions of the $1^{\pm} \leftarrow X^1\Sigma^+$ type are intense in one-photon and weak in two-photon spectra [23] of HI). REMPI and other two-photon spectra also are used for quantitative characterizations of photoinduced excited state distributions and reaction processes [25].

Other types of multiphoton spectroscopy used to probe the same electronic energy regions as one-photon VUV spectroscopies involve multiwavelength pump-probe or multiphoton (four or more) techniques. Although often so, a pumping step need not be optical. Sometimes electron pulse pumping of the excited state to be probed is more efficient than optical pumping, as is the case for the laser-induced fluorescence (LIF) spectrum of Ar_2 in Fig. 4. The spectrum in Fig. 4 was obtained [26] by probing $a\,^3\Sigma_u^+$ levels produced in a pulsed discharge with radiation from a scanning dye laser system while detecting atomic emission produced when the probed Ar_2 levels predissociate. Regardless of how populated, pumped states now provide convenient platforms from which

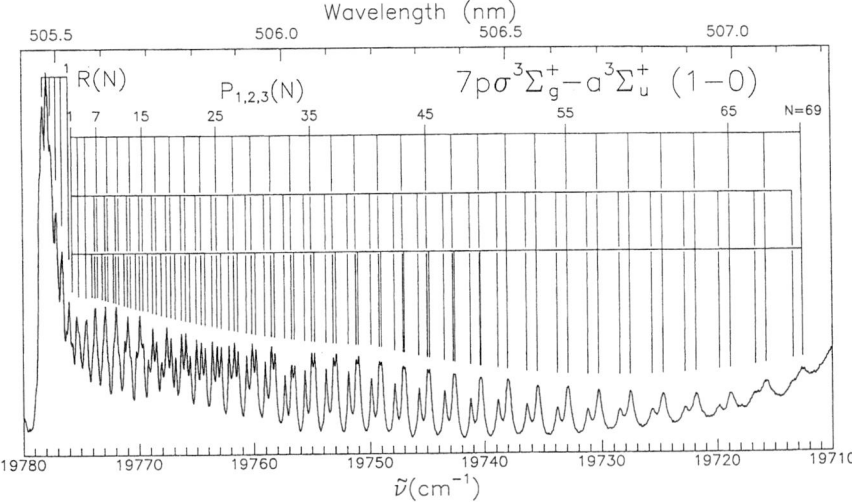

FIG. 4. Laser-induced fluorescence spectrum of the $7p\sigma\,^3\Sigma_g^+ - a\,^3\Sigma_u^+$ (1-0) band of Ar_2 obtained by probing a highly excited Rydberg state from an electronically pumped platform state. Reprinted from [26] with permission of the American Institute of Physics.

to probe levels within a few eV of ionization, an energy region mainly the province of VUV grating spectroscopy just two decades ago.

The VUV spectroscopy of highly energetic processes such as inner shell [27] and ionic [28] spectra continue to be of major importance. Figure 5 shows an absorption spectrum of ionized krypton in the 142–157 Å region [27], taken using a pulsed gas nozzle, a laser-produced plasma light source (LPPLS) and a 1.5-m grazing incidence monochromater with a diode-array multichannel detector [29]. Light from the LPPLS was split, the major part ionizing the gas and the other part time-delayed to provide the background continuum radiation probing the ionized gas. The vertical lines in Fig. 5 indicate calculated [27] positions and relative gf-values for lines from the transition arrays $3d^{10}4s^23p^3 - 3d^94s^24p^4$, $3d^{10}4s^24p^4 - 3d^94s^24p^5$, and $3d^{10}4s^24p^5 - 3d^94s^24p^6$ for Kr IV, Kr III, and Kr II, respectively. Specific level and transition assignments [27] of the spectral lines have been omitted to simplify the figure.

VUV emission spectroscopies also are important in both laboratory and extraterrestrial applications. In some cases, extraterrestrial observations provide extensions to laboratory observations, such as the identification of high-lying

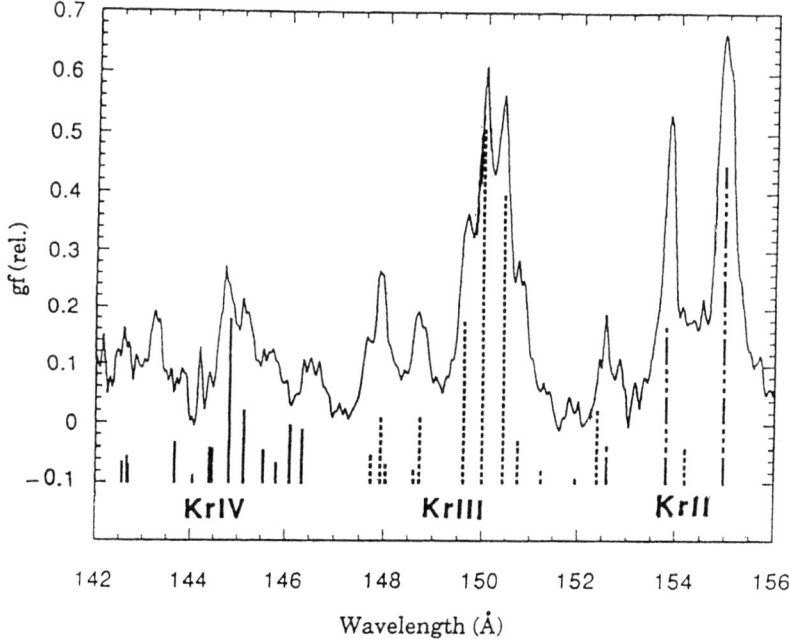

FIG. 5. Absorption spectrum of ionized krypton in the 142–157 Å region. Vertical lines indicate calculated (see text) transitions for Kr IV (solid), Kr III (dotted), and Kr II (dash-dot), respectively. Spectrum provided by Professor T. J. McIlrath [27].

Rydberg states in C I using HR solar spectra [30], although the converse [31] still is more common. Emission spectra are especially important to the characterization and/or identification [32] of highly ionized species that abound in high-voltage electrical discharges, high-temperature plasmas, and stellar coronaspheres.

12.3 Intensity Measurements and Cross Sections

12.3.1 Absolute Photoabsorption Cross Sections

The Beer–Lambert law

$$\ln \frac{I_0(\nu)}{I(\nu)} = \sigma N \tag{1}$$

expresses the ratio of initial, $I_0(\nu)$, and absorbed, $I(\nu)$, intensities in terms of the absorption cross sections σ in cm^2 mol^{-1} and the column density in N mol cm^{-2}. The quantity σN is called *optical depth* (or opacity or optical thickness). Two other expressions also are used for absorption measurements: $I(\nu) = I_0(\nu)e^{-kpl}$ and $I(\nu) = I_0(\nu)10^{-\varepsilon cl}$. The absorption coefficients k and ε are given in units of atm^{-1} cm^{-1} and dm^3 mol^{-1} cm^{-1}, respectively. In the previous expressions, l is the path length in units of cm, p is pressure in units of torr or atm, and c is the concentration in units of mol dm^{-3}. Conversion factors between these absorption coefficients (cross sections) are given in Table A-3 of [33].

Photoabsorption cross-section (coefficient) measurements in the VUV region were performed for many molecules during the period 1960 to 1980. Results for small molecules (up to 1977) were summarized by Okabe [33]. Unfortunately, as pointed out by Hudson and Carter [34], most of those cross-section measurements for molecular bands with fine structures are severely distorted by the instrumental bands widths. The effects of instrumental width on measured cross sections [35] are demonstrated in Fig. 6 for three cases: ratios of line to instrumental widths of $\alpha = 3.6$, 1.8, and 0.7. The ratio of the measured to true cross sections, σ_m/σ_r, is less than 1 even for α values of 3.6 (line widths ~4 times the instrument widths). On the other hand, the integrated cross sections are less affected by the instrumental widths, and approach the true cross sections as optical depth approaches zero. Therefore, the true integrated cross section can be obtained from series of data taken with different column densities.

The measured integrated cross sections are converted to the band oscillator strengths (f-values) according to

$$f(\nu', \nu'') = \frac{mc^2}{\pi e^2} \frac{1}{\tilde{N}} \int_{\text{band}} \sigma(\nu) \, d\nu \tag{2}$$

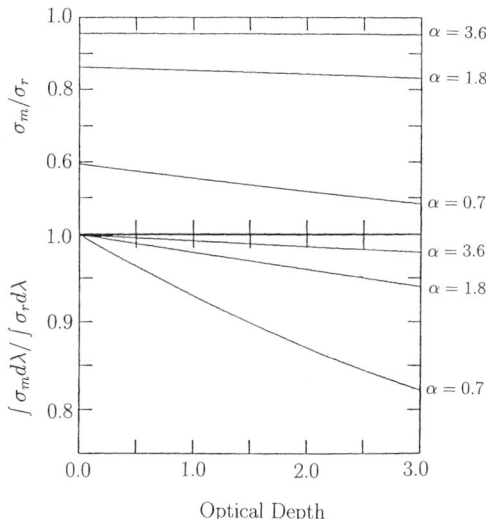

FIG. 6. Ratio of synthetic measured/true peak cross sections vs. true optical depth in the top for α = linewidth/slit function. Ratio of synthetic measured/true integrated cross sections vs. true optical depth in the bottom with the same α.

in which $\tilde{N}$ is the fractional Boltzmann population of the absorbing vibrational level, and the integration of the cross section $\sigma(\nu)$ is performed over all of the rotational lines belonging to the (v', v'') band. The constants m, e, and c are the electron mass, the electron charge, and the velocity of light, respectively. In Eq. (2), $\sigma(\nu)\,d\nu$ can be replaced by $\sigma(\lambda)\,d\lambda$. The rotational line oscillator strengths can be derived from the band oscillator strengths and Hönl–London factors. The cross sections at any temperature could be obtained from the line oscillator strengths and the Boltzmann factors with the known linewidths and/or instrumental widths.

12.3.2 Application to Atmospheric Science

In recent years, high-resolution absorption cross sections of simple molecules have become of increasing interest in atmospheric applications. Solar radiation in the wavelength range 240–175 nm plays a critical role in the photochemistry of the stratosphere, mesosphere, and lower thermosphere. At altitudes of 60–90 km, the predissociation of the Schumann–Runge (S–R) bands of O_2 is an important source of oxygen atoms, and hence of ozone formation. At altitudes of 30–100 km, the solar flux transmitted through the S–R bands is available to dissociate minor atmospheric constituents and pollutants such as nitrogen, hydrogen, and halogenated species, all of which may participate in the catalytic destruction of ozone.

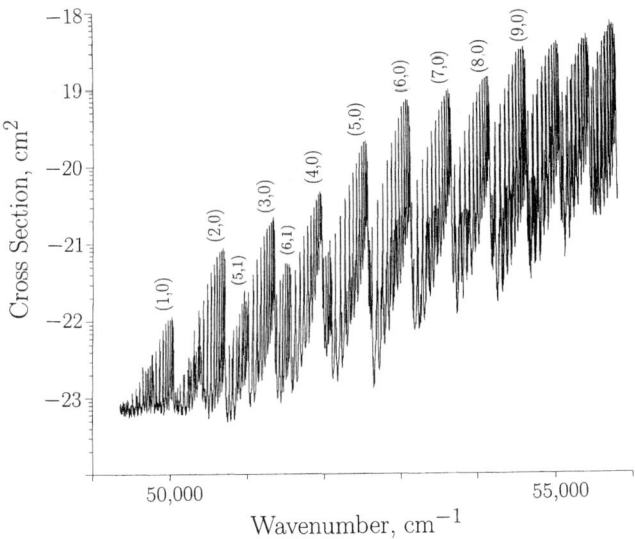

FIG. 7. The absorption cross sections of the Schumann–Runge band at 300 K. The flattening of the cross sections around 50,000 cm^{-1} is caused by the underlying Herzberg continuum.

Recent, high wavelength-resolution measurements [36, 37] have provided absolute absorption cross sections of the S–R bands of O_2 at 300 K (see Fig. 7). Large changes in peak and valley cross sections make possible the significant penetration of the solar radiation between rotational lines. However, molecular cross sections are temperature dependent, and the penetration becomes larger at the altitude of interest because of the lower temperature. Existing temperature-dependent polynomial expressions [38], which are based on the original absorption measurements [37], provide an accurate and efficient means of expressing the cross sections of the S–R bands for temperatures 130 K < T < 500 K.

The absorption cross sections of minor species also are important because many can be dissociated by solar radiation. Many simple molecules (HCl, NO, CCl_4, etc.) have continuous or broad-band absorption spectra below 200 nm. The cross sections of some of these molecules have been measured, but often without accurately calibrated wavelengths. On the other hand, in some cases like NO most of the bands in the wavelength region 200–160 nm consist of many sharp rotational lines whose linewidths are close to the Doppler widths, ~0.1 cm^{-1}.

The best traditional grating spectrometer with a resolution of around 0.3 cm^{-1} cannot measure reliable cross sections unless the spectral features are broadened in some manner. However, the Fourier transform (FT) spectrometer recently developed for use in the VUV region (see Fig. 1) has a resolution of 0.025 cm^{-1} which is sufficient to measure Doppler-limited spectra [8]. To make FT absorption measurements with VUV radiation requires a stable and bright background continuum source and a narrow bandpass filter. The combination of a synchrotron source with a zero-dispersion predisperser and the VUV FT spectrometer from Imperial College was used at the Photon Factory, KEK, Japan, to make Doppler-limited absorption cross-section measurements for the S–R bands of O_2 with $v' \geq 12$ and NO bands in the wavelength region 190–160 nm [39].

12.3.3 Application to Interstellar Molecules

Many interstellar molecules have been observed by radio detection of molecular emission lines. In the VUV region, observations of molecules in diffuse clouds have been carried out by optical measurements of absorption lines in the spectra of the background stars. Early observations were limited to techniques using rocket-borne instruments, and later to satellite experiments. The Lyman bands of H_2 were observed for the first time [40] by sounding rocket instruments, followed by the first observation of electronic transitions of CO [41]. Observations of CO molecules in the VUV region [42–46] were extended with the instruments on the satellites, *Copernicus* and *International Ultraviolet Explorer*. However, recently, observations at high spectral resolution have become available from the Goddard High Resolution Spectrograph (GHRS) on the *Hubble Space Telescope* [47, 48].

Quantitative analyses of such astronomical spectra require accurate and reliable wavelength and band oscillator strength data. As pointed out earlier, the band oscillator strengths can be obtained from the integrated cross sections. The rotational line oscillator strengths can also be obtained from the band oscillator strengths and Hönl–London factors. However, because the levels related to VUV transitions are highly excited, they are often more or less perturbed by other levels. Intensity borrowing from perturbing levels leads to quite different line strengths than those obtained from calculations based on the band oscillator strengths. Moreover, the very low temperature of molecular clouds limit observation to only a few rotational lines, that is, with $J = 0 - 3$ depending on temperature. Thus, the spectroscopic data needed for interstellar absorption studies are accurate rotational line positions and oscillator strengths for lines with $J = 0 - 3$.

Carbon monoxide is the most abundant interstellar molecule after hydrogen. The Fourth Positive bands of CO $(A - X)$ are well known and, in absorption, extend from 154 nm to 110 nm [49], which are the most convenient wavelengths for GHRS observation. The band oscillator strengths (f-value) of the $A - X$ system have been obtained from many different techniques: photoabsorption,

lifetime measurements, various electron impact methods, and theory. The
f-values have been summarized by Chan et al. [50] and the values agree well for
the low vibrational levels. The f-values of higher v can be separated into
two groups. The measured values of Eidelsberg et al. [51] supported by De
Leon [52] are higher than those of Chan et al. [50] supported by Kirby and
Cooper [53]. Smith et al. [54] measured the f-values of the same system with
$v' = 11 - 14$ with high resolution ($\lambda/\Delta\lambda \approx 150{,}000$), and support the lower
f-values of Chan et al.

Basically, the absorption cross sections of molecular bands, at any temperature, can be obtained from their f-values. However, most of the bands in the VUV region overlap with other bands and are sometimes also perturbed. As a consequence, synthetic absorption spectra of molecules at the low temperatures that prevail in interstellar clouds, are uncertain. CO molecules have been cooled in the laboratory down to ~25 K by a supersonic expansion technique, and cross sections of selected bands of CO have been observed at high resolution [55–57]. For example, the cross sections of the $K\,^1\Sigma^+ - X\,^1\Sigma^+$ band have been measured at both 295 K and 20 K. At the bottom of Fig. 8, strong R(0), R(1), and R(2) lines of the $K - X$ band at 20 K are observed with P(1), P(2), P(3), and R(3) lines having medium intensities. In this case, the low-temperature spectrum could be derived from the f-values of high-temperature spectra. Figure 9 presents observation of the diffuse band, $J\,^1\Sigma^+ - X\,^1\Sigma^+$. The spectrum at 295 K appears with two broad peaks, the R- and P-branches. The spectrum at 20 K has a single peak with a shoulder at longer wavelengths, which are assigned to R- and P-branches, respectively, with very large predissociated linewidths.

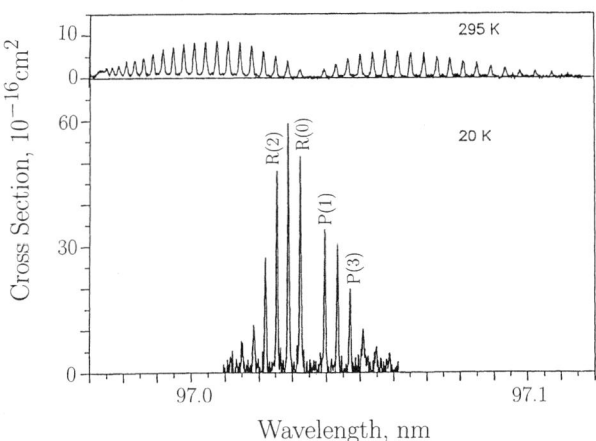

FIG. 8. The absorption cross sections of the $K(0) - X(0)$ band of CO at ~20 K and 295 K. Because of instrumental bandwidth effects, the peak absorption cross sections must be interpreted as lower limits.

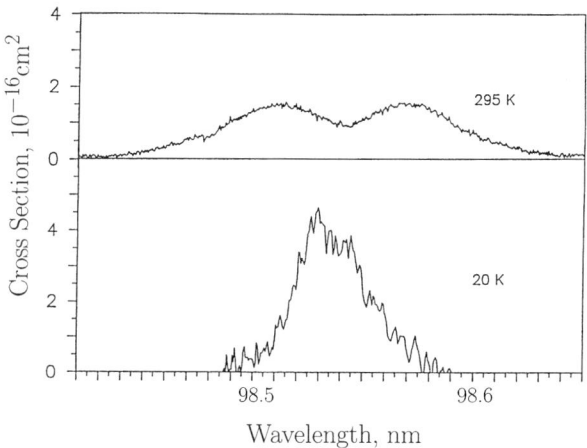

FIG. 9. The absorption cross sections of the $J(0) - X(0)$ band of CO at $\sim$20 K and 295 K. Because of the broadened structures, the absorption cross sections are absolute at both temperatures.

This spectrum could not be derived from the f-values obtained at higher temperature.

Acknowledgments

The authors would like to thank Drs. B. Lewis (Australian National University) and T. McIlrath (University of Maryland) for contributing figures and other information prior to publication, and Drs. D. S. Ginter and W. H. Parkinson for their comments and critical reading of the manuscript. We also would like to thank the National Science Foundation, the National Air and Space Administration, the University of Maryland, and the Harvard College Observatory for their support.

References

1. M. L. Ginter, D. S. Ginter, and C. M. Brown, *Appl. Opt.* **19**, 4015–4020 (1980).
2. M. L. Ginter, *Nucl. Instrum. Methods.* **A249**, 474–484 (1986).
3. F. B. Orth, K. Ueda, T. J. McIlrath, and M. L. Ginter, *Appl. Opt.* **25**, 2215–2217 (1986).
4. R. Mahon, T. J. McIlrath, V. P. Myerscough, and D. W. Koopman, *IEEE J. Quantum Electron.* **15**, 444–451 (1979).
5. R. H. Lipson, S. S. Dimov, J. Y. Cai, P. Wang, and H. A. Bascal, *J. Chem. Phys.* **102**, 5881–5889 (1995).
6. A. Mellinger, C. R. Vidal, and Ch. Jungen, *J. Chem. Phys.* **104**, 8913–8921 (1996).

7. S. B. Kim, D. J. Kane, and J. G. Eden, *Phys. Rev. Lett.* **68**, 1311–1314 (1992).
8. J. E. Murray, K. Yoshino, J. R. Esmond, W. H. Parkinson, Y. Sun, A. Dalgarno, A. P. Thorne, and G. Cox, *J. Chem. Phys.* **101**, 62–79 (1994).
9. J. P. England, B. R. Lewis, and M. L. Ginter, *J. Chem. Phys.* **103**, 1727–1731 (1995).
10. J. T. Vanderslice, S. G. Tilford, and P. G. Wilkinson, *Astrophys. J.* **141**, 395–402 (1965).
11. D. S. Ginter and M. L. Ginter, *J. Molec. Spectrosc.* **90**, 177–196 (1981).
12. P. G. Burke and K. A. Bersington, *Atomic and Molecular Processses: An R-Matrix Approach*, Inst. Phys. Publ., New York, 1993.
13. Ch. Jungen, *Quantum Defect Theory*, Inst. Phys. Publ., New York, 1996.
14. D. S. Ginter and M. L. Ginter, *J. Chem. Phys.* **86**, 1437–1444 (1987).
15. C. M. Brown, S. G. Tilford, and M. L. Ginter, *J. Opt. Soc. Am.* **67**, 584–606 (1977).
16. D. S. Ginter, M. L. Ginter, and C. M. Brown, *J. Chem. Phys.* **85**, 6530–6535 (1986).
17. D. S. Ginter and M. L. Ginter, *J. Chem. Phys.* **85**, 6536–6543 (1987) and references therein.
18. G. Miecznik and C. H. Greene, *J. Opt. Soc. Am.* **B13**, 244–256 (1996) and references therein.
19. J. E. Murphy, E. Friedman-Hill, and R. W. Field, *J. Chem. Phys.* **103**, 6459–6466 (1995) and references therein.
20. J. Stephens and C. H. Greene, *J. Chem. Phys.* **103**, 5470–5475 (1995) and references therein.
21. D. S. Ginter and M. L. Ginter, *J. Chem. Phys.* **88**, 3761–3774 (1988) and references therein.
22. P. G. Wilkinson and E. T. Byram, *Appl. Opt.* **5**, 581–594 (1965).
23. S. T. Pratt and M. L. Ginter, *J. Chem. Phys.* **102**, 1882–1888 (1995).
24. R. J. Yokelson, R. J. Lipert, and W. A. Chupka, *J. Chem. Phys.* **97**, 6144–6152 (1992).
25. P. J. Dagdigian, D. F. Varley, R. Liganage, R. J. Gordon, and R. W. Field, *J. Chem. Phys.* **105**, 10251–10262 (1996).
26. C. M. Herring, S. B. Kim, J. G. Eden, and M. L. Ginter, *J. Chem. Phys.* **101**, 4561–4571 (1994).
27. T. J. McIlrath provided Fig. 5 from data in: M. H. Sher, U. Mohideen, H. W. K. Tom, O. R. Wood II, G. D. Aumiller, R. R. Freeman, D. L. Windt, W. K. Waskiewicz, J. Sugar, and T. J. McIlrath, *Soft x-ray Spectroscopy of Multiphoton Ionized Krypton* (In preparation).
28. S. S. Churilov and Y. N. Joshi, *J. Opt. Soc. Am.* **B13**, 11–28 (1996).
29. M. H. Sher, U. Mohideen, H. W. K. Tom, O. R. Wood II, G. D. Aumiller, T. J. McIlrath, and R. R. Freeman, *Opt. Lett.* **18**, 646–648 (1993).
30. U. Feldman, C. M. Brown, G. A. Doschek, C. E. Moore, and F. D. Rosenberg, *J. Opt. Soc. Am.* **66**, 853–859 (1976).
31. C. E. Moore, R. Tousey, G. D. Sandlin, C. M. Brown, M. L. Ginter, and S. G. Tilford, *Astrophys. and Space Sci.* **38**, 359–364 (1975).
32. H. M. Milchberg, C. G. Durfee III, and J. Lynch, *J. Opt. Soc. Am.* **B14**, 731–737 (1995).
33. H. Okabe, *Photochemistry of Small Molecules*, John Wiley & Sons, New York, 1978.
34. R. D. Hudson and V. L. Carter, *J. Opt. Soc. Am.* **58**, 227 (1968).

35. G. Stark, K. Yoshino, P. L. Smith, K. Ito, and W. H. Parkinson, *Astrophys. J.* **369**, 574–580 (1991).
36. K. Yoshino, D. E. Freeman, J. R. Esmond, and W. H. Parkinson, *Planet. Space Sci.* **31**, 339–353 (1983).
37. K. Yoshino, J. R. Esmond, A. S.-C. Cheung, D. E. Freeman, and W. H. Parkinson, *Planet. Space Sci.* **40**, 185–192 (1992).
38. K. Minschwaner, G. P. Anderson, L. A. Hall, and K. Yoshino, *J. Geophys. Res.* **97**, 10,103–10,108 (1992).
39. K. Yoshino, P. L. Smith, W. H. Parkinson, A. P. Thorne, and K. Ito, *Rev. Scient. Instr.* **66**, 2122–2124 (1995).
40. G. R. Carruthers, *Astrophys. J.* **161**, L81–L85 (1970).
41. A. M. Smith and T. P. Stecher, *Astrophys. J.* **164**, L43–L47 (1971).
42. E. B. Jenkins, J. F. Drake, D. C. Morton, J. B. Rogerson, L. Spitzer, and D. G. York, *Astrophys. J.* **181**, L122–L127 (1973).
43. D. C. Morton, *Astrophys. J.* **197**, 85–115 (1975).
44. S. R. Federman and A. E. Glassgold, *Astrophys. J.* **242**, 545–559 (1980).
45. M. M. Hanson, T. P. Snow, and J. H. Black, *Astrophys. J.* **392**, 571-581(1992).
46. D. E. Welty and J. R. Fowler, *Astrophys. J.* **393**, 193–205 (1992).
47. S. R. Federman, J. A. Cardelli, Y. Sheffer, D. L. Lambert, and D. C. Morton, *Astrophys. J.* **432**, L139–L142 (1994).
48. D. C. Morton and L. Noreau, *Astrophys. J. Suppl. Ser.* **95**, 301–343 (1994).
49. S. G. Tilford and J. D. Simmons, *Phys. Chem. Ref. Data* **1**, 147–188 (1972).
50. W. F. Chan, G. Cooper, and C. E. Brion, *Chem. Phys.* **170**, 123–138 (1993).
51. M. Eidelsberg, F. Rostas, J. Breton, and B. Thieblemont, *J. Chem. Phys.* **96**, 5585–5590 (1992).
52. R. L. DeLeon, *J. Chem. Phys.* **89**, 20–24 (1988).
53. K. Kirby and D. L. Cooper, *J. Chem. Phys.* **90**, 4895–4902 (1989).
54. P. L. Smith, G. Stark, K. Yoshino, and K. Ito, *Astrophys. J.* **431**, L143–L145 (1994).
55. P. L. Smith, G. Stark, K. Yoshino, K. Ito, and M. H. Stevens, *Astron. Astrophys.* **252**, L13–L15 (1991).
56. G. Stark, K. Yoshino, P. L. Smith, J. R. Esmond, K. Ito, and M. H. Stevens, *Astrophys. J.* **410**, 837–842 (1993).
57. K. Yoshino, G. Stark, J. R. Esmond, P. L. Smith, K. Ito, and M. Matsui, *Astrophys. J.* **438**, 1013–1016 (1995).

13. SOFT X-RAY FLUORESCENCE SPECTROSCOPY

Thomas A. Callcott

Department of Physics and Astronomy
University of Tennessee
Knoxville, Tennessee

13.1 Introduction

X-ray fluorescence spectra are produced when x-rays are used to excite characteristic x-ray emission spectra of a particular element. Very low yields for radiative transitions in the soft x-ray spectral range made soft x-ray fluorescence (SXF) impractical before the advent of powerful synchrotron sources. Consequently, the history of SXF spectroscopy is provided by the measurement of soft x-ray emission (SXE) spectra excited by energetic electrons. Even with electron beam excitation, soft x-ray spectra are weak so that their measurement greatly benefits from the modern instrumentation described elsewhere in this volume. This chapter will describe the characteristic features of SXF spectroscopy, specialized spectrometers and detectors that have been developed for measuring the very weak spectra, and some typical experimental results.

Historically, the soft x-ray spectroscopy of interest here and hard x-ray spectroscopy have developed along largely independent paths. At photon energies above about 2 KeV, x-ray spectroscopists took advantage of higher radiative yields, longer absorption lengths, and wavelengths matched to crystalline lattice spacings to develop a rich science based on the use of crystal spectrometers, intense and efficient e-beam excited x-ray sources, and experimental systems with sources, samples, and detectors separated by windows transparent to hard x-radiation. In the 1950s, the range of these measurements was extended to 1000 eV and below using natural crystals of large lattice spacing and liquid crystals of even larger lattice spacing [1]. However, strong absorption, wavelengths longer than typical lattice spacings, and the susceptibility of large lattice spacing crystals to radiation damage make the use of crystal spectrometers impractical at lower energies. In the spectral range below 1000 eV, the most versatile and useful spectrometers use reflection gratings in grazing incidence.

In the choice of spectrometer and many other ways, SXF studies must conform to the constraints imposed by the spectral region. Soft x-rays are strongly absorbed requiring windowless operation from source to sample to spectrometer and detector, and the use of reflective optics. Normal incidence reflectance is very low, requiring the use of grazing incidence optics, and the accommodation or correction of the large aberrations that this implies. The single most important constraint on SXF spectroscopy, however, is the low radiative yield, which results from the fact that holes in shallow core levels are filled more efficiently by nonradiative than by radiative processes [2]. This fact prevented the development of intense conventional sources using e-beam excitation, and limited the development of SXE spectroscopies before the advent of modern synchrotron sources.

E-beam excited SXE spectroscopy using grating spectrometers was pioneered in the 1930s by T. H. Osgood [3], M. Siegbahn [4], and H. W. B. Skinner [5] and was actively pursued by many scientists through the 1950s. Review articles by Tomboulian in 1957 [6] and Parratt in 1959 [7] summarize this earlier work devoted mostly to light elements and light element compounds from the second and third rows of the periodic table. The early work used photographic detection, with a resulting nonlinear response to intensity so that these early spectra often differ in detail from those obtained later with scanning spectrometers and photoelectric detection.

At the time of these reviews, the soft x-ray emission (and absorption) spectra of the light elements were already being interpreted in terms of the one electron density of states of filled (and empty) bands calculated with the recently developed methods of the band theory of solids. These theoretical methods, as well as the use of different instrumentation, set the soft x-ray community apart from hard x-ray spectroscopists, who typically used theoretical descriptions derived from atomic or molecular orbital theory as better suited to the localized systems that they were studying. This separation has long been symbolized by the existence of separate triannual international conference series on Vacuum Ultraviolet Radiation Physics [8] and X-ray Physics and Inner Shell Ionization [9]. In recent years, the advent of synchrotron sources as the best source for both x-ray regions has greatly reduced the barriers between the two communities.

The instrumentation in use in the late 1960s was described in the predecessor to the current volume [10]. State-of-the-art instruments were scanning monochomators of the Seya/Namioka design for energies below about 100 eV, and grazing incidence Rowland circle monochromators for higher energies. Detection primarily used phosphor-coated or open-faced photomultipliers, later supplemented by channeltron and channelplate electron multipliers.

The proceedings of two conferences held at Strathclyde, Scotland, in September of 1967 and 1971, and reported in proceedings edited by D. J. Fabian [11] and Fabian and Watson [12], effectively summarized the state of the theoretical and experimental investigations in the late 1960s. The interpretation of light element spectra in terms of densities of states derived from one electron band structure calculations was well developed, and some many-body effects such as plasmon satellites and spectator holes in d-band metals were understood. Applications to metals, alloys, oxides, and compounds of elements through the transition metals were all reported at these conferences.

In another conference on electronic processes in the x-ray and extreme ultraviolet regions, held in Paris in 1971, physicists from Universitat Munchen in Germany, including G. Wiech, who remains active in the field today, reported fluorescence excitation of the B, Be, and C soft x-ray spectra using synchrotron radiation from the 7.5 GeV synchrotron at DESY [13]. This is, to my knowledge, the first report of soft x-ray fluorescence excitation using synchrotron radiation. About fifteen years would elapse before the technique would come into general use with the development of dedicated synchrotron sources and emission spectrometers.

In subsequent years, but before the advent of modern synchrotron excited SXF studies, several additional advances were made in the interpretation of soft x-ray spectra. An anomalous peak located at the Fermi edge of both emission and absorption spectra of simple metals and associated with the many body response of simple metals to the creation of a core hole was predicted by Mahan [14], and described in further detail by Mahan [15], Nozieres, and DeDominicus [16]. The MND anomaly, named after these authors, was identified and analyzed by several experimental groups for light metals such as Na and Al [17, 18]. An anomalously broadened Fermi edge in the SXE spectrum of Li was explained in terms of lattice relaxation occurring on the same time scale as the core hole lifetime by Mahan [19] and Almbladh [20]. These phonon relaxation effects were subsequently measured and analyzed for Li and Na [21].

Finally, many body theoretical studies by Mahan [22] and von Barth and Grossman [23] for light metals helped resolve the question of whether SXE spectra should be interpreted using densities of states calculated for the neutral atom or for a core excited atom. For nearly free electron metals, these studies established the "final state rule," which holds that densities of states should be calculated using the final state of the electronic process. Thus, for absorption spectra, spectra should be calculated for the core excited atom, while for emission spectra, spectra should be calculated for a filled core level. Although unproved in detail for more localized systems such as transition metals, insulators, and ionic compounds, the final state rule is widely used as a rule of thumb for the interpretation of spectra. It should be noted that the final state of an SXE transition contains a valence hole. In s-p bonded materials with metallic and covalent bonding, the valence hole is delocalized and spectra have been very successfully interpreted using "band structure" calculations with ground state wavefunctions having the symmetry of the lattice. In d- and f-band materials where the valence hole may be localized on a single site, and in wide bandgap insulators supporting valence exciton states, the existence of the valence hole may strongly modify observed spectra.

Photoemission and angular resolved photoemission measurements developed in the 1960s and 1970s largely supplanted SXE spectroscopy as the principal means of obtaining electronic density of states information for solids. Already in 1965, a conference devoted to the electronic structure of metals and alloys contained more papers on photoemission than on SXE spectroscopy [24]. At about the same time, electronic "core level" spectroscopies such as ESCA (Electron Spectroscopy for Chemical Analysis) and Auger spectroscopy came into routine use for the chemical characterization of complex materials [25, 26]. SXE and SXF spectroscopies retain, however, several unique characteristics, which, combined with the development of intense sources and new spectrometers, have resulted in renewed interest in these techniques as valuable complements to various electron spectroscopies. These special characteristics are discussed in the next section.

13.2 Characteristics of the Soft X-Ray Fluorescence Spectra of Solids

13.2.1 Soft X-Ray Emission (SXE) Spectra

Normal emission spectra are produced when electrons make radiative transition into a core hole from a higher energy state. The simplest approximation for the intensity of soft x-ray emission is given by the expression

$$I(\omega) \propto \omega \sum_v |\langle v|\mathbf{p}|c\rangle|^2 \cdot \delta(\hbar\omega - E_v + E_c) \qquad (1)$$

where the momentum matrix element is between core and valence states [27]. Within the dipole approximation, converting from momentum to dipole matrix elements, assuming electronic states in bands indexed by reciprocal lattice vectors **G** and that matrix elements are approximately constant for transitions to a particular band, this becomes

$$I(\omega) \propto \omega^3 \sum_\mathbf{G} N_\mathbf{G}(E_v) |\langle v_{\mathbf{G},l\pm 1}|\mathbf{r}|c_l\rangle|^2 \qquad (2)$$

with $E_v = E_c + \hbar w$. Here $N_\mathbf{G}$ is the density of states for the **G**th band at energy E_v, and the angular momentum selection rule for dipole transitions has been made explicit in the matrix element.

Transitions from valence levels characteristic of the solid to a narrow core level produce broad spectra that provide some measure of the electronic density of states (DOS) of the solid. Because core levels are compact and have distinct binding energies, the spectra are characteristic of, and selective for, chemical species. Also, the dipole selection rules governing one electron transition assure that the spectra are selective for angular momentum state. Thus, normal emission spectra provide a measure of the local partial density of states (LPDOS) of the valence states of a solid that is selective for both chemical species and angular momentum. This selectivity gives the technique special power for analyzing the electronic properties and bonding of complex solids.

The selection rules are particularly simple for x-ray spectra, being just those appropriate for a one-electron atom. This follows from the circumstance that the core hole is excited in a filled shell, so that the quantum numbers characterizing the hole are the negative of those of the missing electron. Those selection rules are: Δn is unconstrained; $\Delta l = \pm 1$; $\Delta s = 0$; $\Delta j = \pm 1, 0$; $\Delta m_j = \pm 1, 0$ except $m_j = 0 \rightarrow m_j = 0$ if $\Delta j = 0$, where n, l, s, m, and j are the principal, orbital angular momentum, spin, magnetic, and total angular momentum quantum numbers, respectively [27].

In x-ray spectroscopy, it is customary to designate spectra using a notation associated with the location of the core hole, where K, L, M, N, ... spectra result from a hole in the n = 0, 1, 2, 3, ... shell. A subscript to this letter increases with decreasing binding energy of the level and hence with increases in the l and j quantum numbers. This notation is summarized in Table I and compared with the common notation of atomic spectroscopy in which principle quantum numbers are followed by letters designating the orbital angular momenta (s, p, d, f, ...) and subscripts giving the j quantum numbers. Both notations are commonly used in the description of experimental results in x-ray physics.

To describe inter-core spectral lines, authors usually designate both the initial and final states of the transition, or use a traditional notation in which Greek subscripts to the x-ray level designations identify spectral lines of increasing energy [28]. For example, for a heavy element, K_α and K_β lines are produced by inner core dipole transitions $L_{2,3} \rightarrow K$ and $M_{2,3} \rightarrow K$, respectively.

13.2.2 The SXE Spectra of Solids

Valence spectra, which will be of principal interest to us here, are identified using the x-ray designation for the core hole alone. As an example of emission

TABLE I. Notations Used in the Description of X-Ray Spectra

X-Ray Notation	Atomic Notation	Quantum Numbers
K	$1s_{1/2}$	n = 1, l = 0, j = 1/2
L_1	$2s_{1/2}$	n = 2, l = 0, j = 1/2
$L_{2,3}$	$2p_{1/2, 3/2}$	n = 2, l = 1, j = 1/2, 3/2
M_1	$3s_{1/2}$	n = 3, l = 0, j = 1/2
$M_{2,3}$	$3p_{1/2, 3/2}$	n = 3, l = 1, j = 1/2, 3/2
$M_{4,5}$	$3d_{3/2, 5/2}$	n = 3, l = 2, j = 3/2, 5/2
N_1	$4s_{1/2}$	n = 4, l = 0, j = 1/2
$N_{2,3}$	$4p_{1/2, 3/2}$	n = 4, l = 1, j = 1/2, 3/2
$N_{4,5}$	$4d_{3/2, 5/2}$	n = 4, l = 2, j = 3/2, 5/2
$N_{6,7}$	$4f_{5/2, 7/2}$	n = 4, l = 3, j = 5/2, 7/2

spectra excited by both energetic electrons and photons, Fig. 1a shows the S-$L_{2,3}$ spectra of CdS produced by transitions from valence states to a hole in the $2p_{1/2,3/2}$ levels of S, which provides a measure of the $(s + d)$-LPDOS at the S site in this compound [29]. These spectra map Cd d-bands (150–153 eV), which lie in a subband gap between an upper valence band (155–159 eV) and a lower valence band (145–150 eV). The lower valence band is derived from S $3s$ states and the upper valence band from mixed covalent/ionic bonding derived mostly from Cd $4s$ and S $3p$ states. The spectral features associated with overlap transitions from the Cd $4d$ states are much sharper in photon than in electron excited spectra, probably because of broadening of the d-spectra by the presence of satellites associated with spectator holes excited by the energetic electrons.

The two d-band peaks and the shoulder on the upper edge of the S-$3s$ peak are produced by the superposition of L_2 and L_3 spectra associated with transitions to S $2p_{1/2}$ and $2p_{3/2}$ core levels. We see later that photon excitation of emission spectra makes possible the selective excitation of the L_3 spectrum alone.

13.2.3 Normal Soft X-Ray Fluorescence (SXF) Spectra

Selectivity for chemical species and angular momentum state are common features of all core level spectroscopies. The use of photons to excite emission spectra in SXF spectroscopy has a number of important advantages, which however could be realized only with the development of synchrotron sources. These include energy selection, polarization selection, bremsstrahlung suppression, and reduced damage.

13.2.3.1 Energy Selection. Monochromatized photons can be used to selectively excite chosen core levels to simplify spectra in complex compounds. With excitation just above an x-ray threshold, deeper levels are not excited. Moreover, because absorption associated with core-hole creation reaches a maximum within a few eV of an x-ray threshold and decreases sharply to higher energies, shallower core levels are excited with reduced efficiency. In contrast, core hole excitation using energetic electrons rises from zero at an energy equal to the x-ray threshold and reaches a maximum at many times the threshold energy.

The value of selective excitation for revealing hidden structure is illustrated by the photon excitation of the S-$L_{2,3}$ spectrum in CdS shown in Fig. 1b. With excitation at photon energies above 165 eV, both L_2 and L_3 spectra are excited with a separation of 1.15 eV. With 163.8 eV excitation between the L_2 and L_3 thresholds, a clean L_3 spectrum is obtained, which reveals an additional d-band peak ($d3$) that had previously been buried by the S-L_2 spectrum.

In Fig. 1c, the S-L_3 spectrum is compared with CdS spectra from other spectroscopies and with theory. All spectra are plotted on a common energy

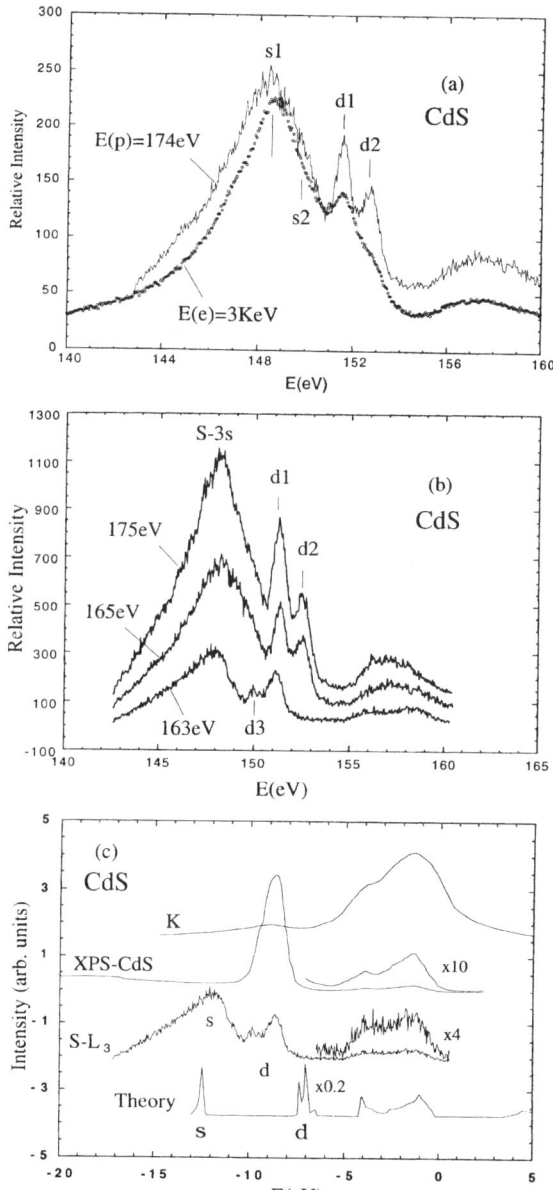

FIG. 1. The S-$L_{2,3}$ emission spectra of CdS. (a) Spectra excited by 3 KeV electrons and 174 eV photons. (b) Spectra excited by photons near threshold showing suppression of L_2 spectrum with excitation between the L_2 and L_3 thresholds. (c) Comparison of L_3 spectrum with other experimental spectroscopies and with theory [29].

scale with zero taken at the top of the upper valence band. The e-beam excited S-K spectrum provides a measure of the p-LPDOS [30] at the S site and sees mostly the upper valence band derived from these states. The XPS spectrum is dominated by the Cd d-states and resolves the upper valence band, but is blind to the lower valence band [31]. The S-L$_3$ spectrum resolves all valence electron features, clearly resolving both the top, bottom, and some features of the upper valence band and locating the Cd d states with respect to both upper and lower valence bands. The principle difference between experiment and theory is in the location of the d-bands in the subband gap. No calculations to date properly locate the d-bands in the subband gap [29, 32]. It seems likely that this discrepancy is produced by an energy shift associated with a valence hole in the final state of both XPS and SXF spectra. In effect, experiments measure $3d^9$ final states while band calculations are made for $3d^{10}$ states.

13.2.3.2 Polarization Selection. Light from synchrotron sources is polarized, providing another potential means of selectively exciting material systems. This selectivity has not yet been significantly utilized in SXF spectroscopy studies, but may have important future applications. These effects require that the excited system retain a "memory" of the polarization effects imposed by the exciting light, and thus must generally be described using the scattering formalism discussed later, rather than as a "normal" fluorescence effect in which emission is decoupled from the excitation process. Here, we briefly describe two examples of the use of polarization in fluorescence spectroscopy.

In materials containing localized d and f orbitals, circularly polarized light and the selection rules on magnetic quantum number can be used to excite particular members of an atomic multiplet. This selectivity has had important applications in Magnetic Circular Dichroism (MCD) studies of magnetic materials using soft x-ray absorption spectroscopy [33, 34]. MCD effects in emission have also been demonstrated in a few cases, but have not moved much beyond the proof of principle stage [35]. Currently available sources of circularly polarized light, which utilize the relatively weak off axis-light from bending magnets, are generally too weak to be practical excitation sources for SXF spectra. Such studies will become more important as elliptical undulators or wigglers are put into service to provide much more intense sources of elliptically polarized light (Volume 31, Section 1.3).

Linear polarization of the exciting radiation is often useful if the E-vector of radiation can be selectively oriented with respect to a fixed axis in a system. It has been particularly useful, for example, in photoemission and SX absorption studies of molecular interactions with surfaces, as described further in Section 13.2.4. In emission spectroscopy, measurements of angular resolved x-ray emission from gaseous molecular systems have been reported [36]. In these studies, the E-vector of the incident light serves to define a preferred axis in the gaseous system, which imposes an angular distribution on subsequent emission that occurs before subsequent rotational or vibrational motion eliminates the preferred axis. Another application of angular resolved fluorescence to molecules on surfaces is described in Section 13.2.4. There, however, the preferred axis is not defined by the incident light, but by the physical geometry of the excited system.

13.2.3.3 Bremsstrahlung Suppression. Another major advantage of photon excitation of emission spectra is that it eliminates the bremsstrahlung background radiation associated with the scattering of energetic electrons. This greatly facilitates the detection of very weak spectra, such as those from thin surface layers, impurities, and the M and N spectra of heavier elements. With e-beam excitation, emission spectra from impurities are buried in the bremsstrahlung background and are not usually measurable at concentrations of less than a few percent. For example, the N K-spectra from N_2 adsorbed at submonolayer coverages on Ni described in Section 13.2.4 would be buried in the bremsstrahlung from the underlying Ni if excited by energetic electrons. As another example of measurements made possible by the suppression of bremsstrahlung, the spectra of boron impurities have been measured to concentrations of 10^{-4} in Si and are detectable at still lower levels [37].

13.2.3.4 Reduced Damage. Finally, photon excitation usually produces less damage in fragile materials than e-beam excitation. With near threshold photon excitation, the incident power is much more efficiently converted to core hole formation than for energetic electrons, thus reducing the power loads for a given excitation rate. Moreover, because exciting photons have energies above an absorption edge, and emitted photons have energies below the absorption edge, the absorption depths are less than escape depths, assuring that emitted electrons escape with high probability.

We note in passing that this disparity in absorption length between exciting and emitted x-rays makes fluorescence yield an unreliable tool for the measurement of absorption spectra near threshold. For the excitation region above a threshold where escape depth is less than absorption depth, the fluorescent yield saturates and provides little information about the true absorption.

13.2.4 Angular Emission

As noted previously, the linear polarization of synchrotron light is commonly used to selectively excite orbitals of particular symmetry in systems where there is one or more well-defined axes. The prototypical example was provided by photoemission studies of CO adsorbed on Ni, in which the molecular axis is oriented perpendicular to the surface [38]. By exciting in s-polarization, with electric field E-vectors parallel to the surface, only π orbitals of CO are excited, whereas excitation with p-polarized light at non-normal angles excites both σ and π orbitals. Similarly, molecular orbital information can be obtained for this

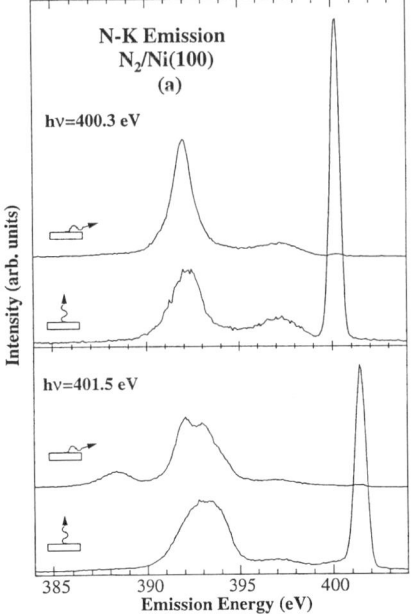 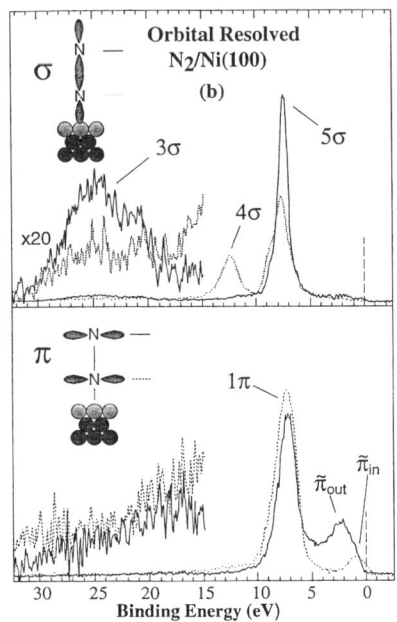

FIG. 2. The N-K spectra from a monolayer of N_2 on Ni(100). (a) Spectra recorded at different exit angles and excitation energies. The 400.3 eV excitation yields contributions from outer N atom only and normal emission yields contributions from π orbitals only. (b) Decomposed spectra in terms of symmetry and atomic site. The upper panel displays states of σ symmetry and the lower panel states of π symmetry.

system in emission spectroscopy by measuring SXF spectra as a function of angle from the surface normal. Nilsson and coworkers have pioneered such studies using a very versatile SXF spectroscopy system in which the emission spectrometer can be rotated around the exciting photon beam. They have reported work not only on the CO on Ni(100) system [39] but also for N_2 on Ni(100) [40].

In adsorption of N_2 on Ni, the molecular axis is oriented perpendicular to the surface so that σ orbitals are oriented perpendicular and π orbitals parallel to the surface. Since orbitals of both π and σ symmetry are derived from N p-states, measuring the N K-spectrum provides a measure of the LPDOS for each symmetry. If emission is measured normal to the surface, the LPDOS for π orbitals are mapped; with grazing emission, the LPDOSs from both symmetries are superimposed. In addition, these authors were able to use the fact that adsorption produces a chemical shift of the 1s level of the N atom adjacent to the Ni substrate to separate contributions of the inner and outer atoms of the adsorbed N_2 molecule [41]. Their measurements are presented in Fig. 2a. In the upper panel, only the outer atom is excited, while in the lower panel, both inner and outer atoms are excited. The normal emission curves, indicated by icons on the figures, contain contributions only from π orbitals, while the grazing incidence spectra contain contributions from orbitals of both symmetry.

In Fig. 2b, these spectra are decomposed in terms of both symmetry and atomic site, and plotted on a binding energy scale. The principle peak in the π spectra at a binding energy of 7.5 eV is derived from the 1π orbital and the intensity at lower binding energies as modifications of the 1π and possible $2\pi^*$ orbitals by interactions with the substrate. For σ symmetry, the peaks at binding energies of 7.6 eV are derived from the 5σ orbital and the peak at 12.3 eV with the 4σ orbital, which only has intensity on the inner atom. Readers are referred to the original paper for a more detailed discussion of these results. It is clear, however, that SXF spectra can provide unique and very detailed information about modification of electronic structure by interactions of molecules with a substrate.

These spectra effectively summarize several useful features of photon-excited SXF spectra. With angular detection, certain orbital symmetries are resolved. With energy selection, contributions from the inner and outer N atoms are resolved. In addition, the suppression of bremsstrahlung makes it possible to detect the emission spectra from submonolayer coverages of N_2 that would be buried in background radiation from Ni using e-beam excitation.

13.2.5 Resonant Inelastic X-ray Scattering

In "normal" SXF spectra, the excitation and de-excitation processes are treated as independent dipole transitions. A principal result of SXF studies has been the demonstration that, for excitation near threshold, excitation and emission processes are often strongly coupled and must be described as a single inelastic scattering process. The scattering process is an x-ray version of the electronic resonant Raman process well known in optics. Several names have been used in the literature, among them resonant inelastic x-ray scattering (RIXS), which we adopt here as being both accurate and descriptive.

The cross section for inelastic scattering, which determines the intensity of inelastically scattered x-rays, has been treated theoretically in some detail for atomic systems by Tulkki and Aberg, but their theory can be applied to electronic excitations in any system [42, 43]. It is described for a single initial (i) and final (f) state, and multiple intermediate (m) states using second-order perturbation theory by an equation

$$\left(\frac{d^2\sigma(\omega_{in})}{d\omega_{out}d\Omega}\right)_{fi} \propto \sum_f \left| \sum_m \frac{\langle f|\mathbf{p} \cdot \mathbf{A}|m\rangle\langle m|\mathbf{p} \cdot \mathbf{A}|i\rangle}{E_m - E_i - \hbar\omega_{in} - i\Gamma_m/2} \right|^2 \cdot \delta(\hbar\omega_{in} - \hbar\omega_{out} - E_f + E_i) \quad (3)$$

Many features of inelastic scattering are implied by this equation. A very important result, represented by the delta function, is that the energy losses involve only the energy difference between initial and final states, independent of the intermediate states. Thus, for a single electron excitation, the energy loss is just the energy needed to create an electron-hole pair in the final state so that

$$\hbar\omega_{in} - \hbar\omega_{out} = E_f - E_i = (E_e - E_h)_f \qquad (4)$$

where $(E_e - E_h)_f$ is the energy of the electron-hole pair in the final state.

The shape and energy position of the Electronic Raman spectra are determined solely by the distribution of energy levels in the final state of the system. In applications to materials with delocalized electrons in which both initial and final states can be described as Bloch sums having the symmetry of the lattice, the scattering process conserves crystalline momentum. It thus produces an electron-hole pair in states with correlated crystalline momentum, so that, in the reduced zone scheme

$$\mathbf{k}_e + \mathbf{q}_{in} = \mathbf{k}_h + \mathbf{q}_{out} \qquad (5)$$

where $\mathbf{k}_e$ and $\mathbf{k}_h$ are wavevectors of the electron and hole states and $\mathbf{q}_{in}$ and $\mathbf{q}_{out}$ the wavevectors of the incident and emitted photons. For the soft x-ray range, where x-ray wavelengths are much larger than atomic spacings, the $q \ll k$, so that these can be treated as a "vertical" excitation for which $\mathbf{k}_e \cong \mathbf{k}_h$. For harder x-rays, the momentum of the x-ray photons must be considered in the interpretation of data, as has been demonstrated for the Si-K spectrum by Y. Ma [44]. This k-conserving scattering has been used in the interpretation of spectra from crystalline Si [45], diamond [46], and graphite [47]. The K spectra from diamond were the first to be accurately interpreted using k-conserving inelastic scattering by Y. Ma.

The spectra for graphite shown in Fig. 3 demonstrate the dispersion of spectral features that is characteristic of k-conserving transitions between bands. The diagram to the left illustrates the correlation of electron-hole pairs in k-space. Varying the excitation energy excites core electrons to conduction band states in regions of k-space near the K and M points. Emission occurs from valence bands in the same region of k-space.

In atomic systems and in insulators, the electronic excitation associated with the inelastic energy can occur between localized states rather than between k-conserving band states. In such solids, localization of the electron near a specific atomic site by a valence hole in the final state can eliminate the k-conservation effects and make a localized description more useful. Spectra from hexagonal boron nitride (h-BN), which provides a simple model of such a system, will be described later.

Although energy losses are determined solely by the final state energies, other important features of the scattering are determined by the intermediate state, which enters through the dipole matrix elements and in the resonant energy denominator. The resonant denominator is affected by both the energy (E_m) and lifetime broadening (Γ_m) of the intermediate state. Though all scattering amplitudes are increased by excitation to energies near the real intermediate state E_m, it is useful to distinguish between *near-resonance* excitation where $|\hbar\omega_{in} - E_m + E_i| > \Gamma_m/2$, and *on-resonance* excitation where $|\hbar\omega_{in} - E_m + E_i| < \Gamma_m/2$. For the near-resonance case, the RIXS process dominates the fluorescence and scattering weakens rapidly as one moves away from the resonant energy. For on-resonance scattering, the intermediate state coincides with an accessible real state of the system, so that the scattering process is dominated by excitation energies within $\Gamma_m/2$ of $E_m - E_i$.

It is frequently useful to think of scattering processes in a time domain rather than the energy domain of spectroscopy. In near-resonance scattering, inelastic scattering is the only relevant process and the scattering occurs in a time related to the energy difference $|\hbar\omega_{in} - E_m + E_i|$ by the Heisenberg uncertainty relation. For on-resonance scattering, other scattering processes in the intermediate

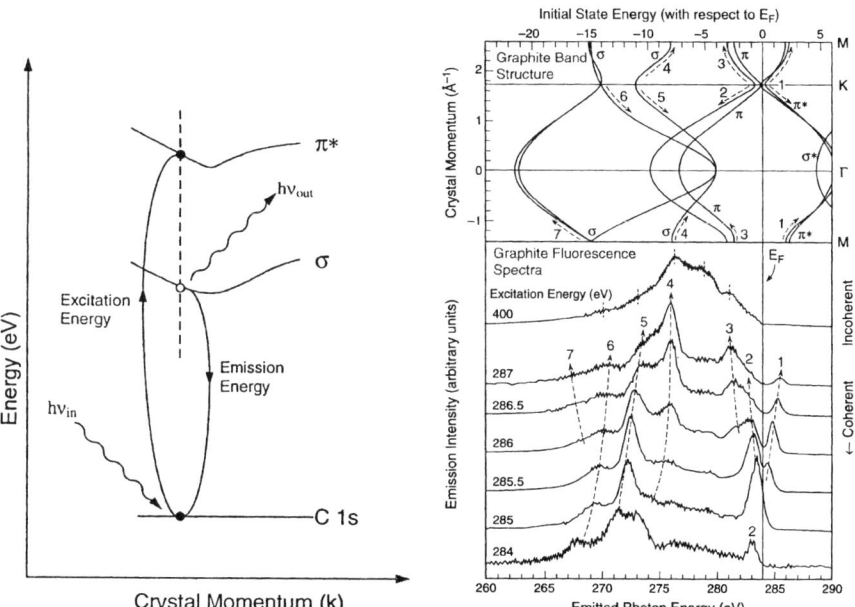

FIG. 3. The C-K spectra of graphite excited near threshold. The SXF spectra are generated by an inelastic scattering process in which the electron-hole pair is correlated in k-space. The process is diagrammed at the left. Spectra are labeled with excitation energies. Spectral features show dispersion along the bands as indicated by the numbers [47].

state—including non-radiative de-excitation and electron-electron and electron-phonon scattering—limit the lifetime of the intermediate state.

The difference between near-resonance and on-resonance scattering have important implications for spectroscopy. Near-resonance scattering is a pure RIXS process, so the selection rules implicit in the resonant Raman formalism are rigidly observed. Moreover, the scattering time is related directly to the energy difference from resonance, giving a potential means of measuring time-dependent electronic processes in the sub-femtosecond range. For on-resonance scattering, Γ_m is determined by all processes that limit the lifetime of the intermediate state. The processes responsible for limiting lifetime may also destroy the coherence of the intermediate state and open additional channels for refilling the core hole. Thus, on-resonance RIXS spectra are typically superimposed on "normal" fluorescence spectra produced by radiative de-excitation of the core hole that follows electron or phonon scattering in the intermediate state.

Some of these features of inelastic scattering in more localized systems are illustrated by the boron K-spectra of hexagonal BN excited near threshold illustrated in Fig. 4 [48]. Hexagonal BN has a layered structure similar to that of graphite and a similar band structure; the principle difference is that h-BN has a large band gap and is an insulator, which permits h-BN to support a localized exciton state with an energy level within the band gap. The presence of this exciton state produces remarkably different effects in its inelastic scattering spectra. The spectra are labeled with the excitation energies. These spectra show an extraordinarily strong localized excitonic resonance at 192 eV, which is bound about 1.5 eV below the conduction band threshold. The high energy "elastic" peak marked E reaches a maximum amplitude about 500× larger than other spectral features at the resonance energy. Theoretical studies indicate that this core exciton is derived from π^* states near the conduction band minimum at the M point of the 2D hexagonal Brillouin zone and is localized within about two lattice spacings from the core excited atom [49].

These studies further indicate that the final state of the inelastic scattering process is a similarly localized *valence* exciton state on which an electron is bound to a valence hole. The inelastic spectra excited at photon energies below this resonance and between the resonance and the conduction band threshold are near resonance spectra. The peaks labeled A and B exactly track the elastic peak at energies approximately 16 and 11 eV below the elastic peak. These spectra are essentially a map of the σ-p density of states, produced by energy loss to electronic Raman excitations from the σ-p valence bands to a localized valence exciton derived from π^*-p conduction band states. The π-p valence to π^* states are forbidden by symmetry in the RIXS process. It should be emphasized that the excitons associated with the intermediate and final states of this process are different, the strong resonance in the intermediate state being associated with a core exciton and the final state having a localized valence exciton.

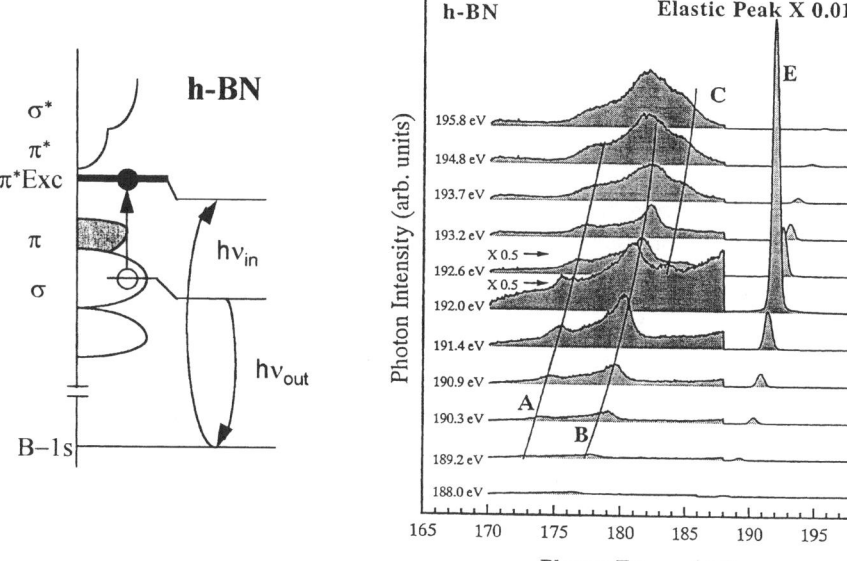

FIG. 4. The B-K spectra of h-BN excited near threshold. The SXF spectra are generated by an inelastic process that produces electronic excitation between valence band states of σ symmetry and a localized π* exciton state. The process is diagrammed to the left. Spectra are labeled with excitation energies. Peak E is a strong elastic scattering peak associated with scattering through a core exciton state. Below the exciton resonance at 192 eV, peaks A and B result from an electronic Raman scattering process. Peak C observed on resonance and in normal fluorescence spectra excited above 194 eV is symmetry forbidden in the inelastic scattering process [48].

For excitation on the excitonic resonance and above the conduction band, the spectra shows a third peak C associated with transitions from the π-p states, characteristic of normal fluorescence spectra. This is a clear example of how alternative decay channels in on-resonance excitation relax selection rules of the RIXS process.

13.3 Instrumentation for SXF Spectroscopy

Soft x-ray fluorescence spectroscopy at photon excitation energies below about 1000 eV became a marginally useful spectroscopy with the development of bending magnet sources on electron storage rings, and has become a versatile and truly practical spectroscopy with the advent of undulator sources. These sources and the grating monochromators that deliver their output at high resolutions to an experimental chamber are described elsewhere in this volume.

In standard configurations presently in use, they are capable of delivering fluxes of 10^{14} photons/sec into a focused spot of 200 µm diameter with a resolving power of 1000 over the energy range of 100–1000 eV.

13.3.1 Rowland Circle Spectrometers

The other critical equipment development for fluorescence was the design of new emission spectrometers with greatly enhanced measuring efficiencies at the University of Tennessee [50] and the Uppsala University in Sweden [51]. These grazing incidence spectrometers are of Rowland circle design with fixed input slit and scanned detectors, but dramatically improve performance in several ways. They improve light collection by moving the entrance slits to within a few mm of the entrance slit so that gratings can be completely filled through the entrance slit. They use multiple interchangeable grating to permit coverage of the energy range from <50 eV to >1000 eV in a single instrument with excellent resolution. In more recent versions of these instruments, the detector is scanned on the Rowland circle by a precision X/Y/Θ drive that permits different Rowland circles to be utilized for high- and low-energy spectra [52, 53].

These SXF spectrometers are compact enough to permit them to be mounted so that they can be rotated about an axis to record emission as a function of angle [54]. When combined with sophisticated means of sample manipulation, preparation, and characterization, these instruments make possible sophisticated measurements such as those described in Section 13.2.4. The state-of-the-art instrumentation used for these studies is shown in Fig. 5. It incorporates a sample preparation and characterization chamber with sample transfer to an analysis chamber that mounts a Nordgren-type SXF spectrometer and a high-resolution electron analyzer that may be rotated about an axis coincident with the output beam from a synchrotron source.

13.3.2 Area Detectors

The greatest single improvement in the performance of modern SX spectrometers is obtained by replacing the output slit by large, photon-counting area detectors, which record a broad spectrum of ≥100 resolution elements. These detectors, mounted tangent to the Rowland circle, are programmed to record at high resolution in the dispersion direction to resolve the spectrum. In the transverse direction, the long curved astigmatic image is imaged and sliced into 10–20 slices. These slices can be translated and added to compensate for curvature and thus to maintain resolution, while collecting as much dispersed radiation as possible.

Several different area detectors have been utilized including CCD detectors with phosphor wavelength shifting [55, 56], a CCD detector with microchannel plate (MCP) preamplification [50], and a detector consisting of 3–5 stacked

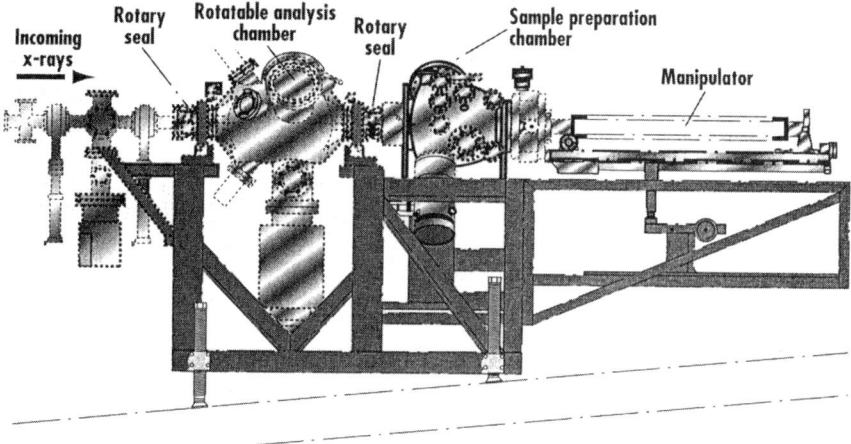

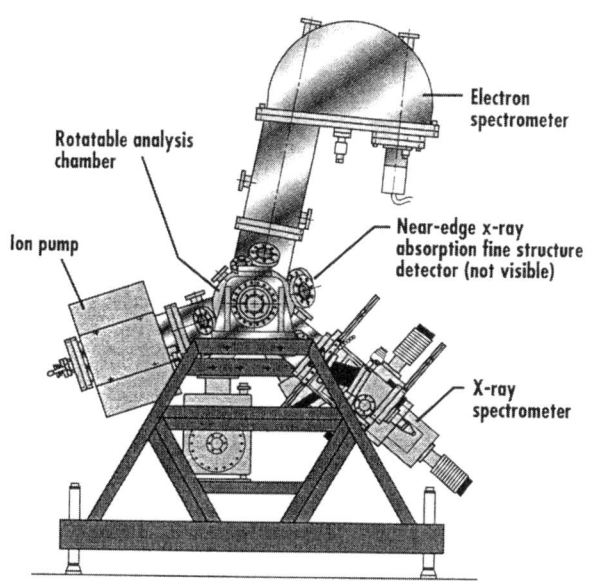

FIG. 5. A diagram of an experimental apparatus providing for *in situ* sample preparation and characterization, followed by transfer to an analysis chamber with SXF and electron spectrometers that can be rotated about the x-ray beam from a synchrotron light source. This apparatus provides maximum versatility for measuring soft x-ray fluorescence spectra excited by a polarized synchrotron source.

MCPs output to a shaped resistive sheet [51, 57]. The latter detector is currently used by most researchers. It has the virtue that it is a true photon-counting detector with good stability and excellent discrimination against noise pulses so that uncertainties in the spectra are limited by statistical noise in the recorded signal. Its performance is frequently enhanced by overcoatings such as CsI on the input plate of the MCP stack, and by the use of a retarding potential screen that drives emitted electrons back onto the MCP surface. The major drawbacks of this detector are the relatively low quantum yields of its detecting surface for soft x-ray photons and its limited dynamic range for high incident fluxes. The limited dynamic range is seldom a problem in practice, because of the very low intensities of typical SXF spectra.

The much improved performance, reliability, and availability of cooled CCDs, may mandate a new look at the possibility of direct recording of SX spectra with CCD detectors. Very low-noise (thermal noise $<$1e per hour per pixel, readout noise $<$5e per pixel), back thinned CCDs are now available for soft x-ray imaging. They offer quantum efficiencies of 30 to 90 percent and detection gains (electrons per detected photon) of 30 to 300 in the energy range between 100 and 1000 eV. The major drawback would be an expected slow increase in noise over time due to radiation damage. An alternative is to use wavelength shifting with high gain phosphors [55]. A practical geometry would use a cooled, back thinned CCD packaged with a fiber optics faceplate on which the phosphor is deposited.

13.3.3 VLS Spectrometers

Most SXF spectrometers currently in use utilize the traditional Rowland circle geometry with extreme grazing incidence at both the grating and the detector. In order to maintain resolution and calibration, this design requires that very high accuracy and stability be maintained in scanning the detector over a long curved path along the focal plane on the Rowland circle. As detailed in Chapter 3, design work by many groups has shown that varied line space (VLS) gratings have many virtues for soft x-ray spectroscopy, among them the ability to shift the focal plane away from the Rowland circle. One design study specifically addressed to the problem of designing a compact, lightweight SXF spectrometer has demonstrated that excellent resolution can be obtained using a VLS grating on a concave grating and scanning the detector along a short path nearly normal to the exit path of the dispersed light [58]. Other designs rotate a VLS grating while keeping the input slit and detector fixed.

For maximum resolution, these VLS designs require that higher-order corrections to linear variations of line spacing be used. In practice, this requires the use of ruled gratings. Currently available ruled gratings are quite satisfactory for use in the low-energy photon energy range below a few hundred volts, but ruled gratings used at higher energies may be compromised by scattering problems. If future technical developments succeed in reducing scattering from ruled gratings, possibly by combining ruling and etching steps, it seems likely that many future SXF spectrometers will use VLS designs.

13.4 Survey of Recent SXF Spectroscopy Research

We will very briefly provide reference here to some of the more recent SXF studies as a means of giving the reader a sense of the types of applications for which SXF spectroscopy has proved useful. Work by most of the more active groups currently using the technique is cited, but the list is not intended to be comprehensive or complete.

Swedish scientists from Uppsala University, using spectrometers designed by Joseph Nordgren, have been among the most active of SXF researchers. Nordgren and Wassdahl have led a large group that has carried out research at various synchrotron sources in the United States and Europe. Their studies have covered a wide range of materials and topics including high Tc superconductors and related compounds [59–63], diamond, graphite, and the fullerenes [48, 64–68], free molecules and surface layers [69–73], transition metals, rare earths and their compounds [74–78], and magnetic circular dichroism [35, 79]. Working with Y. Ma, this group first identified k-conservation in inelastic scattering as an important process in near threshold SXF spectroscopy [48]. Another Uppsala group including A. Nilsson, N. Wassdahl, N. Martensson, collaborating with J. Stohr of IBM–Almaden has concentrated on surface studies such as those described in Section 13.2.4 [39, 40] and on magnetic multilayers [80].

A collaboration of scientists from the University of Tennessee, Tulane University, Lawrence Berkeley National Laboratory, Lawrence Livermore National Laboratory, and the National Institute for Standards and Technology has also been very active. They have used spectrometers designed by T. Callcott and D. Ederer and have worked at the NSLS, ALS, and CAMD synchrotron sources. This group has published broadly on light elements and their compounds, including silicides [81–83], oxides [84–86], sulfides [29, 87], and borides, as well as transition metals and their compounds [88, 89], and high Tc superconductors and related compounds [90–92]. This group has also utilized the bulk sensitivity of SXF spectroscopy to study multilayers and buried structures [93–95]. In addition, they have studied inelastic scattering in several systems [47–49].

The work of several other groups should be noted. G. Wiech and A. Simunek in Munchen Germany have reported SXE and SXF spectroscopy studies of light element compounds, and particularly of oxides and silicides [95–100]. As noted earlier, G. Wiech was a coauthor of the first reported studies of SXF spectroscopy using synchrotron radiation. Another German group, led by W. Eberhardt in Jülich, have added SXF spectroscopy to EXAFS, XANES, and photoelectron spectroscopies they use for synchrotron-based studies of solids. They have reported SXF studies of transition metal compounds [101, 102], and of intercore transitions in CaF and BaF$_2$ [103, 104] and potassium halides [105].

Japanese scientists, using a spectrometer designed by S. Shin [52], have reported studies on a variety of light elements and light element compounds [106–109]. These studies continue a strong tradition of x-ray spectroscopy studies in Japan extending from the pioneering work of M. Sawada of Osaka University in the 1930s and continuing in the work of T. Sagawa and coworkers at Tohuku University in the 1960s.

References

1. B. L. Henke and P. A. Jaanimagi, *Rev. Sci. Instrum.* **56**, 1537 (1985).
2. M. O. Krause, *J. Phys. Chem. Ref. Data* **8**, 307 (1979).
3. T. H. Osgood, *Phys. Rev.* **44**, 517 (1933).
4. M. Siegbahn, *Ergebn. Exakt. Naturw.* **16**, 104 (1937).
5. H. W. B. Skinner, *Phil. Trans. Roy. Soc. Lon., Ser.* A **239**, 95 (1940).
6. D. H. Tomboulian, *Hanbuch der Physik,* S. Flugge and Marburg, eds. (Springer-Verlag, Berlin, 1957) Vol. 30, pp. 246–304.
7. L. G. Parratt, *Rev. Mod. Phys.* **31**, 616 (1959).
8. Twelve International Conferences on Vacuum Ultraviolet Radiation Physics have been held since 1965, the most recent being VUV 12 in San Francisco in 1998.
9. Seventeen International Conferences on X-ray Physics and Inner Shell Ionization have been held, the most recent being X-96 in Prague, Czechoslovakia in 1996.
10. J. A. R. Samson, *Techniques of VUV Spectroscopy.* (John Wiley, New York, 1967).
11. D. J. Fabian, ed. *Soft X-ray Band Structure.* (Academic Press, London, 1968).
12. D. J. Fabian and L. M. Watson, eds. *Band Structure Spectroscopy of Metals and Alloys.* (Academic Press, London, 1973).
13. K. Feser et al., *J. de Physique* **32**(10-suppl.), C4–331 (1971).
14. G. D. Mahan, *Phys. Rev.* **163**, 612 (1967).
15. G. D. Mahan and W. Pardee, *Phys. Rev. Letters* **45**, 117 (1973).
16. P. Nozieres and C. T. DeDominicus, *Phys. Rev.* **178**, 1097 (1969).
17. T. A. Callcott et al., *Phys. Rev.* B **18**, 6622 (1978); *Jap. J. Appl. Phys* **17**, suppl. 2, 149 (1978).
18. J. D. Dow and D. R. Ranceschetti, *Phys. Rev. Letters* **34**, 1320 (1975); P. H. Citrin et al., *Phys. Rev. Letters* **35**, 885 (1975).
19. G. D. Mahan, *Phys. Rev.* B **15**, 4587 (1977).
20. C. O. Almbladh, *Phys. Rev.* B **16**, 4343 (1977).
21. T. A. Callcott et al., *Phys. Rev.* B **16**, 5185 (1977); *Phys. Rev. Letters* **38**, 442 (1977).
22. G. D. Mahan, *Phys. Rev.* B **21**, 1421 (1980).
23. U. Von Barth and G. Grossman, *Solid State Commun.* **32**, 645 (1979).
24. F. Abeles, ed. *Optical Properties and Electronic Structure of Metals and Alloys.* (North-Holland, Amsterdam, 1966).

25. A comprehensive review of early work with XPS and ESCA is found in E. W. Plummer and W. Eberhard, *Chem. Phys.* **49**, 533 (1982).
26. Bi-annual reviews of XPS and ESCA studies are found in articles by N. H. Turner and J. A. Schreifels, *Anal. Chem* **68**, 309R (1996); **66**, 163R (1994); **64**, 302R (1992); **62**, 113R (1990).
27. R. B. Leighton, *Principles of Modern Physics,* (McGraw-Hill, New York, 1959), p. 228.
28. F. K. Richtmeyer, E. H. Kennard, and T. Lauritsen, *Introduction to Modern Physics,* (McGraw-Hill, New York, 1955), pp. 365–374.
29. L. Zhou et al., *Phys. Rev.* B **55**, 5051 (1997).
30. C. Sugiura et al., *J. Phys. Soc.* Jpn **31**, 1784 (1971).
31. L. Ley et al., *Phys. Rev.* B **9**, 600 (1974)
32. P. Schroer et al., *Phys. Rev.* B **47**, 6971 (1993).
33. G. Shutz et al., *Phys. Rev. Letters* **58**, 737 (1987); C. T. Chen et al., *Phys. Rev.* B **42**, 7262 (1990).
34. B. T. Thole et al., *Phys. Rev. Letters* **68**, 1943 (1992); P. Carra et al., *Phys. Rev. Letters* **70**, 694 (1993).
35. L. C. Duda et al., *Phys. Rev.* B **50**, 16758 (1994).
36. R. Mayer et al., *Phys. Rev.* A **43**, 235 (1991).
37. J. A. Carlisle et al., *Appl. Phys. Letters* **67**, 34 (1995).
38. R. J. Smith et al., *Phys. Rev. Letters* **37**, 1081 (1976).
39. A. Nilsson et al., *Phys. Rev. B* **51**, 10244 (1995).
40. P. Bennich et al., *Phys. Rev. Letters* **78**, 2847 (1997).
41. A. Sandell et al., *Phys. Rev. Letters* **70**, 2000 (1993).
42. J. Tulkki and T. Aberg, *J. Phys. B* **15**, L435 (1982).
43. J. Tulkki, *Phys. Rev.* A **27**, 3375 (1983).
44. Y. Ma et al., *Phys. Rev. Letters* **74**, 478 (1995).
45. J.-E. Rubensson et al., *Phys. Rev. Letters* **64**, 1047 (1990); K. E. Miyano et al., *Phys. Rev.* B **48**, 1918 (1993-I).
46. Y. Ma et al., *Phys. Rev. Letters* **69,** 2598 (1992).
47. J. A. Carlisle et al., *Phys. Rev. Letters* **74**, 1234 (1995).
48. J. J. Jia et al., *Phys. Rev. Letters* **76**, 4054 (1996); W. L. O'Brien et al., *Phys. Rev. Letters* **70**, 238 (1993).
49. E. L. Shirley in procedings of workshop on *Raman Emission by X-ray Scattering,* D. L. Ederer and J. H. Mcguire, eds. (World Scient. Publ., Singapore, 1996) p. 71.
50. T. A. Callcott et al., *Rev. Sci. Instrum.* **57**, 2680 (1986).
51. J. Nordgren et al., *Rev. Sci. Instrum.* **60**, 1690 (1989).
52. S. Shin et al., *Rev. Sci. Instrum.* **66**, 1584 (1995).
53. T. Callcott et al., unpublished.
54. T. Studt, *Res. and Dev. Magazine* **36**(11), 105 (1994).
55. R. D. Carson et al., *Rev. Sci. Instrum.* **55**, 1973 (1984).
56. M. Lampton and F. Paresce, *Rev. Sci. Instrum.* **45**, 1098 (1974).
57. T. A. Callcott et al., *Nuc. Instrum. Methods Phys. Res. A* **266**, 578 (1988).
58. K. Osborn et al., *Rev. Sci. Instrum.* **66**, 3131 (1995).
59. J.-H. Guo et al., *J. Phys. Chem. Solids* **54**, 1203 (1993).
60. J.-H. Guo et al., *Phys. Rev.* B **49**, 1376 (1994).
61. S. M. Butorin et al., *J. Phys. Condens. Matter* **6**, 9267 (1994).
62. S. Butorin et al., *Phys. Rev.* B **51**, 11915 (1995).
63. L. D. Finkelstein et al., *Solid State Communications* **95**, 503 (1995).

64. Y. Ma, P. Skytt, *et al.*, *Phys. Rev. Letters* **71**, 3725 (1993).
65. P. Skytt *et al.*, *Phys. Rev.* B **50**, 10457 (1994).
66. J.-H. Guo *et al.*, *Chem. Phys. Letters* **235**, 152 (1995).
67. J.-H. Guo *et al.*, *Phys. Rev.* B **52**, 10681 (1995).
68. Y. Luo *et al.*, *Phys. Rev.* B **52**, 14479 (1995).
69. N. Wassdahl *et al.*, *Phys. Rev. Letters* **69**, 812 (1992).
70. P. Skytt *et al.*, *Phys. Rev.* A **52**, 3730 (1995).
71. Y. Luo *et al.*, *Phys. Rev.* A **52**, 3730 (1995).
72. P. Glans *et al.*, *Phys. Rev. Letters* **76**, 2448 (1996).
73. P. Skytt *et al.*, *Phys. Rev. Letters* (submitted).
74. M. G. Samant *et al.*, *Phys. Rev. Letters* **72**, 1112 (1993).
75. Y. Ma *et al.*, *Phys. Rev.* B **48**, 2109 (1993).
76. S. Eisebitt *et al.*, *Phys. Rev.* B **48**, 5042 (1993).
77. D. C. Mancini *et al.*, *Vacuum* **46**, 1165 (1995).
78. K. Lawmiczak-Jablonska *et al.*, *Physica* B **217**, 78 (1996).
79. L. C. Duda *et al.*, *Vacuum* **46**, 1125 (1995).
80. A. Nilsson *et al.*, *Phys. Rev.* B **54**, 2917 (1996).
81. J. J. Jia *et al.*, *Phys. Rev.* B **43**, 4863 (1991); *Phys. Rev.* B **46**, 9446 (1992).
82. J. J. Jia *et al.*, *J. Appl. Phys.* **69**, 7800 (1991).
83. J. J. Jia *et al.*, *Phys. Rev.* B **52**, 4904 (1995).
84. W. L. O'Brien *et al.*, *Phys. Rev.* B **44**, 1013 (1991).
85. W. L. O'Brien *et al.*, *Phys. Rev.* B **45**, 3882 (1992).
86. W. L. O'Brien *et al.*, *Phys. Rev.* B **47**, 15482 (1993).
87. L. Zhou *et al.* (transition metal sulfides—to be published).
88. D. R. Mueller *et al.*, *Phys. Rev.* B **54**, 1 (1996).
89. D. R. Mueller *et al.*, *Phys. Rev.* B **52**, 9702 (1995).
90. C. H. Zhang *et al.*, *Phys. Rev.* B **39**, 4796 (1989).
91. D. R. Mueller *et al.*, *Physica Scripta* **41**, 979 (1990).
92. D. R. Mueller *et al.*, *Phys. Rev.* B **52**, 9702 (1995).
93. R. C. C. Perera *et al.*, *J. Appl. Phys.* **66**, 3676 (1989).
94. J. A. Carlisle *et al.*, *Appl. Phys. Letters* **67**, 34 (1995).
95. J. A. Carlisle *et al.*, *Phys. Rev.* B (Rapid Com.) Fe/Si.
96. A. Simunek *et al.*, *J. Phys. Cond. Matter* **5**, 867 (1993).
97. A. Simunek and G. Wiech, *Z. Phys.* B **93**, 51 (1993).
98. G. Wiech and H.-O. Feldhutter, *Phys. Rev.* B **47**, 6981 (1993).
99. G. Wiech and A. Simunek, *Phys. Rev.* B **49**, 5398 (1994).
100. A. Simunek, M. Polcik, and G. Wiech, *Phys. Rev.* B **52**, 11865 (1995).
101. S. Eisibitt *et al.*, *Phys. Rev.* B **48**, 5042 (1993).
102. Y. Ma *et al.*, *Phys Rev.* B **48**, 2109 (1993).
103. J. E. Rubensson *et al.*, *Phys. Rev.* B **50**, 9035 (1994).
104. J. E. Rubensson *et al.*, *Phys. Rev.* B **51**, 13856 (1995).
105. J. Luning *et al.*, *Phys. Rev.* B **51**, 10399 (1995).
106. S. Shin *et al.*, *Phys. Rev.* B **52**, 11853 (1995).
107. S. Shin *et al.*, *Phys. Rev.* B **52**, 15082 (1995).
108. A. Agui *et al.*, *Phys. Rev.* B **55**, 2073 (1997).
109. S. Shin *et al.*, *Phys. Rev.* B **55**, 2623 (1997).

INDEX

A

Aberration
  astigmatic curvature of focal lines, 32–33
  grazing incidence optics, 234
  in off-plane Eagle spectrographs, 8
  in spherical grating monochromator, 37–38
  ray aberration, 30–32
Aberration-limited line width, 41
Absolute flux measurement, 177–190
  accuracy, 178–179
  in photoionization chambers, 180–182
  in photoionization quantometer, 182–185
  transfer detector standards, 186–189
Absolute intensities in ionization chambers, 108
Absolute photoabsorption cross sections, 270–271
Absorption cross section measurement, 270–275
Advanced Light Source, 9
Advantages of interferometry, 74–76
Airy intensity distribution, 77
All-reflection FTS, 97–99
Amplifying detectors, 148–162
Angular emission of SXF, 287–289
Anode voltage selection
  photoemission photodiodes, 123
  windowless photoemissive photodiodes, 125
Aperture-based spectromicroscopes, 235
Applications
  of spherical-grating monochromators, 22
  VUV FTS, 91–93
  x-ray microscopy, 239–245
Area detectors
  SXF spectroscopy, 294–296
Astigmatism
  calculation, 5
  correction in uniformly spaced grating instruments, 55–56
  of focal lines, 32–33
  in Seya-Namioka monochromator, 2
Asymmetry of interferogram, 81

Atmospheric science, 271–273
Avalanche photodiodes, 133–134

B

Backgammon encoder, 168–169
Background characteristics of microchannel plates
Bakeout, 196–198
Bandgap vs. energy of incident photons, 126
Beam direction, outgoing, 29
Beamline-experiment chamber isolation, 198–199
Beamlines
  high flux undulator, 9
  high-resolution undulator, 9
Beamline vacuum, 198–200
Beamsplitter, transmission grating, 101–102
Beer-Lambert law, 270
Bremsstrahlung suppression, 287

C

Calibration
  with electron spin resonance, 185–186
  primary and secondary standards, 178–179
  self-calibration procedure, 187
  technological improvements, 177–178
  transfer detector standard, 177–189
Carbon contamination, cleaning, 202–203
Chamber
  beamline, 198–200
  categories, 193
  cost estimate, 198
  gas phase, 193–196
  isolation, 198–199
  photoionization, 180–182
  solid state, 196–198
Channel electron gain, 150–151
Channel electron multipliers, 149–152
Charged coupled device (CCD), 162

detectors in SXF spectroscopy, 294–296
Charge division centroiding schemes, 166–170
Chemical dynamics study, 9
Chemical mapping of microstructured thin films, 244
CODACON position readout system, 164–165
Commercial interferometers, 87–88
Composites, 253
Concave monochromators, 1–17
Constant deviation angle, 66–69
Corrosion study, 252
Crossed delay lines (XDL), 171
Crossed multiwire array position sensor system, 168
Cross sections of photoionization, 183
Cryopumps, 195

D

Damage to materials, 287
Daresbury Synchrotron Radiation Source, 114
Dark noise in photocathodes, 145–146
Dark spots in microchannel plate images, 160
Delay line centroiding scheme, 170–172
Demagnifying exit slit, 67
Demagnifying plane VLS design, 67
Design
 beamline vacuum chambers, 200
 Fabry–Perot interferometry, 74
 FTS spectrometer, 74
 ionization chambers, 107–109
 Michelson interferometer, 88–91
 microchannel plates, 152
 semiconductor photodiodes, 127–128
 soft x-ray FTS, 101–102
 undulator-based spherical grating monochromator, 33–43
 vacuum environment, 193–200
Detection efficiency
 microchannel plates, 153
 of photocathodes, 144–145
Detectors
 amplifying, 148–162
 calibration with electron spin resonance, 185–186
 standards for absolute flux measurement, 186–189
 see also Photocathodes
Development

of soft x-ray spectroscopy, 279–282
of spectromicroscopy, 225–227
Diffraction
 efficiency, 46–47
 and microprobe formation, 230–238
Direction of outgoing beam, 29
Discrete electronic position sensors, 164–166
Discrete optical position sensors, 162–164
Dispersive spectrometry vs. interferometric spectrometry, 73

E

Eagle-type monochromators, 7–9
E-beam vs. photon excitation, 287
Efficiency
 photoemission photodiodes, 123
 semiconductor diode, 127
 silicon photodiodes, 130
Electron bombarded CCD (EBCCD), 166
Electronic Raman spectra, 290
Electron optics microscopy, 236–237
Electron storage ring calibration, 178–179
Energy levels and wavelength measurements, 265–270
Energy selection in SXF spectroscopy, 284–285
Evolution of spectromicroscopy, 229–230
Exit slit, demagnifying, 67
Exposure time XRL system, 216

F

Fabrication of photocathodes, 141
Fabry–Perot interferometry, 77–80
Far ultraviolet photocathodes
 dark noise, 145–146
Flanges, 197
Flat field spectrographs, 62–65
Flexibility in spherical monochromators, 22
Flux, see Absolute flux
Focused-mesh multipliers, 148–149
Focusing in variable-induced-angle monochromators, 43–46
Fourier transform spectrometry, 80–86
 grating spectrometer vs., 86–87
F–P spectrometry, see Fabry–Perot spectrometry

INDEX

Free spectral range
  FTS, 83–84
  in spatially heterodyned spectrometers, 95
Full width at half maximum
  microchannel plates, 156
  position error, 147
  semi-transparent photocathodes, 147
FWHM, see Full width at half maximum

G

GaAsP Schottky diode, 186
Gain
  channel electron multipliers, 150
  microchannel plates, 153–158
Gallium arsenide phosphide photodiode, 134
Gallium phosphide photodiodes, 134
Gas
  choice ionization chambers, 111
  detectors, 107–115
  disposal, 196
Gas phase experiments, 194–196
Gaussian image point location, 29–30
Geological samples, 254–256
Geometry for grating theory, 24
Grating spectrometers vs. FTS, 86–87
Grating theory, 23–32
Grating types in Seya–Namioka monochromators, 3
Gratings
  holographically recorded, 70
  non-Rowland, 28–29
  plane, 17–19
  Rowland, 23–28
  spherical, 63–65
  uniformly spaced, 55–56
  VLS, 56–58, 65–66
Grazing incidence optics in microprobe formation, 233–234

H

Hamamatsu S1337 silicon photodiode, 186
Harada-style focusing, 58–60
Harmonics filter in high-resolution monochromator, 11
Health considerations and gas phase chamber design, 195–196

Hettrick-style focusing, 60–62
High-resolution spectrographs, 62–63
High-flux undulator beamline, 9
High-order corrections in Harada-style focusing, 60
High-resolution monochromator
  beamline, 13–17
  undulator beamline, 9–13
High resolution spectral data, 267–268
Holographically recorded gratings, 70
  in Seya–Namioka monochromators, 3

I

Image formation
  projection printing technique, 210–214
  proximity printing, 210
Image magnification
  magnetic projection microscope, 236
Image projection
  process, 213–214
Images
  dark spots in, 160
  types of paraxial focusing, 57
Imaging
  characteristics in microchannel plates, 159–160
  effects in photocathodes, 147–148
  microscopes, 235–238
Incidence angle effects for photodiodes in VUV, 135, 136
Infrared FTS spectrometry, 74
In-plane Eagle spectrograph, 8
Instrumentation for SXF spectroscopy, 293–294
Integrated circuit production, 205–206
Intensified charge injection devices, 163
Intensity levels for photodiodes in VUV, 135
Intensity measurements, 270–275
Interaction region of vacuum chamber, 194
Interferometric spectrometry vs. dispersive spectrometry, 73
Interstellar absorption studies, 273–275
Interstellar molecule study, 273–275
Ionization chambers, 107–111

K

Kevlar® fiber study, 244–245

L

Leak detection, 200–201
Light path function in mirror/VLS grating system, 65–66
Linear dichroism microscopy of Kevlar® fibers, 244–245
Line width, aberration-limited, 41
Lithography, 205–221
Lithography process, 206–208
Littrow mount, 18
 in Eagle-type monochromators, 7–9
Low-temperature spectrum values, 274–275

M

Mach–Zehnder interferometer, 100
Magnetic fields for photodiodes in VUV, 135
Magnetic material study by XMCD spectroscopy, 245–249
Magnetic projection microscopes, 235–237
Manufacture
 photoemissive photodiodes, 122–123
 VLS gratings, 56
Memory chip size, 205
Michelson interferometer
 design, 80–81
 technical considerations, 88–91
Microchannel plate detectors, 152–161
 SXF spectroscopy, 294–296
Microfocused x-ray photoelectron spectroscopy, 22
Microlithography, 206–208
Microprobe optics, 230–238
Microscopy, 47–52
Microspectroscopy vs. spectromicroscopy, 225
Microsphere plates, 161–162
Micro-XPS, *see* Microfocused x-ray photoelectron spectroscopy
Mirror aberration correction by VLS grating, 65
Mirror/VLS grating system, 65–66
Mode switching at 6Vope facility, 14–17
Monk-Gillieson plane grating mount, 18
Monochromators
 Czerny–Turner mount, 17
 design development, 21–22
 Eagle-type, 7–9
 Ebert–Fastie mount, 17
 HERMON, 70
 High-resolution, 9–13
 High-resolution on a bending magnet beamline, 13–17
 Littrow mount, 18
 Monk–Gillieson mount, 18–19
 Plane grating, 17–19
 Seya–Namioka, 1–6
 for synchrotron radiation, 66–70
 6Vope, 13–14, 16
 Wadsworth, 6–7
Multianode microchannel plate array (MAMA), 165–166
Multianode readout schemes, 164–165
Multilayer polymer film characterization, 244
Multiphase polymer phase separation, 241–242
Multiphoton spectroscopy, 268–269

N

Near Edge X-ray Absorption Fine Structure (NEXAFS), 225–227
Negative resist, 207–208
Noise characteristics in spatially heterodyned spectrometers, 96
Non-Rowland grating path function, 28–29
Nonscanning interferometers, 93–97
Notation
 for grating theory, 24
 for soft x-ray spectra, 283
Nyquist frequency, 82

O

Off-plane Eagle spectrographs, 8
Opaque photocathodes, 139
 FWHM position error, 147
Optical centroiding schemes, 166
Optical depth, 270
Optical lithography, 208–214
Optical mask design, 214–215
Optical matching in spherical monochromators, 22
Optical spectroscopy in VUV, 263–275
 applications, 271–275
 cross section measurement, 270–275
 wavelength measurement and energy levels, 265–270

INDEX

Optics cleaning, 202–203
Optimized beam light, 47–52
Optimizing
 imaging performance, 163–164
 PGM, 69
Outgoing beam direction, 29
Overlayers study, 250–251

P

PAPA sensor and imaging, 163
Parallel plate detector, 114
Paraxial, definition, 56
Paraxial focusing equations for VLS gratings, 56–58
Path function
 for non-Rowland gratings, 28–29
 for Rowland grating, 23–27
PC, see Photoionization chamber
PEEM, see Photoemission electron microscope
Penumbra blur, 215
Photocathodes, 139–147
 dark noise, 145–146
 detection efficiency, 141–142
 imaging effects, 147–148
 stability, 143–145
 types, 139, 141
Photodiodes, 117–118
 arrays and imaging, 163
 avalanche, 133–134
 manufacture, 122
 photoelectric, 121–122
 photoemissive, 120–126
 radiation detectors, 117–137
 semiconductor, 126–134
 silicon, 128–133
 types, 119
 use in VUV, 134–137
 windowed photoemissive, 120–121
 windowless photoemissive, 125
Photoelectron emission microscopy, 47–52
Photoemission from semiconductor photodiodes, 136
Photoemission electron microscope, 236–237
Photoemissive photodiodes, 120–126
Photoionization
 chamber, 180–182
 cross-sections, 183, 185
 quantometer, 182–185

 yield calculation, 180
Photomultiplier tubes, 148
Photon centroiding position sensors, 166–171
Photon
 counting with microchannel plates, 155
 detection, 139–148
 position sensing techniques, 162–166
Photon Factory, 13
Photon vs. e-beam excitation, 287
Photoresist in lithography, 206–208
Plane grating monochromators, 17–19
Polarization of fluorescence spectroscopy, 286–287
Polymer study by x-ray microscopy, 239–244
Position sensing techniques, 162–166
Positive resist, 207–208
Practical FTS, 87–91
Primary detector radiometry, 179–180
Primary radiator radiometry, 178–179
Projection lens function, 211–213
Proportional counters, 112–115
Proximity printing technique, 209
Pulse amplitude distribution in microchannel plates, 158–159
Pulse height distribution in microchannel plates, 153–158
Pumps
 beamline vacuum, 200
 cryopumps, 195
 for gas phase experiment, 194–195
 solid state experiments, 197
 TSP pumps, 195

Q

QE, see Resolution
Quantum detection efficiency, 141–145

R

Radiation detectors
 types, 178
 in VUV, 117–118
Radiation sources for XRL system, 216–220
Radiometric standards for photodiodes, 118–119
Ray aberrations, calculation, 30–32
Reflection photocathodes, 139

Requirements of soft x-ray spectroscopy, 279–280
Resistive anode positioning technique, 166–170
Resist types, 208
Resolution
 crossed delay lines, 171
 crossed multiwire arrays, 168
 effect of lithographic conditions, 215, 215–216
 Fabry–Perot interferometers, 77–79
 FTS, 83–84
 of images using microchannel plates, 159–160
 of intensified charged-coupled device, 163–164
 limitations in interferometry, 74–75
 limitations in VUV, 263
 projection system, 211–212
 resistive anode technique, 168
 in spatially heterodyned spectrometers, 94
 spiral anode, 170
Resonant inelastic x-ray scattering (RIXS), 289–293
Rowland circle spectrometers for SXF spectroscopy, 294
Rowland grating path function, 23–27

S

Sampling theorem of a signal, 82
Scanning
 with auxiliary mirror, 69
 by grating rotation, 66–69
Scanning micro-XPS, 52
Scanning Transmission X-Ray Microscope, 231–233
Scattering
 inelastic, 289–293
 near- vs. on-resonance, 291
Schottky diode, 134
Schwarzschield objectives microscopes, 234–235
Semiconductor detectors, 186–189
Semiconductor integrated circuit production, 205–206
Semiconductor photodiodes, 126–134
Semi-transparent photocathodes, 139
 FWHM position error, 147

Seya–Namioka monochromator, 1–5
SGM, see Spherical grating monochromator
SHS, see Spatially heterodyned spectrometer
Signal
 Michelson interferometer, 88
 sampling theorem, 82
Signal-to-noise ratio in FTS, 85–86
Sign conventions for paraxial focusing, 57
Silicon oxidation states, 249
Silicon photodiodes, 128–130
 as detector standards, 187
 with integral filters, 132–133
 with nitrided oxide, 131–132
6Vope facility, 13
SNR, see Signal-to-noise ratio
Soft x-ray
 design, 21
 emission (SXE) spectra, 282–283
 fluorescence spectroscopy (SXF), 279–298
 projection lithography, 220–221
 spectra of solids, 283–284
 VLS monochromator, 58
Soft x-ray FTS by grazing reflection, 99–105
Solid state experiments, 196–198
Spatially heterodyned spectrometer, 93–97
Spatial distribution of polymers, 239–244
Spectra
 interpretation of soft x-ray, 281–282
 normal soft x-ray fluorescence, 284–287
 soft x-ray of solids, 283–284
 lineshapes in SGM, 37–47
 one- vs. two-photon, 268
 purity for photodiodes in VUV, 135
Spectral resolution for interferometry, 74–75
Spectromicroscopy
 history, 229–230
 x-ray, 225–257
Spectroscopy
 optical, 263–275
 soft x-ray vs. hard x-ray, 280
SPEM (Scanning Photoemission Microscope), 229, 232–233
Spherical grating monochromator, undulator based, 33–43
Spherical grating mount in Wadsworth monochromator, 6–7
Spherical grating spectrometers, 63–65
Spherical grating/VLS monochromators, 69–70
Spinodal decomposition in polymer thin films, 242–243

Spiral anode (SPAN), 170
Spot diagrams in Seya–Namioka monochromators, 4
Stability
  of photocathodes, 144–145
  of photoemissive photodiodes, 122
Streak cameras, 62
STXM, see Scanning Transmission X-Ray Microscope
Surface contaminant study, 249–250
SX700 system, 45–46, 69
SXF spectroscopy
  instrumentation, 293–294
  research, 297–298
SXR, see Soft x-ray
Synchrotron rings, suitability for XRL, 216–220

T

Tangential focal line curvatures, 32–33
TERAS storage ring, 181
Third generation synchrotron sources, 21
Throughput advantage of interferometers, 75–76
Time response microchannel plates, 160–161
$TiSi_2$ resistivity transformation, 250
Transfer detectors calibration, 178, 179–180
Transfer detector standards, 186–189
Transmission photocathodes, 139
Transmission x-ray microscopes, 237–238
Tribology, 252
TSP pumps, 195
Tunable plane gratings, 62–63

U

Ultrahigh vacuum chambers, 196–198
Undulator-based SGM, 33–43
Uniformly spaced grating instruments, 55–56

V

Vacuum
  condition for photodiodes in VUV, 135
  leak detection, 200–201
  requirements for solid state experiments, 196–198
  techniques, 193–203
Vacuum chamber, see Chamber
Vacuum ultraviolet design, 21
Valves for solid state experiments, 197
Variable induced angle, 69
Varied-line spacing (VLS) gratings
  advantages, 55–56
  in Seya–Namioka monochromator, 3
  SXF spectrometers, 296–297
VUV spectroscopy applicaton, Fabry–Perot interferometry, 79–80

W

Wadsworth monochromator, 6–7
Wavelength
  advantage of FTS, 75
  dispersion in all-reflection FTS, 97
  measurements, 265–270
Wavenumber accuracy in FTS, 84–85
Wedge and strip anode, 169
Windowed photoemissive photodiodes, 120–126
Windowless photoemissive photodiodes, 124–126

X

XMCD spectroscopy, 245–249
X-ray absorption-based photoelectron emission microscopy (XPEEM), 22
X-ray lithographic exposure window, 215–216
X-ray lithography, 214–221
X-ray photoelectron spectroscopy (XPS), 52
X-ray spectromicroscopy, 225–257
  applications, 239–245
  history, 229–230
  other microscopic techniques vs., 228–229

Z

Zone-plate based SPEM, 232–233
Zone plate optics and microprobe formation, 230–238

ISBN 0-12-475979-3

QC 457 .V33 1998

Vacuum ultraviolet
 spectroscopy II

DATE DUE